A Primer on Finite Element Analysis

A PRIMER ON FINITE ELEMENT ANALYSIS

By

Anand V. Kulkarni
Professor
Department of Mechanical Engineering
S.D.M. College of Engineering and Technology
Dharwad
(Karnataka)

Venkatesh K. Havanur
Lecturer
Department of Mechanical Engineering
K.L.S.'s, V.D.R. Institute of Technology
Haliyal
(Karnataka)

UNIVERSITY SCIENCE PRESS
(An Imprint of Laxmi Publications Pvt. Ltd.)
An ISO 9001:2008 Company
BENGALURU • CHENNAI • COCHIN • GUWAHATI • HYDERABAD
JALANDHAR • KOLKATA • LUCKNOW • MUMBAI • RANCHI • NEW DELHI
BOSTON (USA) • NAIROBI (KENYA)

A PRIMER ON FINITE ELEMENT ANALYSIS

Printed and bound in India
Typeset at Sukuvisa Enterprises, Delhi
First Edition : 2011, Reprint : 2014
ISBN 978-93-81159-10-1

Branches

Bengaluru	080-26 75 69 30	
Chennai	044-24 34 47 26,	24 35 95 07
Cochin	0484-237 70 04,	405 13 03
Guwahati	0361-254 36 69,	251 38 81
Hyderabad	040-27 55 53 83,	27 55 53 93
Jalandhar	0181-222 12 72	
Kolkata	033-22 27 43 84	
Lucknow	0522-220 99 16	
Mumbai	022-24 93 12 61	
Ranchi	0651-220 44 64	

PUBLISHED IN INDIA BY

UNIVERSITY SCIENCE PRESS
(An Imprint of Laxmi Publications Pvt.Ltd.)
An ISO 9001:2008 Company
113, GOLDEN HOUSE, DARYAGANJ,
NEW DELHI - 110002, INDIA
Telephone : 91-11-4353 2500, 4353 2501
Fax : 91-11-2325 2572, 4353 2528
www.laxmipublications.com info@laxmipublications.com

C—
Printed at: Repro Knowledgecast Limited, Thane

CONTENTS

PREFACE

Now-a-days FEM has become a common tool for solving Engineering problems in industries for the obvious reasons of its versatility and affordability. To expose an UG student to this powerful method most of the universities have included this subject in the UG curriculum. There are many good books on the subject available in the market.

This book contains complementary/supplementary material to the textbooks covering comprehensively the theoretical aspects of the subject. This book is written primarily to help the students as a gentle introduction to the practice of FEM. This book contains many 1-D and 2-D problems solved by analytical method, by FEM using hand calculations and by using commercial FEM, ANSYS 12 academic teaching software. Results of all the methods have been compared. This helps the students in building confidence in the method.

Chapter 1 contains brief introduction to FEA, theoretical background and its application. Chapter 2 contains the linear static analysis of bars of constant cross-section, tapered cross-section and stepped bar. In each section different variety of exercise problems are given. Chapter 3 contains the linear static analysis of trusses. Trusses problems are also selected in such way that each problem is having different boundary conditions to apply. Chapter 4 contains the linear static analysis of simply supported and cantilever beams. In the chapters 2 to 4 all the problems are considered as one dimensional in nature. Stress analysis of rectangular plate with a circular hole is covered in chapter 5. In this chapter emphasis is given on the concept of exploiting symmetric geometry and symmetric loading conditions. Also stress and deformation plots are given. Chapter 6 contains the thermal analysis of cylinders and plates. Here both one dimensional and two dimensional problems are considered. Chapter 7 contains the problems of potential flow distribution over a cylinder and over an airfoil. Chapter 8 contains the dynamic analysis (modal and transient analysis) of bars and beams. Chapter 9 contains the introduction to ANSYS Workbench.

Though attempts have been made to minimize the errors, some inadvertent errors might have crept in. We will be grateful to those who bring these mistakes to our notice so that they can be corrected in the subsequent editions.

Authors are grateful to their families for their support and sacrifice made during the period of writing this book. Authors are also thankful to ANSYS INC. U.S.A. and ANSYS INC. INDIA for giving permission to use the menus and dialog boxes of ANSYS 12 academic teaching software in the book.

— **AUTHORS**

FOREWORD

I have great pleasure and feel honour in sharing my thoughts with the readers through this foreword, because of the following reasons:

- This is a wonderful primer on Finite Element Analysis which is based on simple and logical approach and hands on procedure defined for the analysis using software.
- The authors have done the solutions using the conventional approach and the same has been compared with the results obtained from the software package. This leads to the clarity in understanding the basic concepts.
- This book can act as a good reference for the students who would like to work on FEA Assignments.

In general, Finite Element Analysis is considered to be a powerful numerical approach, with great amount of mathematical complexities. It is also considered as, an indispensable tool for any industrial product development exercise. It is observed that, if we intend to use the Finite Element Analysis from first principles for any defined problems, more efforts are needed in formulation and programming with the help of a suitable high level programming language. These efforts consume huge amount of time and needs good expertise and understanding of Finite Element Analysis procedure.

However, for the most of industrial applications requiring use the Finite Element Analysis, one would prefer a proven and validated Finite Element Analysis Software such as Ansys, Nastran etc. These packages are good but have certain limitations in understanding and controlling the behaviour of the governing equations and the associated variable, properties, etc. In spite of these issues, these software platforms have been approved by the practicing Engineers, due to realistic and reliable results within the acceptable limits. Also, the total procedure is simple and can be completed in minimum time.

Most of the students and the professionals use these packages without having much understanding of the basics and procedure followed in Finite Element Analysis. The total exercise is done in a typical 'Mechanical' mode. The present book would address these requirements in a most effective manner.

The book illustrates the Finite Element Analysis procedure for few simple exercises using manual approach and followed by total click-by-click procedure with software platform.

I wish that this book will be a seed for better understanding Finite Element Analysis and ultimately bring out the Engineers with more confidence and clarity in product development assignments.

Dr. A.S. Deshpande,
PRINCIPAL
Gogte Institute of Technology,
Belgaum.

ACKNOWLEDGEMENT

We honour Dr. D.H. Rao, Director, of Jain Group of Institutions Belgaum, Dr. A.S. Deshpande, Principal, Gogte Institute of Technology Belgaum, Dr. G.R. Udupi, Principal, V.D.R. Institute of Technology Haliyal, Dr. B.T. Achyutha, Principal, Bapuji Institute of Engineering and Technology, Davanagere and Karnataka Law Society, Belgaum for inspiring us to carry out this work. We take this opportunity to express our deep sense of gratitude to them.

We express our gratitude to Shri Jayant Deshpande, Project Manager, Goodrich Inc, U.S.A. for helping in getting the permission from ANSYS Inc, U.S.A. and ANSYS Inc., India.

We also express our gratitude to Dr. S. M. Kulkarni, Professor, National Institute of Technology, Surathkal, to Dr. S. Subrahmanya Swamy, Professor, Bapuji Institute of Engineering and Technology, Davanagere, to Dr. D. Ramesh Rao HOD-Mechanical Engineering department, Bapuji Institute of Engineering and Technology, Davanagere, for reviewing this book.

We also express our gratitude to Shri Shreekant S. Deshpande of Tata Consultancy Services, Bangalore, to Shri Gangadhar Patil of Hubli and to Shri Anand J. Kulkarni of Wipro Limited, Bangalore for their support.

We sincerely acknowledge with thanks to Shri Belur Anand, Country Manager, ANSYS Inc., India and also to Shri Kiran Kumar, Senior technology specialist ANSYS Inc., India for their support.

Authors are grateful to their families and parents for their support and sacrifice made during the period of writing this book.

Authors are also thankful to ANSYS INC., U.S.A. specially to Mr. Paul Lethbridge and Mr. Fran Hensler and ANSYS INC., INDIA for giving permission to use the menus, dialog boxes and plots of ANSYS 12 academic teaching software in the book.

We acknowledge with thanks to M/s Laxmi Publications for providing whole hearted cooperation in publishing this book.

We also thank to those people who have helped us directly or indirectly during this work.

— **AUTHORS**

Chapter 1

Introduction to Finite Element Analysis

1.1 METHODS OF SOLVING ENGINEERING PROBLEM

There are three methods to solve any engineering problem:

1. Analytical method,
2. Numerical method,
3. Experimental method.

1.1.1 Analytical Method

This is classical approach. The method gives closed form solutions. Results obtained with this method are accurate within the assumptions made. This method is applicable only for solving problems of simple geometry and loading, like cantilever and simply supported beams etc.

1.1.2 Numerical Method

This approximate method is resorted to when analytical method fails. This method is applicable to real life problems of complex nature. Results obtained by this method cannot be believed blindly and must be carefully assessed against experience and judgment of the analyst. Examples of this method are Finite Element Method, Finite Difference Method etc.

1.1.3 Experimental Method

This method involves actual measurement of the system response. This method is time consuming and needs expensive set up. This method is applicable only if physical prototype is available. Results obtained by this method cannot be believed blindly and minimum three to five prototypes must be tested. Examples of this method are strain gauge, photo elasticity etc.

1.2 PROCEDURE OF FINITE ELEMENT ANALYSIS (RELATED TO STRUCTURAL PROBLEMS)

STEP (*i*). *Discretization of the structure*

This first step involves dividing the structure or domain of the problem into small divisions or elements. The analyst has to decide about the type, size and arrangement of the elements.

STEP (*ii*). *Selection of a proper interpolation (or displacement) model*

A simple polynomial equation (linear/quadratic/cubic) describing the variation of state variable (*e.g.*, displacement) within an element is assumed. This model generally is of the interpolation/shape function type. Certain conditions are to be satisfied by this model in order that the results are meaningful and converging.

STEP (*iii*). *Derivation of element stiffness matrices and load vectors*

Response of an element to the loads can be represented by element equation of the form

$$[k]\ \{q\} = \{Q\}.$$

where, $[k]$ = Element stiffness matrix,

$\{q\}$ = Element response matrix or element nodal displacement vector,

$\{Q\}$ = Element load matrix or element nodal load vector.

From the assumed displacement model, the element properties, namely stiffness matrix and the load vector are derived. Element stiffness matrix $[k]$ is characteristic property of element and depends on geometry as well as material. There are three approaches for deriving element equations. They are,

(*a*) Direct approach,

(*b*) Variational approach (Piece-wise Rayleigh–Ritz method),

(*c*) Weighted residual approach (*e.g.*, Galerkin method).

(*a*) **Direct approach:** In this method, direct physical reasoning is used to establish the element properties (stiffness matrices and load vectors) in terms of pertinent variables. Although this approach is limited to simple types of elements, it helps to understand the physical interpretation of the finite element method.

(*b*) **Variational approach:** This approach can be adopted when variational theorem (extrenum principle) that governs the physics of the problem is available. The method involves minimizing a scalar quantity known as functional that is typical of the problem at hand (*e.g.*, potential energy in stress analysis problems).

(*c*) **Weighted residual approach:** This approach is more general in the sense that it is applicable to all situations where governing differential equation of the problem is available. The method involves minimizing error resulting from substituting trial solution in to the differential equation.

STEP (*iv*). *Assembling of element equations to obtain the global equations*

Element equations obtained in step (*iii*) are assembled to form global equations of the form $[K]\ \{r\} = \{R\}$ that describes the behaviour of entire structure.

where, $[K]$ is the global stiffness matrix,

$\{r\}$ is the vector of global nodal displacements and

$\{R\}$ is the global load vector of nodal forces for the complete structure.

STEP (*v*). *Solution for the unknown nodal displacements*

The global equations are to be modified to account for the boundary conditions of the problem. After specifying the boundary conditions, the equilibrium equations can be expressed as,

$$[K_1]\{r_1\} = \{R_1\}.$$

For linear problems, the vector $\{r_1\}$ can be solved very easily.

STEP (*vi*). *Computation of element strains and stresses*

From the known nodal displacements $\{r_1\}$, the element strains and stresses can be computed by using predefined equations for structure.

The terminology used in the previous six steps has to be modified if we want to extend the concept to other fields. For example, field variable in place of displacement, characteristic matrix in place of stiffness matrix and element resultants in place of element strains.

1.3 METHODS OF PRESCRIBING BOUNDARY CONDITIONS

There are two methods of prescribing boundary conditions.

1.3.1 Elimination Method

This method is useful while performing hand calculations and poses difficulties in implementing in software. This method has been used in this book for solving the problems by finite element method using hand calculations. This method results in reduced sizes of matrices thus making it suitable for hand calculations. The method is explained below in brief. Consider following set of global equations,

$$\begin{bmatrix} k_{11} & k_{12} & k_{13} & k_{14} \\ k_{21} & k_{22} & k_{23} & k_{24} \\ k_{31} & k_{32} & k_{33} & k_{34} \\ k_{41} & k_{42} & k_{43} & k_{44} \end{bmatrix} \begin{Bmatrix} u_1 \\ u_2 \\ u_3 \\ u_4 \end{Bmatrix} = \begin{Bmatrix} P_1 \\ P_2 \\ P_3 \\ P_4 \end{Bmatrix}$$

Let u_3 be prescribed *i.e.*, $u_3 = s$. This condition is imposed as follows:

(*i*) Eliminate the row corresponding to u_3 (3rd row).

(*ii*) Transfer the column corresponding to u_3 (3rd column) to right hand side after multiplying it by 's'. These steps result in the following set of modified equations,

$$\begin{bmatrix} k_{11} & k_{12} & k_{14} \\ k_{21} & k_{22} & k_{24} \\ k_{41} & k_{42} & k_{44} \end{bmatrix} \begin{Bmatrix} u_1 \\ u_2 \\ u_4 \end{Bmatrix} = \begin{Bmatrix} P_1 \\ P_2 \\ P_4 \end{Bmatrix} - s \begin{Bmatrix} k_{13} \\ k_{23} \\ k_{43} \end{Bmatrix}$$

This set of equation now may be solved for non trivial solution.

1.3.2 Penalty Method

This is the method used in most of the commercial software because of the reason that this method facilitates prescribing boundary conditions without changing the sizes of the matrices involved. This makes implementation easier. This method is not used in this book.

1.4 PRACTICAL APPLICATIONS OF FINITE ELEMENT ANALYSIS

- Analysis of new design
- Optimization projects
- Failure analysis

1.5 FINITE ELEMENT ANALYSIS SOFTWARE PACKAGE

There are three main steps involved in solving an engineering problem using any commercial software:

(*i*) Pre processing
(*ii*) Processing (or solving)
(*iii*) Post processing

STEP (*i*). *Pre processing*

In this step, CAD model of the system (component) is prepared and is meshed (discretized). Boundary conditions (support conditions and loads) are applied to the meshed model.

STEP (*ii*). *Processing*

In this step software internally calculates element stiffness matrices, element load vectors, global stiffness matrix, global load vector and solves after applying boundary conditions for primary unknowns (*e.g.*, displacements/temperatures/....), and secondary unknowns (*e.g.*, stress/strain/heat flux/....).

STEP (*iii*). *Post processing*

Post processing involves sorting and plotting the output to make the interpretation of results easier.

Chapter 2

FINITE ELEMENT ANALYSIS OF AXIALLY LOADED MEMBERS

2.1 INTRODUCTION

Bar Element is used in the analysis of rod-like axially loaded members. There are two popular bar elements.

2-node bar element

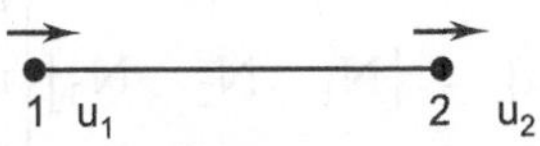

Figure 2.1

This element has two end nodes and each node has one degree of freedom namely translation along its length. Its formulation is based on linear interpolation. It gives accurate results only if loads are applied at nodes and area is constant over the element. However required accuracy for practical purposes in other cases can be obtained by taking more number of smaller elements. The interpolation equation, element stiffness matrix, strain-displacement matrix, element strain and element stress for 2-node (linear) bar element are given by,

$$\{u\} = \begin{bmatrix} N_1 & N_2 \end{bmatrix} \begin{bmatrix} u_1 \\ u_2 \end{bmatrix}$$

$$= \begin{bmatrix} \dfrac{(x_2 - x)}{L} & \dfrac{(x - x_1)}{L} \end{bmatrix}$$

$$[k] = \frac{AE}{L} \begin{bmatrix} 1 & -1 \\ -1 & 1 \end{bmatrix}$$

$$[B] = \frac{1}{L}[-1 \quad 1]$$

$$\{\varepsilon\} = [B]\{q\}$$
$$\{\sigma\} = E[B]\{q\}$$

Uniformly distributed load per unit length w, acting on the element can be converted into equivalent loads using,

$$\{W\} = wL\begin{Bmatrix} \frac{1}{2} \\ \frac{1}{2} \end{Bmatrix}$$

Thermal loads due to change in temperature ΔT can be converted into equivalent nodal loads using $\{Q\} = EA\alpha(\Delta T)\begin{Bmatrix} -1 \\ 1 \end{Bmatrix}$

3-node bar element

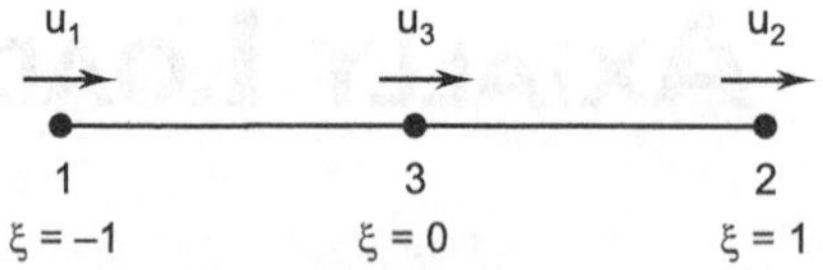

Figure 2.2

This element has a mid-side node in addition to two end nodes and each node has one degree of freedom namely translation along its length. Its formulation is based on quadratic interpolation and this element gives accurate results even with distributed loads and linearly varying cross-sectional area. Coarse mesh with fewer of these elements can give the desired accuracy as compared to fine mesh of 2-node bar element. The interpolation equation, element stiffness matrix, strain-displacement matrix, element strain and element stress for quadratic bar element are given by,

$$[u] = \begin{bmatrix} N_1 & N_2 & N_3 \end{bmatrix}\begin{Bmatrix} u_1 \\ u_2 \\ u_3 \end{Bmatrix} = \begin{bmatrix} -\frac{\xi}{2}+\frac{\xi^2}{2} & \frac{\xi}{2}+\frac{\xi^2}{2} & 1-\xi^2 \end{bmatrix}$$

$$[k] = \frac{AE}{3L}\begin{bmatrix} 7 & 1 & -8 \\ 1 & 7 & -8 \\ -8 & -8 & 16 \end{bmatrix}$$

$$[B] = \frac{2}{L}\begin{bmatrix} -\frac{1-2\xi}{2}, & \frac{1+2\xi}{2}, & -2\xi \end{bmatrix}$$

$$\{\varepsilon\} = [B]\{q\}$$
$$\{\sigma\} = E[B]\{q\}$$

Uniformly distributed load per unit length w, acting on the element can be converted into equivalent loads using,

$$\{W\} = w\frac{L}{6}\begin{Bmatrix} 1 \\ 1 \\ 4 \end{Bmatrix}$$

2.2 BARS OF CONSTANT CROSS-SECTION AREA

PROBLEM 2.1. *Consider a 2 m long steel bar of 50 mm^2 cross-sectional area as shown in figure 2.3. Use two element mesh to model this problem. Find nodal displacements, element stresses and reaction. Take Young's modulus, E = 2 × 10^5 N/mm^2, P = 100 N.*

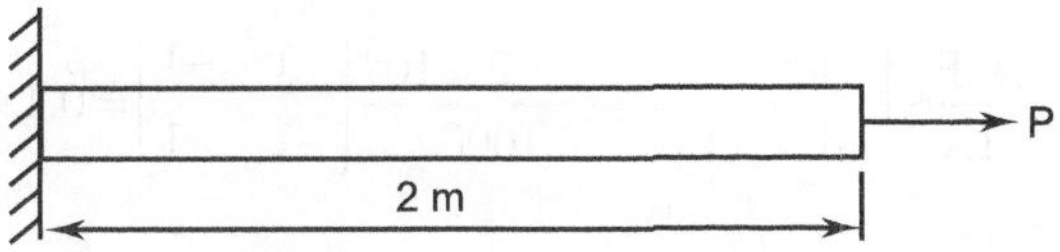

Figure 2.3. Bar with tip load

SOLUTION. (I) Analytical method. [Refer figure 2.3(*a*)]

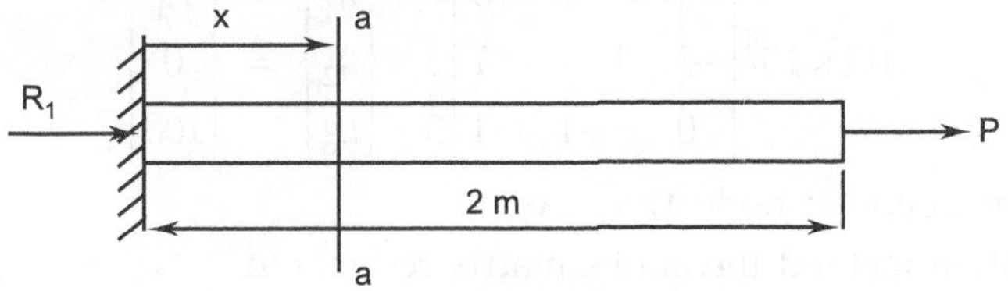

Figure 2.3(a)

Displacement calculation

$$\text{Displacement at section } a\text{-}a\ \delta = \frac{Px}{AE} = \frac{100 \times x}{50 \times 2 \times 10^5} = 1 \times 10^{-5}x$$

$$\text{Displacement at node 2}\quad \delta_{x=1000} = 1 \times 10^{-5} \times 1000 = 0.01 \text{ mm}$$

$$\text{Displacement at node 3}\quad \delta_{x=2000} = 1 \times 10^{-5} \times 2000 = 0.02 \text{ mm}$$

Stress calculation

$$\text{Maximum stress in the bar} = \frac{P}{A} = \frac{100}{50} = 2 \text{ N/mm}^2 \text{ (Constant)}$$

Reaction calculation

For reaction calculation, $\Sigma F_x = 0$

$$R_1 + 100 = 0$$

$$\therefore \quad R_1 = -100 \text{ N (direction is leftwards)}$$

(II) FEM by hand calculations

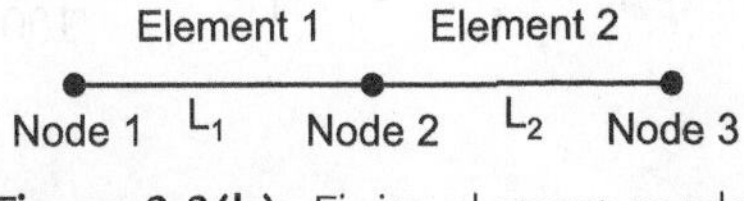

Figure 2.3(b). Finite element model

$$L_1 = L_2 = 1000 \text{ mm}$$

$$A = A_1 = A_2 = 50 \text{ mm}^2$$

$$E = E_1 = E_2 = 2 \times 10^5 \text{ N/mm}^2$$

Stiffness matrix for element 1 is,

$$[k_1] = \frac{A_1E_1}{L_1}\begin{bmatrix} 1 & -1 \\ -1 & 1 \end{bmatrix} = \frac{50 \times 2 \times 10^5}{1000}\begin{bmatrix} 1 & -1 \\ -1 & 1 \end{bmatrix} = 0.1 \times 10^5 \begin{array}{c} \begin{array}{cc} 1 & 2 \end{array} \\ \left[\begin{array}{cc} 1 & -1 \\ -1 & 1 \end{array}\right]\begin{array}{c} 1 \\ 2 \end{array} \end{array}$$

$$[k_2] = \frac{A_2E_2}{L_2}\begin{bmatrix} 1 & -1 \\ -1 & 1 \end{bmatrix} = \frac{50 \times 2 \times 10^5}{1000}\begin{bmatrix} 1 & -1 \\ -1 & 1 \end{bmatrix} = 0.1 \times 10^5 \begin{array}{c} \begin{array}{cc} 2 & 3 \end{array} \\ \left[\begin{array}{cc} 1 & -1 \\ -1 & 1 \end{array}\right]\begin{array}{c} 2 \\ 3 \end{array} \end{array}$$

Global equation is,

$$[K]\,\{r\} = \{R\}$$

$$0.1 \times 10^5 \begin{array}{c} \begin{array}{ccc} 1 & 2 & 3 \end{array} \\ \left[\begin{array}{ccc} \cancel{1} & \cancel{-1} & \cancel{0} \\ \cancel{-1} & 1+1 & -1 \\ \cancel{0} & -1 & 1 \end{array}\right]\begin{array}{c} 1 \\ 2 \\ 3 \end{array} \end{array} \begin{Bmatrix} \cancel{u_1} \\ u_2 \\ u_3 \end{Bmatrix} = \begin{Bmatrix} \cancel{R_1} \\ 0 \\ 100 \end{Bmatrix} \qquad ...(i)$$

Boundary conditions are, at node 1, $u_1 = 0$

By using elimination method the above matrix reduces to,

$$0.1 \times 10^5 \begin{bmatrix} 2 & -1 \\ -1 & 1 \end{bmatrix} \begin{Bmatrix} u_2 \\ u_3 \end{Bmatrix} = \begin{Bmatrix} 0 \\ 100 \end{Bmatrix}$$

By matrix multiplication, we get

$$0.1 \times 10^5(2 \times u_2 - u_3) = 0 \qquad ...(ii)$$

$$0.1 \times 10^5(-u_2 + u_3) = 100 \qquad ...(iii)$$

By solving equations (*ii*) and (*iii*),

we get

$$u_2 = 0.01 \text{ mm}$$

$$u_3 = 0.02 \text{ mm}$$

Stress (σ) calculation

Stress for element 1 is,

$$\{\sigma_1\} = \frac{E}{L_1}[-1 \quad 1]\begin{Bmatrix} u_1 \\ u_2 \end{Bmatrix} = \frac{2 \times 10^5}{1000}[-1 \quad 1]\begin{Bmatrix} 0 \\ 0.01 \end{Bmatrix} = 2 \text{ N/mm}^2$$

Stress for element 2 is,

$$\{\sigma_2\} = \frac{E}{L_2}[-1 \quad 1]\begin{Bmatrix} u_2 \\ u_3 \end{Bmatrix} = \frac{2 \times 10^5}{1000}[-1 \quad 1]\begin{Bmatrix} 0.01 \\ 0.02 \end{Bmatrix} = 2 \text{ N/mm}^2$$

Reaction calculation

From equation (*i*)

$$0.1 \times 10^5(u_1 - u_2) = R_1$$

$$0.1 \times 10^5(0 - 0.01) = R_1$$

$$R_1 = -100 \text{ N}$$

(III) Software results

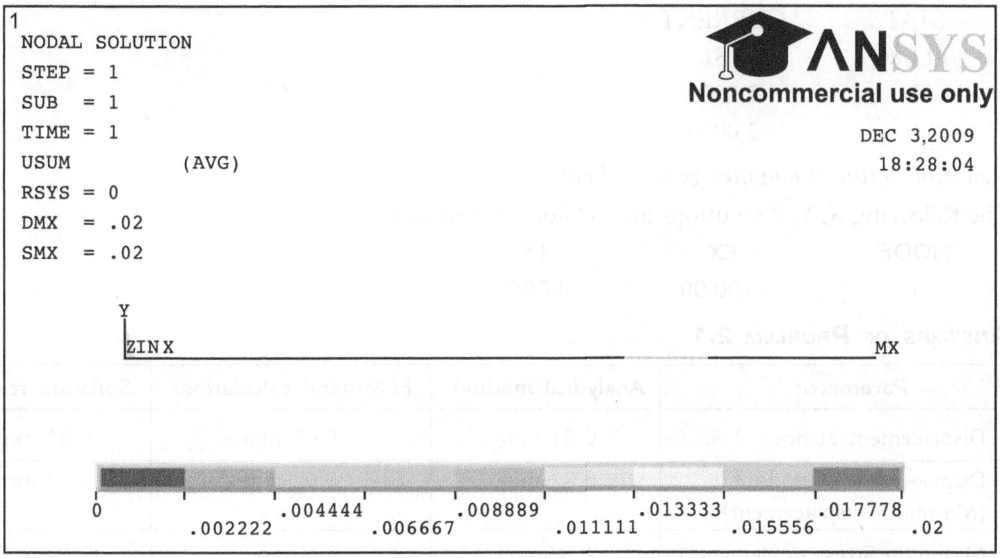

Figure 2.3(c). Deflection pattern for a bar (Refer to appendix for colour figures)

Deflection values at nodes *(Computer generated output)*

The following degree of freedom results are in Global coordinates

NODE	UX	UY	UZ	USUM
1	0.0000	0.0000	0.0000	0.0000
2	0.10000E-01	0.0000	0.0000	0.10000E-01
3	0.20000E-01	0.0000	0.0000	0.20000E-01

Maximum absolute values

NODE	3	0	0	3
VALUE	0.20000E-01	0.0000	0.0000	0.20000E-01

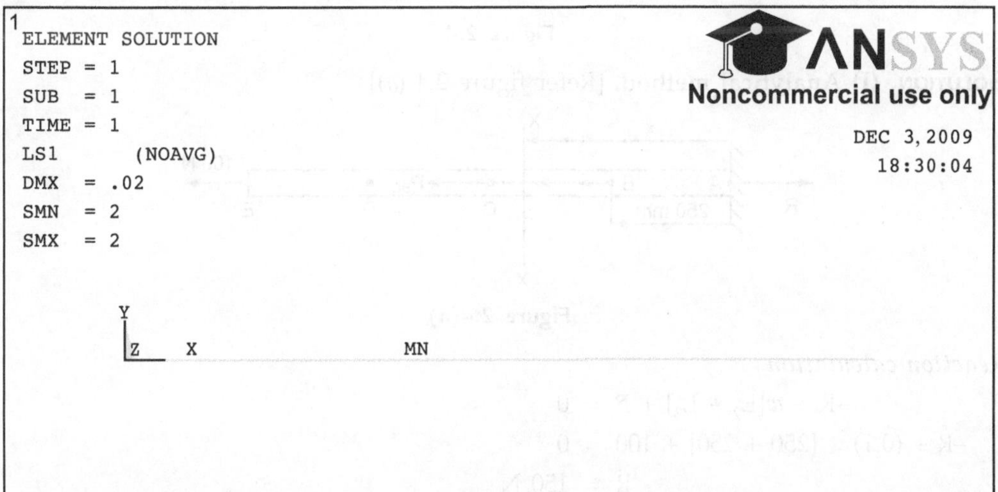

Figure 2.3(d). Stress pattern for a bar (Refer to appendix for colour figures)

Stress values at elements (Computer generated output)

STAT	CURRENT
ELEM	LS1
1	2.0000
2	2.0000

Reaction value (Computer generated output)

The following X, Y, Z solutions are in Global coordinates

NODE	FX	FY
1	-100.00	0.0000

ANSWERS OF PROBLEM 2.1

Parameter	Analytical method	FEM-hand calculations	Software results
Displacement at node 2	0.01 mm	0.01 mm	0.01 mm
Displacement at node 3 (Maximum displacement)	0.02 mm	0.02 mm	0.02 mm
Maximum Stress in element 1	2 N/mm^2	2 N/mm^2	2 N/mm^2
Maximum Stress in element 2	2 N/mm^2	2 N/mm^2	2 N/mm^2
Reaction at fixed end	–100 N	–100 N	–100 N

PROBLEM 2.2. *Bar under distributed and concentrated forces. Consider the bar shown in figure 2.4 subjected to loading as shown below. Use four element mesh model and find nodal displacements, element stresses and reaction at fixed end. Take* $E = 2 \times 10^5$ *N/mm*^{2}*,* $A = 50$ *mm*^{2}*,* $P = 100$ *N.*

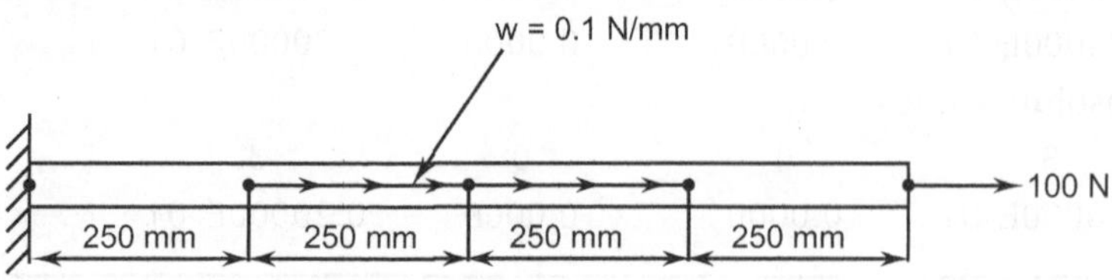

Figure 2.4

SOLUTION. (I) Analytical method. [Refer figure 2.4 (*a*)]

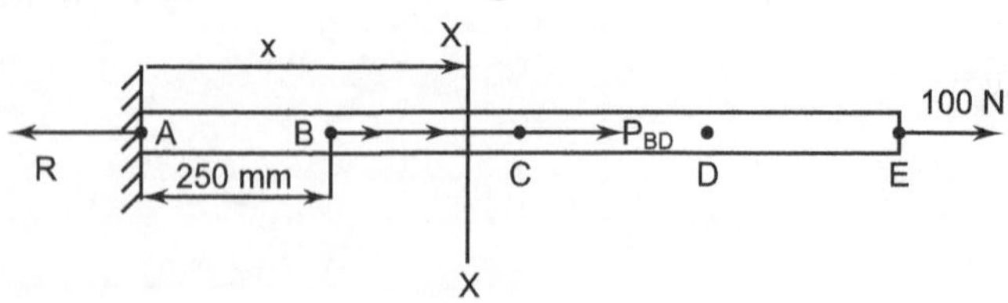

Figure 2.4(a)

Reaction calculation

$$-R + w[L_2 + L_3] + P = 0$$

$$-R + (0.1) \times [250 + 250] + 100 = 0$$

$$R = 150 \text{ N}$$

Stress calculation

$$\sigma_{AB} = \frac{R}{A} = \frac{150}{50} = 3\ \text{N/mm}^2$$

$$\sigma_{DE} = \frac{P}{A} = \frac{100}{50} = 2\ \text{N/mm}^2$$

To find σ_{BD}, consider section XX

$$\sigma_{XX} = \frac{P_{BD}}{A_{BD}} = \frac{150-(x-250)\times 0.1}{50} = \frac{150-0.1x+25}{50} = \frac{175-0.1x}{50}$$

$$\sigma_{XX}\Big|_{\text{at B}} = \sigma_{XX}\Big|_{x=250} = \frac{175-0.1\times 250}{50} = 3\ \text{N/mm}^2$$

$$\sigma_{XX}\Big|_{\text{at C}} = \sigma_{XX}\Big|_{x=500} = \frac{175-0.1\times 500}{50} = 2.5\ \text{N/mm}^2$$

$$\sigma_{XX}\Big|_{\text{at D}} = \sigma_{XX}\Big|_{x=750} = \frac{175-0.1\times 750}{50} = 2\ \text{N/mm}^2$$

Displacement calculation: Displacement at E,

$$\delta_E = \Delta_{AB} + \Delta_{BD} + \Delta_{DE}$$

$$\delta_E = \frac{\sigma_{AB}\,L_{AB}}{E} + \int_{x=250}^{750} \frac{(175-0.1\,x)\,dx}{AE} + \left(\frac{\sigma_{DE}\,L_{DE}}{E}\right)$$

$$\delta_E = \frac{3\times 250}{2\times 10^5} + \frac{1}{50\times 2\times 10^5}\left[175x - 0.1\frac{x^2}{2}\right]_{250}^{750} + \frac{2\times 250}{2\times 10^5}$$

$$= 0.00375 + 0.00625 + 0.0025 = 0.0125\ \text{mm}$$

Displacement at B,

$$\delta_B = \Delta_{AB} = \frac{3\times 250}{2\times 10^5} = 0.00375\ \text{mm}$$

Displacement at D,

$$\delta_D = \Delta_{AB} + \Delta_{BD} = 0.00375 + 0.00625 = 0.01\ \text{mm}$$

Displacement at C,

$$\delta_C = \Delta_{AB} + \int_{x=250}^{500} \frac{(175-0.1x)\,dx}{AE} = 0.00375 + \frac{1}{50\times 2\times 10^5}\left[175\,x - \frac{0.1x^2}{2}\right]_{250}^{500}$$

$$= 0.0072\ \text{mm}$$

(II) FEM by hand calculations

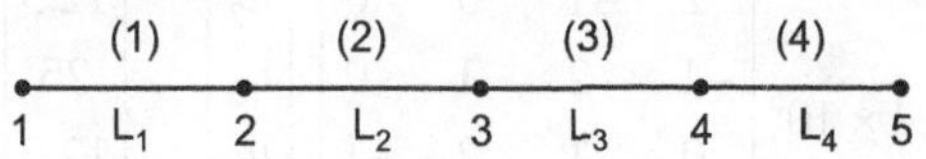

Figure 2.4(b). Finite element model

$$L_1 = L_2 = L_3 = L_4 = 250\ \text{mm}$$

$$A = A_1 = A_2 = A_3 = A_4 = 50\ \text{mm}^2$$

$$E = E_1 = E_2 = E_3 = E_4 = 2\times 10^5\ \text{N/mm}^2$$

Stiffness matrix for elements is,

$$[k_1] = \frac{A_1E_1}{L_1}\begin{bmatrix} 1 & -1 \\ -1 & 1 \end{bmatrix} = \frac{50 \times 2 \times 10^5}{250}\begin{bmatrix} 1 & -1 \\ -1 & 1 \end{bmatrix} = 0.4 \times 10^5 \begin{array}{c} \begin{matrix} 1 & \;\;2 \end{matrix} \\ \begin{bmatrix} 1 & -1 \\ -1 & 1 \end{bmatrix} \end{array}\begin{matrix} 1 \\ 2 \end{matrix}$$

$$[k_2] = \frac{A_2E_2}{L_2}\begin{bmatrix} 1 & -1 \\ -1 & 1 \end{bmatrix} = \frac{50 \times 2 \times 10^5}{250}\begin{bmatrix} 1 & -1 \\ -1 & 1 \end{bmatrix} = 0.4 \times 10^5 \begin{array}{c} \begin{matrix} 2 & \;\;3 \end{matrix} \\ \begin{bmatrix} 1 & -1 \\ -1 & 1 \end{bmatrix} \end{array}\begin{matrix} 2 \\ 3 \end{matrix}$$

$$[k_3] = \frac{A_3E_3}{L_3}\begin{bmatrix} 1 & -1 \\ -1 & 1 \end{bmatrix} = \frac{50 \times 2 \times 10^5}{250}\begin{bmatrix} 1 & -1 \\ -1 & 1 \end{bmatrix} = 0.4 \times 10^5 \begin{array}{c} \begin{matrix} 3 & \;\;4 \end{matrix} \\ \begin{bmatrix} 1 & -1 \\ -1 & 1 \end{bmatrix} \end{array}\begin{matrix} 3 \\ 4 \end{matrix}$$

$$[k_4] = \frac{A_4E_4}{L_4}\begin{bmatrix} 1 & -1 \\ -1 & 1 \end{bmatrix} = \frac{50 \times 2 \times 10^5}{250}\begin{bmatrix} 1 & -1 \\ -1 & 1 \end{bmatrix} = 0.4 \times 10^5 \begin{array}{c} \begin{matrix} 4 & \;\;5 \end{matrix} \\ \begin{bmatrix} 1 & -1 \\ -1 & 1 \end{bmatrix} \end{array}\begin{matrix} 4 \\ 5 \end{matrix}$$

Nodal load calculation for element 2 and 3,

$$W_1 = \begin{Bmatrix} \dfrac{wL_2}{2} \\ \dfrac{wL_2}{2} \end{Bmatrix} = \begin{Bmatrix} \dfrac{0.1 \times 250}{2} \\ \dfrac{0.1 \times 250}{2} \end{Bmatrix} = \begin{Bmatrix} 12.5 \\ 12.5 \end{Bmatrix}\begin{matrix} 2 \\ 3 \end{matrix}$$

$$W_2 = \begin{Bmatrix} \dfrac{wL_3}{2} \\ \dfrac{wL_3}{2} \end{Bmatrix} = \begin{Bmatrix} \dfrac{0.1 \times 250}{2} \\ \dfrac{0.1 \times 250}{2} \end{Bmatrix} = \begin{Bmatrix} 12.5 \\ 12.5 \end{Bmatrix}\begin{matrix} 3 \\ 4 \end{matrix}$$

Global equation is,

$$[K]\{r\} = \{R\}$$

$$0.4 \times 10^5 \begin{array}{c} \begin{matrix} 1 & \quad 2 & \quad 3 & \quad 4 & \quad 5 \end{matrix} \\ \begin{bmatrix} 1 & -1 & 0 & 0 & 0 \\ -1 & 1+1 & -1 & 0 & 0 \\ 0 & -1 & 1+1 & -1 & 0 \\ 0 & 0 & -1 & 1+1 & -1 \\ 0 & 0 & 0 & -1 & 1 \end{bmatrix} \end{array}\begin{matrix} 1 \\ 2 \\ 3 \\ 4 \\ 5 \end{matrix} \begin{Bmatrix} u_1 \\ u_2 \\ u_3 \\ u_4 \\ u_5 \end{Bmatrix} = \begin{Bmatrix} R_1 \\ 12.5 \\ 12.5+12.5 \\ 12.5 \\ 100 \end{Bmatrix} \quad ...(i)$$

Boundary conditions are, at node 1, $u_1 = 0$

By using elimination method the above matrix reduces to,

$$0.4 \times 10^5 \begin{bmatrix} 2 & -1 & 0 & 0 \\ -1 & 2 & -1 & 0 \\ 0 & -1 & 2 & -1 \\ 0 & 0 & -1 & 1 \end{bmatrix}\begin{Bmatrix} u_2 \\ u_3 \\ u_4 \\ u_5 \end{Bmatrix} = \begin{Bmatrix} 12.5 \\ 25 \\ 12.5 \\ 100 \end{Bmatrix}$$

By solving the above matrix and equations, we get

$$u_2 = 0.0038 \text{ mm}$$

$$u_3 = 0.0072 \text{ mm}$$
$$u_4 = 0.01 \text{ mm}$$
$$u_5 = 0.0125 \text{ mm}$$

Stress (σ) calculation

$$\{\sigma_1\} = \frac{E}{L_1}[-1 \quad 1]\begin{Bmatrix} u_1 \\ u_2 \end{Bmatrix} = \frac{2 \times 10^5}{250}[-1 \quad 1]\begin{Bmatrix} 0 \\ 0.0038 \end{Bmatrix} = 3.04 \text{ N/mm}^2$$

$$\{\sigma_2\} = \frac{E}{L_2}[-1 \quad 1]\begin{Bmatrix} u_2 \\ u_3 \end{Bmatrix} = \frac{2 \times 10^5}{250}[-1 \quad 1]\begin{Bmatrix} 0.0038 \\ 0.0072 \end{Bmatrix} = 2.72 \text{ N/mm}^2$$

$$\{\sigma_3\} = \frac{E}{L_3}[-1 \quad 1]\begin{Bmatrix} u_3 \\ u_4 \end{Bmatrix} = \frac{2 \times 10^5}{250}[-1 \quad 1]\begin{Bmatrix} 0.0072 \\ 0.01 \end{Bmatrix} = 2.24 \text{ N/mm}^2$$

$$\{\sigma_4\} = \frac{E}{L_4}[-1 \quad 1]\begin{Bmatrix} u_4 \\ u_5 \end{Bmatrix} = \frac{2 \times 10^5}{250}[-1 \quad 1]\begin{Bmatrix} 0.01 \\ 0.0125 \end{Bmatrix} = 2 \text{ N/mm}^2$$

Reaction calculation: From equation (*i*)

$$0.4 \times 10^5(u_1 - u_2) = R_1$$
$$0.4 \times 10^5(0 - 0.0038) = R_1$$
$$R_1 = -152 \text{ N}$$

(III) Software results

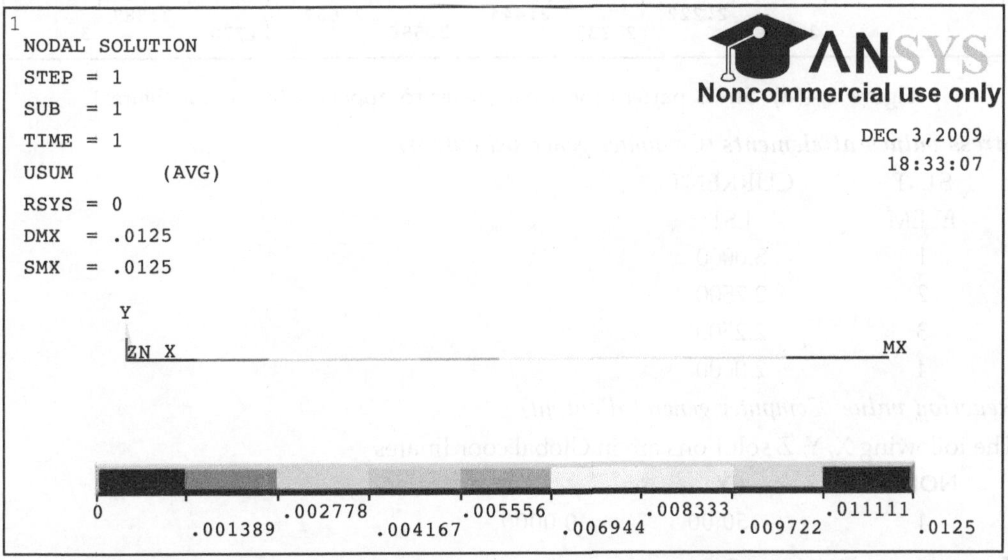

Figure 2.4(c). Deflection pattern for a bar (Refer to appendix for colour figures)

Deflection values at nodes (Computer generated output)

The following Degree of Freedom results are in Global coordinates

NODE	UX	UY	UZ	USUM
1	0.0000	0.0000	0.0000	0.0000
2	0.37500E-02	0.0000	0.0000	0.37500E-02

3	0.71875E-02	0.0000	0.0000	0.71875E-02
4	0.10000E-01	0.0000	0.0000	0.10000E-01
5	0.12500E-01	0.0000	0.0000	0.12500E-01

Maximum absolute values

NODE	5	0	0	5
VALUE	0.12500E-01	0.0000	0.0000	0.12500E-01

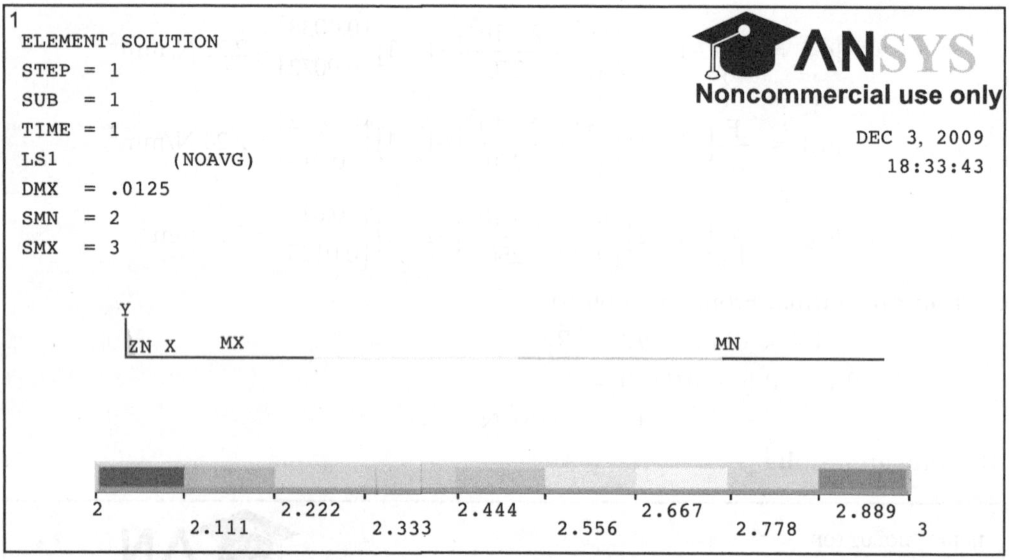

Figure 2.4(d). Stress pattern for a bar (Refer to appendix for colour figures)

Stress values at elements *(Computer generated output)*

STAT ELEM	CURRENT LS1
1	3.0000
2	2.7500
3	2.2500
4	2.0000

Reaction value *(Computer generated output)*

The following X, Y, Z solutions are in Global coordinates

NODE	FX	FY
1	-150.00	0.0000

Answers of Problem 2.2

Parameter	Analytical method	FEM-hand calculations	Software results
Displacement at node 2	0.00375 mm	0.0038 mm	0.00375 mm
Displacement at node 3	0.0072 mm	0.0072 mm	0.00719 mm
Displacement at node 4	0.01 mm	0.01 mm	0.01 mm
Displacement at node 5	0.0125 mm	0.0125 mm	0.0125 mm

Stress in element 1	3 N/mm^2	3.04 N/mm^2	3 N/mm^2
Stress in element 2	3 N/mm^2 to 2.5 N/mm^2	2.72 N/mm^2	2.75 N/mm^2
Stress in element 3	2.5 N/mm^2 to 2 N/mm^2	2.24 N/mm^2	2.25 N/mm^2
Stress in element 4	2 N/mm^2	2 N/mm^2	2 N/mm^2
Reaction at fixed end	-150 N	-152 N	-150 N

PROBLEM 2.3. *A load P = 80 kN is applied as shown in figure 2.5. Determine the nodal displacements, element stresses and support reactions in the bar. Take E = 20 × 10^3 N/mm^2.*

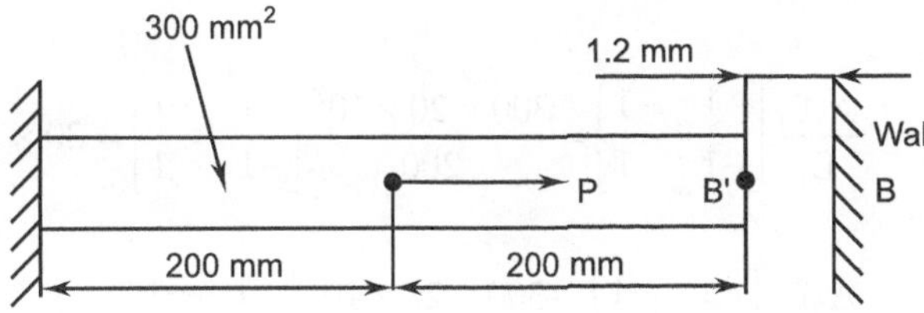

Figure 2.5

SOLUTION. (I) Analytical method. [Refer figure 2.5(*a*)]

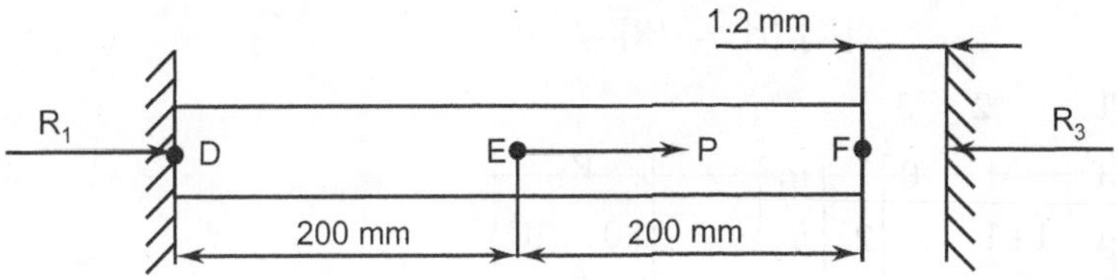

Figure 2.5(a)

Let R_3 be the reaction developed at the wall after contact.

$$\frac{P_{DE}L}{AE} + \frac{P_{EF}L}{AE} = 1.2$$

$$\frac{R_1 \times 200}{300 \times 20 \times 10^3} + \frac{(-R_3) \times 200}{300 \times 20 \times 10^3} = 1.2 \qquad ...(i)$$

$$\Sigma F_x = 0 \Rightarrow R_1 + R_3 = P = 80 \times 10^3 \qquad ...(ii)$$

Solving equations (*i*) and (*ii*)

$$R_1 = 58018 \text{ N}$$

$$R_3 = 21982 \text{ N}$$

Stresses are,
$$\sigma_{DE} = \frac{R_1}{A} = \frac{58018}{300} = 193.39 \text{ N/mm}^2$$

$$\sigma_{EF} = -\frac{R_3}{A} = -\frac{21982}{300} = -73.27 \text{ N/mm}^2$$

Deflections are,
$$\delta_2 = \delta_E = \Delta_{DE} = \frac{\sigma_{DE}L}{E} = \frac{193.39 \times 200}{20 \times 10^3} = 1.934 \text{ mm}$$

$$\delta_3 = 1.2 \text{ mm}$$

(II) FEM by hand calculations

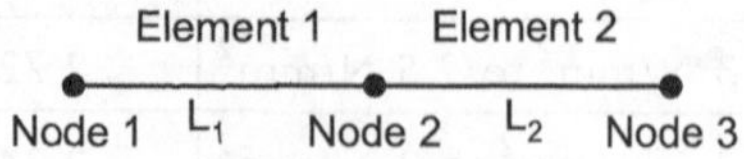

Figure 2.5(b). Finite element model

L_1 = 200 mm

L_2 = 200 mm

First we should check whether contact occurs between the bar and the wall. For this assume that wall does not exist. Then solution to the problem is as below. (Consider two element model)

Stiffness matrices are,

$$[k_1] = \frac{A_1E_1}{L_1}\begin{bmatrix} 1 & -1 \\ -1 & 1 \end{bmatrix} = \frac{300 \times 20 \times 10^3}{200}\begin{bmatrix} 1 & -1 \\ -1 & 1 \end{bmatrix} = 30 \times 10^3 \begin{bmatrix} 1 & -1 \\ -1 & 1 \end{bmatrix}\begin{matrix} 1 \\ 2 \end{matrix} \quad (\text{columns } 1, 2)$$

$$[k_2] = \frac{A_2E_2}{L_2}\begin{bmatrix} 1 & -1 \\ -1 & 1 \end{bmatrix} = \frac{300 \times 20 \times 10^3}{200}\begin{bmatrix} 1 & -1 \\ -1 & 1 \end{bmatrix} = 30 \times 10^3 \begin{bmatrix} 1 & -1 \\ -1 & 1 \end{bmatrix}\begin{matrix} 2 \\ 3 \end{matrix} \quad (\text{columns } 2, 3)$$

Global equation is,

$$[K]\{r\} = \{R\}$$

$$30 \times 10^3 \begin{bmatrix} 1 & -1 & 0 \\ -1 & 1+1 & -1 \\ 0 & -1 & 1 \end{bmatrix}\begin{matrix} 1 \\ 2 \\ 3 \end{matrix} \begin{Bmatrix} u_1 \\ u_2 \\ u_3 \end{Bmatrix} = \begin{Bmatrix} R_1 \\ 80 \times 10^3 \\ 0 \end{Bmatrix} \quad (\text{columns } 1, 2, 3)$$

Boundary conditions are, at node 1, $u_1 = 0$

By using elimination method the above matrix reduces to,

$$30 \times 10^3 \begin{bmatrix} 2 & -1 \\ -1 & 1 \end{bmatrix}\begin{Bmatrix} u_2 \\ u_3 \end{Bmatrix} = \begin{Bmatrix} 80 \times 10^3 \\ 0 \end{Bmatrix}$$

By matrix multiplication, we get

$$30 \times 10^3(2 \times u_2 - u_3) = 80 \times 10^3 \quad ...(iii)$$

$$30 \times 10^3(-u_2 + u_3) = 0 \quad ...(iv)$$

By solving equations (*iii*) and (*iv*),

we get u_2 = 2.67 mm and u_3 = 2.67 mm

Since displacement at node 3 is 2.67 mm (greater than 1.2 mm) we can say that contact does occur. The problem has to be resolved, since the boundary conditions are now different. The displacement at B′ is specified to be 1.2 mm as shown in figure 2.5.

Global element equation is,

$$[K]\{r\} = \{R\}$$

$$30 \times 10^3 \begin{bmatrix} 1 & -1 & 0 \\ -1 & 1+1 & -1 \\ 0 & -1 & 1 \end{bmatrix} \begin{Bmatrix} u_1 \\ u_2 \\ u_3 \end{Bmatrix} = \begin{Bmatrix} R_1 \\ 80 \times 10^3 \\ 0 \end{Bmatrix} \quad ...(v)$$

(Rows and columns of the matrix are labelled 1, 2, 3; row 1 and row 3 and column 1 are struck out.)

Boundary conditions are, at node 1, $u_1 = 0$ and at node 3, $u_3 = 1.2$

By using elimination method the above matrix reduces to,

$$30 \times 10^3[2]\{u_2\} = \{80 \times 10^3\} - 1.2[30 \times 10^3 \times -1]$$
$$30 \times 10^3 \times 2 \times u_2 = 80 \times 10^3 + 36 \times 10^3$$
$$u_2 = 1.933 \text{ mm.}$$

Stress* (σ) *calculation: Stress for element 1 is,

$$\{\sigma_1\} = \frac{E}{L_1}[-1 \quad 1]\begin{Bmatrix} u_1 \\ u_2 \end{Bmatrix} = \frac{20 \times 10^3}{200}[-1 \quad 1]\begin{Bmatrix} 0 \\ 1.933 \end{Bmatrix} = 193.3 \text{ N/mm}^2$$

Stress for element 2 is,

$$\{\sigma_2\} = \frac{E}{L_2}[-1 \quad 1]\begin{Bmatrix} u_2 \\ u_3 \end{Bmatrix} = \frac{20 \times 10^3}{200}[-1 \quad 1]\begin{Bmatrix} 1.933 \\ 1.2 \end{Bmatrix} = -73.3 \text{ N/mm}^2$$

Reaction calculation: From equation (*v*)

$$30 \times 10^3(u_1 - u_2) = R_1$$
$$30 \times 10^3(0 - 1.933) = R_1$$
$$R_1 = -57990 \text{ N (direction is leftwards)}$$

We know that, $R_1 + P + R_3 = 0$

$$-57990 + 80 \times 10^3 + R_3 = 0$$
$$R_3 = -22010 \text{ N (direction is leftwards)}$$

(III) Software results

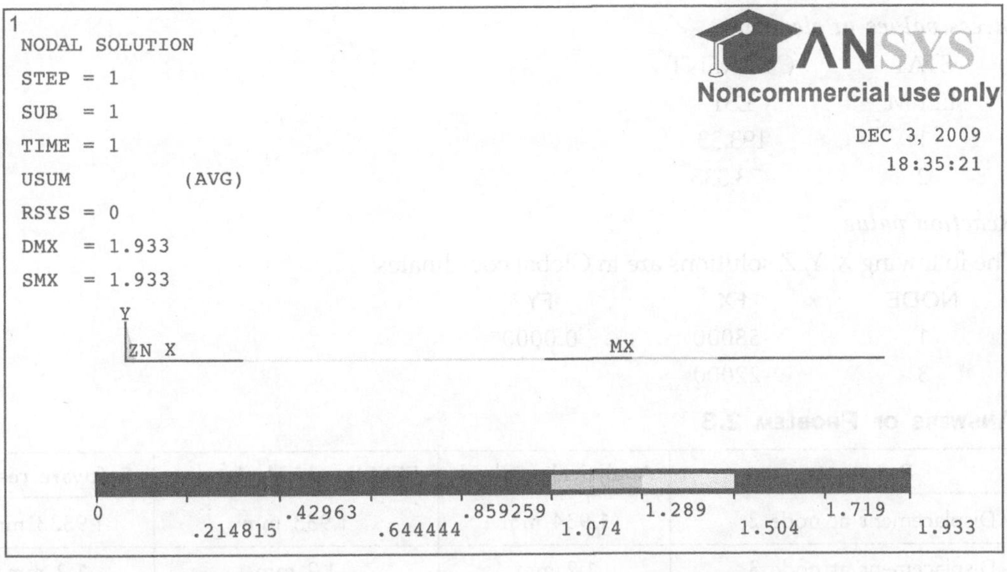

Figure 2.5(c). Deflection pattern for a bar (Refer to appendix for colour figures)

Deflection values at nodes

The following Degree of Freedom results are in Global coordinates

NODE	UX	UY	UZ	USUM
1	0.0000	0.0000	0.0000	0.0000
2	1.9333	0.0000	0.0000	1.9333
3	1.2000	0.0000	0.0000	1.2000

Maximum absolute values

NODE	2	0	0	2
VALUE	1.9333	0.0000	0.0000	1.9333

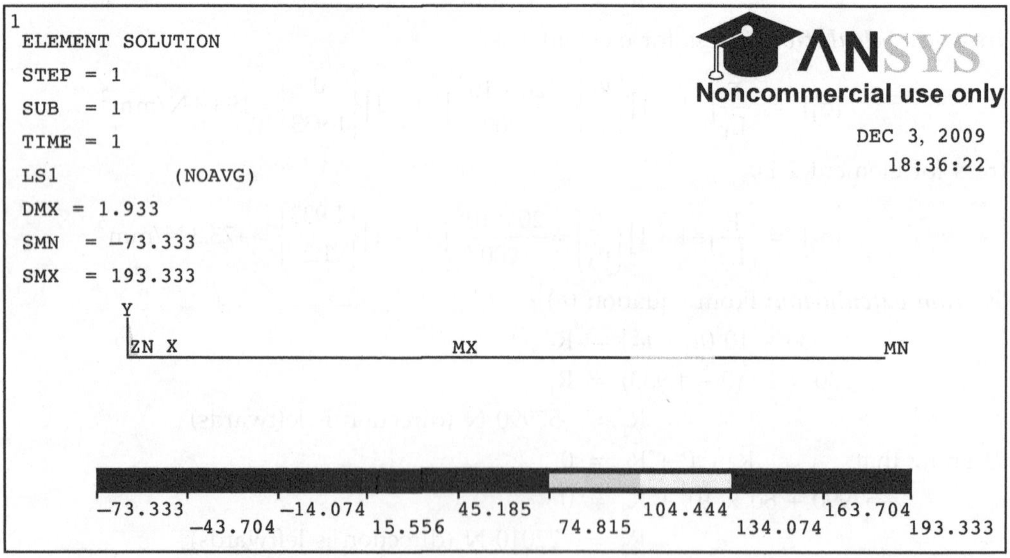

Figure 2.5(d). Stress pattern for a bar (Refer to appendix for colour figures)

Stress values at elements

STAT	CURRENT
ELEM	LS1
1	193.33
2	–73.333

Reaction value

The following X, Y, Z solutions are in Global coordinates

NODE	FX	FY
1	-58000	0.0000
3	-22000	

ANSWERS OF PROBLEM 2.3

Parameter	Analytical method	FEM-hand calculations	Software results
Displacement at node 2	1.934 mm	1.933 mm	1.9333 mm
Displacement at node 3	1.2 mm	1.2 mm	1.2 mm

Stress in element 1	193.39 N/mm^2	193.3 N/mm^2	193.33 N/mm^2
Stress in element 2	−73.27 N/mm^2	−73.3 N/mm^2	−73.333 N/mm^2
Reaction at fixed end	58.02 kN	57.94 kN	−58 kN
Reaction at wall	−21.98 kN	−22.01 kN	−22 kN

PROBLEM 2.4. *A bar subjected to self weight. Determine the nodal displacements for the bar hanging under its own weight as shown in figure 2.6. Use two equal length elements. Let* $E = 2 \times 10^{11}$ *N/m*^{2}*, mass density* $\rho = 7800$ *kg/m*^{3}*, Area* $A = 1000$ *mm*^{2}*. Consider length of rod* $L = 2$ *m.*

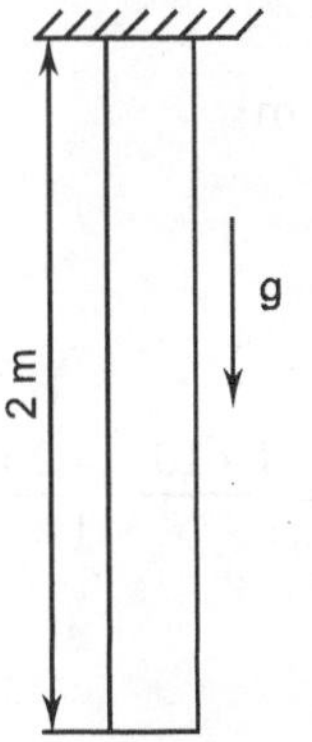

Figure 2.6. Bar under self weight

SOLUTION. (I) Analytical method. [Refer figure 2.6(*a*)]

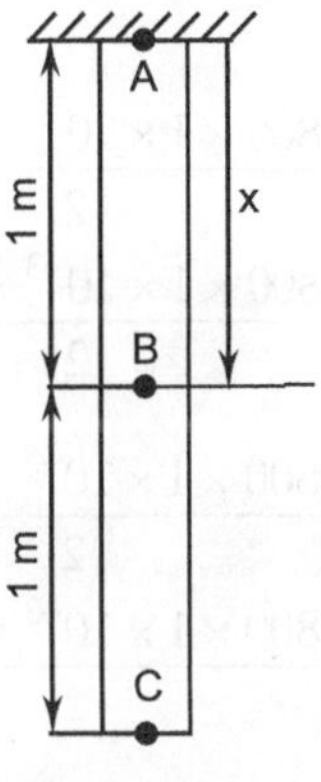

Figure 2.6(a)

$$\delta_3 = \delta_c = \frac{\rho g L^2}{2E} = \frac{7800 \times 9.81 \times (2)^2}{2 \times 2 \times 10^{11}} = 7.6518 \times 10^{-7} \text{m}$$

$$\delta_2 = \delta_B = \int_1^2 \frac{x A \rho g}{AE}\, dx = \frac{\rho g}{E}\left[\frac{x^2}{2}\right]_1^2 = \frac{7800 \times 9.81}{2 \times 10^{11}}\left[\frac{(2)^2}{2} - \frac{(1)^2}{2}\right] = 5.7389 \times 10^{-7} \text{m}$$

(II) FEM by hand calculations

Figure 2.6(b). Finite element model

$$L_1 = L_2 = \frac{L}{2} = \frac{2}{2} = 1 \text{ m}$$

The element stiffness matrices are,

For element 1,

$$[k_1] = \frac{AE}{L_1}\begin{bmatrix} 1 & -1 \\ -1 & 1 \end{bmatrix} = \frac{1 \times 10^{-3} \times 2 \times 10^{11}}{1}\begin{bmatrix} 1 & -1 \\ -1 & 1 \end{bmatrix} = 2 \times 10^{8} \begin{array}{c} \begin{matrix} 1 & 2 \end{matrix} \\ \begin{bmatrix} 1 & -1 \\ -1 & 1 \end{bmatrix} \end{array} \begin{matrix} 1 \\ 2 \end{matrix}$$

For element 2,

$$[k_2] = \frac{AE}{L_2}\begin{bmatrix} 1 & -1 \\ -1 & 1 \end{bmatrix} = \frac{1 \times 10^{-3} \times 2 \times 10^{11}}{1}\begin{bmatrix} 1 & -1 \\ -1 & 1 \end{bmatrix} = 2 \times 10^{8} \begin{array}{c} \begin{matrix} 2 & 3 \end{matrix} \\ \begin{bmatrix} 1 & -1 \\ -1 & 1 \end{bmatrix} \end{array} \begin{matrix} 2 \\ 3 \end{matrix}$$

Nodal load vector due to weight is,

$$F_1 = \begin{Bmatrix} \dfrac{\rho A g L_1}{2} \\ \dfrac{\rho A g L_1}{2} \end{Bmatrix} = \begin{Bmatrix} \dfrac{7800 \times 1 \times 10^{-3} \times 9.81 \times 1}{2} \\ \dfrac{7800 \times 1 \times 10^{-3} \times 9.81 \times 1}{2} \end{Bmatrix} = \begin{Bmatrix} 38.26 \\ 38.26 \end{Bmatrix} \begin{matrix} 1 \\ 2 \end{matrix}$$

$$F_2 = \begin{Bmatrix} \dfrac{\rho A g L_2}{2} \\ \dfrac{\rho A g L_2}{2} \end{Bmatrix} = \begin{Bmatrix} \dfrac{7800 \times 1 \times 10^{-3} \times 9.81 \times 1}{2} \\ \dfrac{7800 \times 1 \times 10^{-3} \times 9.81 \times 1}{2} \end{Bmatrix} = \begin{Bmatrix} 38.26 \\ 38.26 \end{Bmatrix} \begin{matrix} 2 \\ 3 \end{matrix}$$

Global equation is,

$$[K]\ \{r\} = \{R\}$$

$$2 \times 10^{8} \begin{array}{c} \begin{matrix} 1 & 2 & 3 \end{matrix} \\ \begin{bmatrix} 1 & -1 & 0 \\ -1 & 1+1 & -1 \\ 0 & -1 & 1 \end{bmatrix} \end{array} \begin{matrix} 1 \\ 2 \\ 3 \end{matrix} \begin{Bmatrix} u_1 \\ u_2 \\ u_3 \end{Bmatrix} = \begin{Bmatrix} 38.26 \\ 76.52 \\ 38.26 \end{Bmatrix} \qquad ...(i)$$

Boundary conditions are, at node 1, $u_1 = 0$

By using elimination method the above matrix reduces to,

$$2 \times 10^8 \begin{bmatrix} 2 & -1 \\ -1 & 1 \end{bmatrix} \begin{Bmatrix} u_2 \\ u_3 \end{Bmatrix} = \begin{Bmatrix} 76.52 \\ 38.26 \end{Bmatrix}$$

By matrix multiplication, we get

$$2 \times 10^8 (2 \times u_2 - u_3) = 76.52 \quad ...(ii)$$

$$2 \times 10^8 (-u_2 + u_3) = 38.26 \quad ...(iii)$$

By solving equations (*ii*) and (*iii*),

we get

$$u_2 = 5.739 \times 10^{-7} \text{ m}$$

$$u_3 = 7.652 \times 10^{-7} \text{ m}$$

(III) Software results

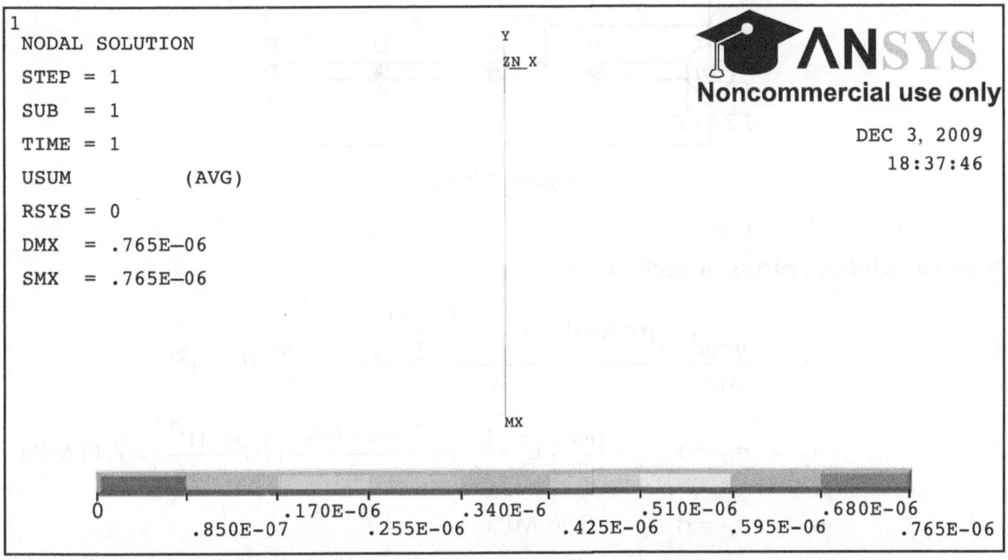

Figure 2.6(c). Deflection pattern for a bar (Refer to appendix for colour figures)

Deflection values at nodes

The following degree of freedom results are in Global coordinates

NODE	UX	UY	UZ	USUM
1	0.0000	0.0000	0.0000	0.0000
2	0.0000	0.57389E-06	0.0000	0.57389E-06
3	0.0000	0.76518E-06	0.0000	0.76518E-06

Maximum absolute values

NODE	0	3	0	3
VALUE	0.0000	0.76518E-06	0.0000	0.76518E-06

ANSWERS OF PROBLEM 2.4

Parameter	Analytical method	FEM-hand calculations	Software results
Displacement at node 2	5.7389×10^{-7} m	5.739×10^{-7} m	5.7389×10^{-7} m
Displacement at node 3	7.6518×10^{-7} m	7.652×10^{-7} m	7.6518×10^{-7} m

PROBLEM 2.5. *A rod rotating at constant angular velocity ω = 45 rad/sec is shown in figure 2.7. Determine the nodal displacements and stresses in the rod. Consider only the centrifugal force. Ignore bending of the rod. Use two quadratic elements. Take A = 350 mm^2, E = 70 GPa, Mass density ρ = 7850 kg/m^3, Length of the rod L = 1 m.*

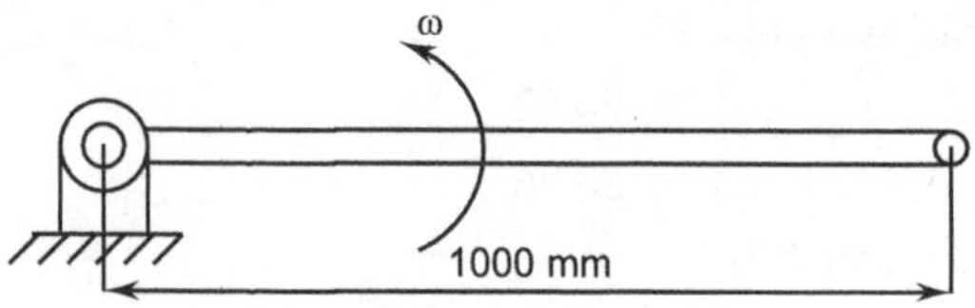

Figure 2.7

SOLUTION. (I) Analytical method [Refer figure 2.7(*a*)]

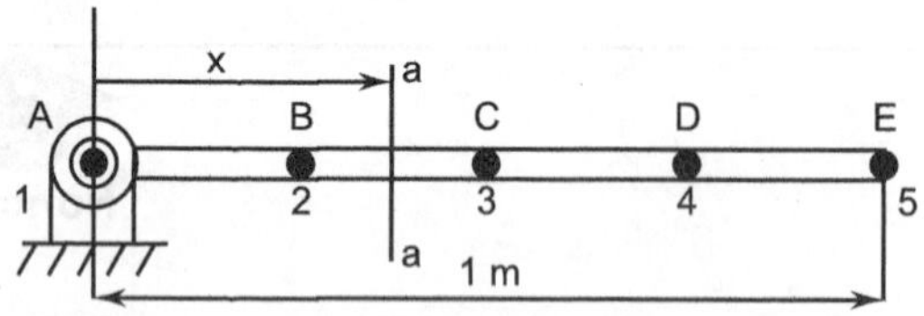

Figure 2.7(a)

$$L = 1 \text{ m}$$

Stress calculation: Stress at section *a-a*

$$\sigma = \frac{mr\omega^2}{\text{Area}} = \frac{\rho \times A \times (L-x)\left[x + \frac{(L-x)}{2}\right]\omega^2}{A} = \frac{\rho\omega^2}{2}[L^2 - x^2]$$

$$\sigma_2 = \sigma_B = \sigma_{x=\frac{L}{4}} = \frac{\rho\omega^2}{2}\left[L^2 - \frac{L^2}{16}\right] = \frac{7850 \times (45)^2}{2}\left[(1)^2 - \frac{(1)^2}{16}\right] = 7.45 \text{ MPa}$$

$$\sigma_3 = \sigma_C = \sigma_{x=\frac{L}{2}} = 5.96 \text{ MPa}$$

$$\sigma_4 = \sigma_D = \sigma_{x=\frac{3L}{4}} = 3.48 \text{ MPa}$$

$$\sigma_5 = \sigma_E = \sigma_{x=L} = 0$$

$$\sigma_1 = \sigma_A = \sigma_{x=0} = 7.95 \text{ MPa}$$

Displacement at section *a-a* = change in length of *x*

$$\Delta x = \int_0^x \frac{P_x}{AE}dx = \int_0^x \frac{A\rho\omega^2(L^2 - x^2)}{2AE}dx = \frac{\rho\omega^2}{2E}\left[L^2 x - \frac{x^3}{3}\right]_0^x$$

$$\delta_1 = \delta_A = \Delta x_{x=0} = 0$$

$$\delta_2 = \delta_B = \Delta x\Big|_{x=\frac{L}{4}} = \frac{\rho\omega^2}{2E}\left[L^2 x - \frac{x^3}{3}\right]_0^{L/4} = \frac{7850 \times (45)^2}{2 \times 70 \times 10^9}\left[L^2 \times \frac{L}{4} - \frac{L^3}{64 \times 3}\right]$$

$$= \frac{7850 \times (45)^2}{2 \times 70 \times 10^9}\left[\frac{(1)^3}{4} - \frac{(1)^3}{192}\right] = 2.78 \times 10^{-5} \text{ m} = 0.0278 \text{ mm}$$

$$\delta_3 = \delta_C = \Delta x\Big|_{x=\frac{L}{2}} = 0.052 \text{ mm}$$

$$\delta_4 = \delta_D = \Delta x\Big|_{x=\frac{3L}{4}} = 0.069 \text{ mm}$$

$$\delta_5 = \delta_E = \Delta x\big|_{x=L} = 0.076 \text{ mm}$$

Reaction calculation

$$\sum F_x = 0$$

$$R_1 + AL\rho\frac{L}{2}\omega^2 = 0$$

$$\therefore \quad R_1 = -\frac{AL^2\rho\omega^2}{2} = -\frac{3.5\times10^{-4}\times(1)^2\times7850\times(45)^2}{2} = -2781.84 \text{ N}$$

(II) FEM by hand calculations

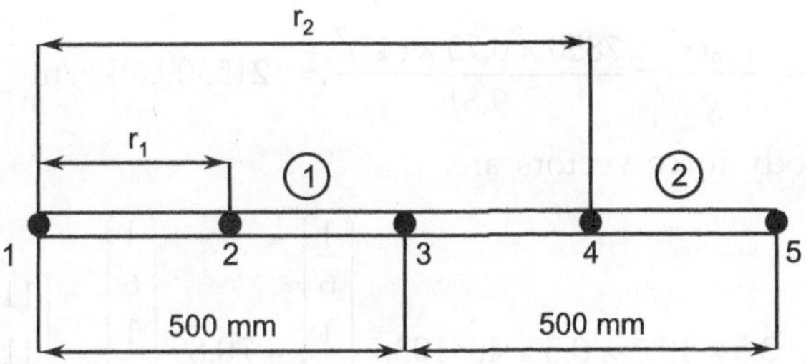

Figure 2.7(b). Finite element model (with 2 quadratic elements)

A finite element model of the rod, with two quadratic elements, is shown in figure 2.7(*b*). The element stiffness matrices are,

$$L = L_1 = L_2 = 0.5 \text{ m}$$

$$[k_1] = \frac{A_1E_1}{3L_1}\begin{bmatrix} 7 & 1 & -8 \\ 1 & 7 & -8 \\ -8 & -8 & 16 \end{bmatrix} = \frac{35\times10^{-4}\times70\times10^{9}}{3\times0.5}\begin{bmatrix} 7 & 1 & -8 \\ 1 & 7 & -8 \\ -8 & -8 & 16 \end{bmatrix}$$

$$= 163.33\times10^5 \begin{array}{c} \begin{matrix} 1 & 3 & 2 \end{matrix} \\ \begin{bmatrix} 7 & 1 & -8 \\ 1 & 7 & -8 \\ -8 & -8 & 16 \end{bmatrix} \end{array} \begin{matrix} 1 \\ 3 \\ 2 \end{matrix}$$

$$[k_2] = 163.33\times10^5 \begin{array}{c} \begin{matrix} 3 & 5 & 4 \end{matrix} \\ \begin{bmatrix} 7 & 1 & -8 \\ 1 & 7 & -8 \\ -8 & -8 & 16 \end{bmatrix} \end{array} \begin{matrix} 3 \\ 5 \\ 4 \end{matrix}$$

Thus, Global stiffness matrix is,

$$[K] = 163.33 \times 10^5 \begin{bmatrix} 7 & -8 & 1 & 0 & 0 \\ -8 & 16 & -8 & 0 & 0 \\ 1 & -8 & 14 & -8 & 1 \\ 0 & 0 & -8 & 16 & -8 \\ 0 & 0 & 1 & -8 & 7 \end{bmatrix} \begin{matrix} 1 \\ 2 \\ 3 \\ 4 \\ 5 \end{matrix} \quad (\text{columns } 1\ 2\ 3\ 4\ 5)$$

The centrifugal force or body force F_c (kg/m^3) is given by,

$$F_c = \frac{\rho r \omega^2}{g}$$

Note that F is a function of the distance r from the pin. Taking average values of F over each element, we have,

$$F_1 = \frac{\rho r_1 \omega^2}{g} = \frac{7850 \times 0.25 \times (45)^2}{9.81} = 405103.2 \text{ kg/m}^3$$

$$F_2 = \frac{\rho r_2 \omega^2}{g} = \frac{7850 \times 0.75 \times (45)^2}{9.81} = 1215309.6 \text{ kg/m}^3$$

Thus, the element body force vectors are,

$$f_1 = A \times L_1 \times F_1 \begin{Bmatrix} \frac{1}{6} \\ \frac{1}{6} \\ \frac{2}{3} \end{Bmatrix} = 3.5 \times 10^{-4} \times 0.5 \times 405103.2 \begin{Bmatrix} \frac{1}{6} \\ \frac{1}{6} \\ \frac{2}{3} \end{Bmatrix} = 70.89 \begin{Bmatrix} \frac{1}{6} \\ \frac{1}{6} \\ \frac{2}{3} \end{Bmatrix} = \begin{Bmatrix} 11.815 \\ 11.815 \\ 47.26 \end{Bmatrix} \begin{matrix} 1 \\ 3 \\ 2 \end{matrix} \leftarrow \text{Global dof}$$

$$f_2 = A \times L_2 \times F_2 \begin{Bmatrix} \frac{1}{6} \\ \frac{1}{6} \\ \frac{2}{3} \end{Bmatrix} = 3.5 \times 10^{-4} \times 0.5 \times 1215309.6 \begin{Bmatrix} \frac{1}{6} \\ \frac{1}{6} \\ \frac{2}{3} \end{Bmatrix} = 212.68 \begin{Bmatrix} \frac{1}{6} \\ \frac{1}{6} \\ \frac{2}{3} \end{Bmatrix} = \begin{Bmatrix} 35.45 \\ 35.45 \\ 141.79 \end{Bmatrix} \begin{matrix} 3 \\ 5 \\ 4 \end{matrix} \leftarrow \text{Global dof}$$

Assembling f_1 and f_2, we obtain,

$$F = \begin{Bmatrix} 11.815 \\ 47.26 \\ 47.26 \\ 141.79 \\ 35.45 \end{Bmatrix} \times 9.81 = \begin{Bmatrix} 115.91 \\ 463.62 \\ 463.62 \\ 1390.96 \\ 347.76 \end{Bmatrix} \text{N}$$

Global equation is,

$$[K]\{r\} = \{R\}$$

$$163.33 \times 10^5 \begin{bmatrix} 7 & -8 & 1 & 0 & 0 \\ -8 & 16 & -8 & 0 & 0 \\ 1 & -8 & 14 & -8 & 1 \\ 0 & 0 & -8 & 16 & -8 \\ 0 & 0 & 1 & -8 & 7 \end{bmatrix} \begin{Bmatrix} u_1 \\ u_2 \\ u_3 \\ u_4 \\ u_5 \end{Bmatrix} = \begin{Bmatrix} 115.91 \\ 463.62 \\ 463.62 \\ 1390.96 \\ 347.76 \end{Bmatrix} \quad ...(i)$$

(Columns and rows numbered 1 2 3 4 5; first row and column struck out.)

Boundary conditions are, at node 1, $u_1 = 0$

By using elimination method the above matrix reduces to,

$$163.33 \times 10^5 \begin{bmatrix} 16 & -8 & 0 & 0 \\ -8 & 14 & -8 & 1 \\ 0 & -8 & 16 & -8 \\ 0 & 1 & -8 & 7 \end{bmatrix} \begin{Bmatrix} u_2 \\ u_3 \\ u_4 \\ u_5 \end{Bmatrix} = \begin{Bmatrix} 463.62 \\ 463.62 \\ 1390.96 \\ 347.76 \end{Bmatrix}$$

By solving the above matrix and equations, we get

$$u_2 = 2.661 \times 10^{-5} \text{ mm} = 0.0266 \text{ mm}$$
$$u_3 = 0.0497 \text{ mm}$$
$$u_4 = 0.0657 \text{ mm}$$
$$u_5 = 0.0709 \text{ mm}$$

The stress at node 1 in element 1 is given by,

$$\sigma_{11} = \frac{2E}{L_1}[-1.5 \quad -0.5 \quad 2]\begin{Bmatrix} u_1 \\ u_3 \\ u_2 \end{Bmatrix} = \frac{2 \times 70 \times 10^3}{500}[-1.5 \quad -0.5 \quad 2]\begin{Bmatrix} 0 \\ 0.0497 \\ 0.0266 \end{Bmatrix} = 7.924 \text{ MPa}$$

The stress at node 2 in element 1 is given by,

$$\sigma_{12} = \frac{2E}{L_1}[-0.5 \quad 0.5 \quad 0]\begin{Bmatrix} u_1 \\ u_3 \\ u_2 \end{Bmatrix} = \frac{2 \times 70 \times 10^3}{500}[-0.5 \quad 0.5 \quad 0]\begin{Bmatrix} 0 \\ 0.0497 \\ 0.0266 \end{Bmatrix} = 6.972 \text{ MPa}$$

The stress at node 3 in element 1 is given by,

$$\sigma_{13} = \frac{2E}{L_1}[0.5 \quad 1.5 \quad -2]\begin{Bmatrix} u_1 \\ u_3 \\ u_2 \end{Bmatrix} = \frac{2 \times 70 \times 10^3}{500}[0.5 \quad 1.5 \quad -2]\begin{Bmatrix} 0 \\ 0.0497 \\ 0.0266 \end{Bmatrix} = 5.992 \text{ MPa}$$

The stress at node 1 in element 2 is given by,

$$\sigma_{21} = \frac{2E}{L_2}[-1.5 \quad -0.5 \quad 2]\begin{Bmatrix} u_3 \\ u_5 \\ u_4 \end{Bmatrix} = \frac{2 \times 70 \times 10^3}{500}[-1.5 \quad -0.5 \quad 2]\begin{Bmatrix} 0.0497 \\ 0.0709 \\ 0.0657 \end{Bmatrix} = 5.992 \text{ MPa}$$

The stress at node 2 in element 2 is given by,

$$\sigma_{22} = \frac{2E}{L_2}[-0.5 \quad 0.5 \quad 0]\begin{Bmatrix} u_3 \\ u_5 \\ u_4 \end{Bmatrix} = \frac{2 \times 70 \times 10^3}{500}[-0.5 \quad 0.5 \quad 0]\begin{Bmatrix} 0.0497 \\ 0.0709 \\ 0.0657 \end{Bmatrix} = 2.968 \text{ MPa}$$

The stress at node 3 in element 2 is given by,

$$\sigma_{23} = \frac{2E}{L_2}[0.5 \quad 1.5 \quad -2]\begin{Bmatrix} u_3 \\ u_5 \\ u_4 \end{Bmatrix} = \frac{2 \times 70 \times 10^3}{500}[0.5 \quad 1.5 \quad -2]\begin{Bmatrix} 0.0497 \\ 0.0709 \\ 0.0657 \end{Bmatrix} = -0.056 \text{ MPa}$$

(III) Software Results

While solving the problem using software, four linear bar elements are taken instead of two quadratic elements.

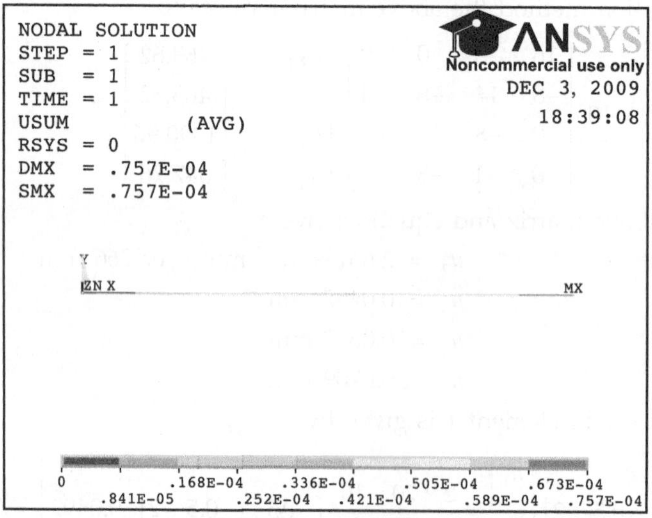

Figure 2.7(c). Deflection pattern for a rod (Refer to appendix for colour figures)

Deflection values at nodes (in m)

The following degree of freedom results are in Global coordinates

NODE	UX	UY	UZ	USUM
1	0.0000	0.0000	0.0000	0.0000
2	0.27795E-04	0.0000	0.0000	0.27795E-04
3	0.52041E-04	0.0000	0.0000	0.52041E-04
4	0.69191E-04	0.0000	0.0000	0.69191E-04
5	0.75696E-04	0.0000	0.0000	0.75696E-04

Maximum absolute values

NODE	5	0	0	5
VALUE	0.75696E-04	0.0000	0.0000	0.75696E-04

Reaction value

The following X,Y,Z solutions are in Global coordinates

NODE	FX	FY
1	-2781.8	-3.3691

ANSWERS OF PROBLEM 2.5

Parameter	Analytical method	FEM – hand calculations	Software results
Displacement at node 2	0.0278 mm	0.0266 mm	0.0278 mm
Displacement at node 3	0.052 mm	0.0497 mm	0.052 mm
Displacement at node 4	0.069 mm	0.0657 mm	0.069 mm
Displacement at node 5	0.076 mm	0.0709 mm	0.076 mm
Stress in node 1 of element 1	7.95 MPa	7.924 MPa	–
Stress in node 2 of element 1	7.45 MPa	6.972 MPa	–
Stress in node 3 of element 1	5.96 MPa	5.992 MPa	–
Stress in node 1 of element 2	5.96 MPa	5.992 MPa	–
Stress in node 2 of element 2	3.48 MPa	2.968 MPa	–
Stress in node 3 of element 2	0 MPa	-0.056 MPa	–
Reaction at fixed end	-2781.84 N	–	-2781.8 N

Procedure for solving the problem using ANSYS® 12.0 academic teaching software

Each problem given in this book is having different procedure for solving using software. For familiarizing, procedure for one problem is given from each chapter using software. Other problems are left to the user to explore the software for solving the problems.

FOR PROBLEM 2.3

PREPROCESSING

1. **Main Menu > Preprocessor > Element Type > Add/Edit/Delete > Add >Structural Link >2D spar 1 > OK>Close**

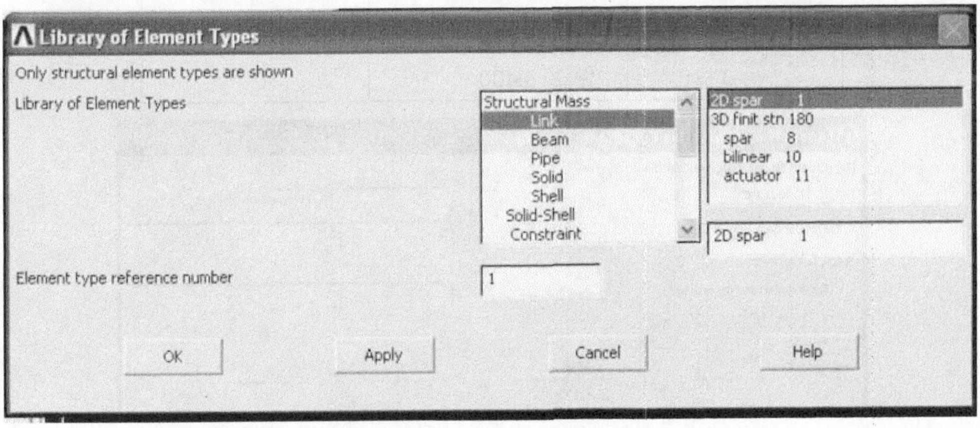

Figure 2.8. Element selection

2. **Main Menu > Preprocessor > Real Constants > Add/Edit/Delete > Add > OK**

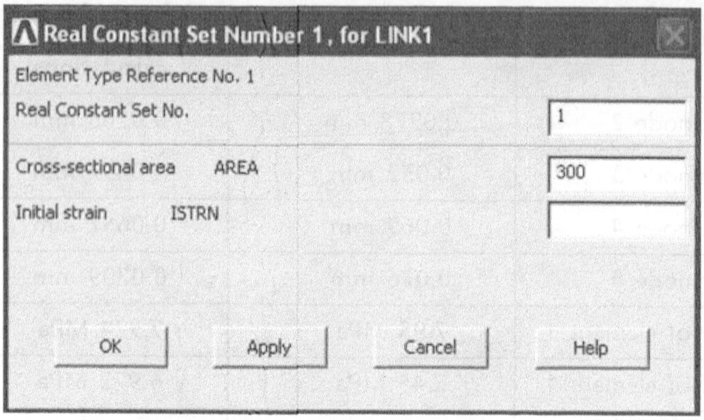

Figure 2.9. Enter the cross-sectional area

Cross-sectional area AREA >**Enter 300 > OK > Close**

Enter the material properties.

3. **Main Menu > Preprocessor > Material Props > Material Models**

Material Model Number 1, click **Structural > Linear > Elastic > Isotropic**

Enter **EX = 2E4 and PRXY = 0.3 > OK**

(**Close** the Define Material Model Behaviour window.)

Create the nodes and elements. Create 3 nodes and 2 elements.

4. **Main Menu > Preprocessor > Modeling > Create > Nodes > In Active CS** Enter the coordinates of node 1 > **Apply** Enter the coordinates of node 2 > **Apply** Enter the coordinates of node 3 > **OK.**

Node locations		
Node number	**X coordinate**	**Y coordinate**
1	0	0
2	200	0
3	400	0

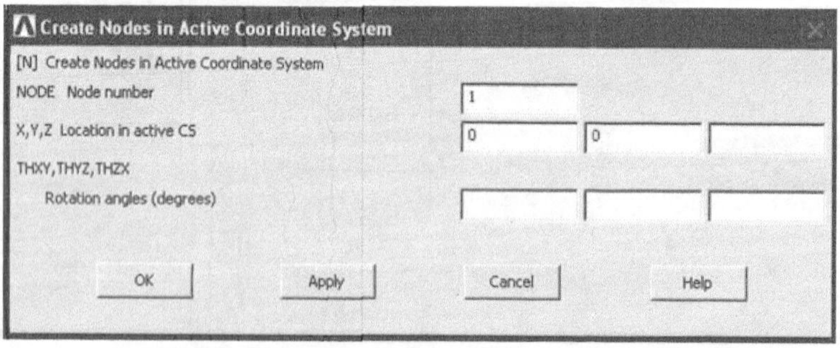

Figure 2.10. Enter the node coordinates

5. **Main Menu > Preprocessor > Modeling > Create > Elements > Auto Numbered > Thru nodes** Pick the 1st and 2nd node > **Apply** Pick 2nd and 3rd node > **OK**

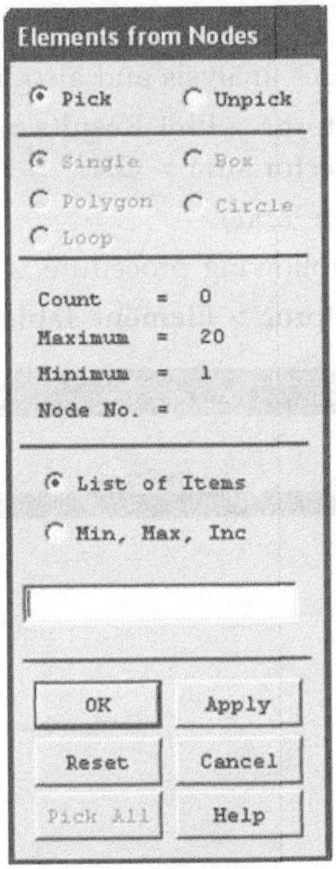

Figure 2.11. Pick the nodes to create elements

Apply the displacement boundary conditions and loads.

6. **Main Menu > Preprocessor > Loads > Define Loads > Apply > Structural >Displacement > On Nodes** Pick the 1st node **> Apply > All DOF = 0 > OK**
7. **Main Menu > Preprocessor > Loads > Define Loads > Apply > Structural >Displacement > On Nodes** Pick the 3rd node **> Apply > Select UX and enter displacement value = 1.2 > OK**
8. **Main Menu > Preprocessor > Loads > Define Loads > Apply > Structural > Force/ Moment > On Nodes** Pick the 2nd node **> OK > Force/Moment value = 80e3 > OK**

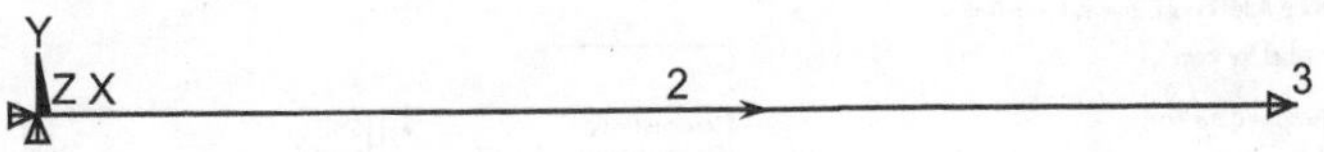

Figure 2.12. Model with loading and displacement boundary conditions

The model-building step is now complete, and we can proceed to the solution. First to be safe, save the model.

SOLUTION.

The interactive solution proceeds.

9. **Main Menu > Solution > Solve > Current LS > OK**

The **/STATUS Command** window displays the problem parameters and **the Solve Current Load Step** window is shown. Check the solution options in the **/STATUS** window and if all is OK, select **File > Close**

In the Solve **Current Load Step** window, Select **OK**, and when the solution is complete, **close** the **'Solution is Done!'** window.

POSTPROCESSING

We can now plot the results of this analysis and also list the computed values.

10. **Main Menu > General Postproc > Plot Results > Contour Plot > Nodal Solu > DOF Solution > Displacement vector sum > OK**

    This result is shown in figure 2.5(*c*).

    To find the axial stress the following procedure is followed.

11. **Main Menu > General Postproc > Element Table > Define Table > Add**

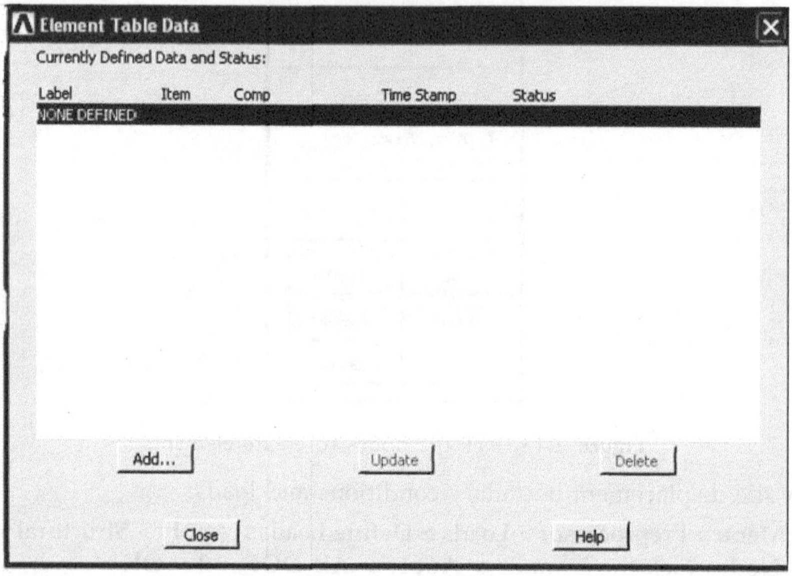

Figure 2.13. Define the element table

Select **By sequence num and LS** and type **1 after LS** as shown in figure 2.14.

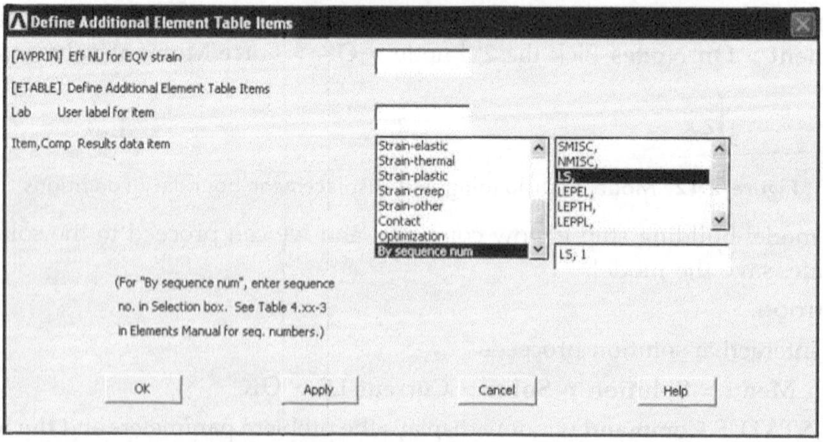

Figure 2.14. Selecting options in element table

> OK>Close

12. **Main Menu > General Postproc > Plot Results > Contour Plot > Elem Table >** Select **LS1 > OK**

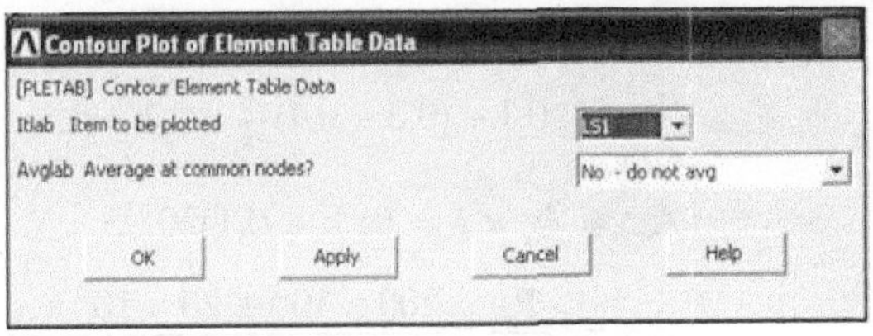

Figure 2.15. Selecting options for finding out axial stress

This result is shown in figure 2.5(*d*).

2.3 BARS OF VARYING CROSS-SECTION AREA

PROBLEM 2.6. *Solve for displacement and stress for the given figure 2.16 using 2 finite elements model. Take Young's modulus E = 200 GPa.*

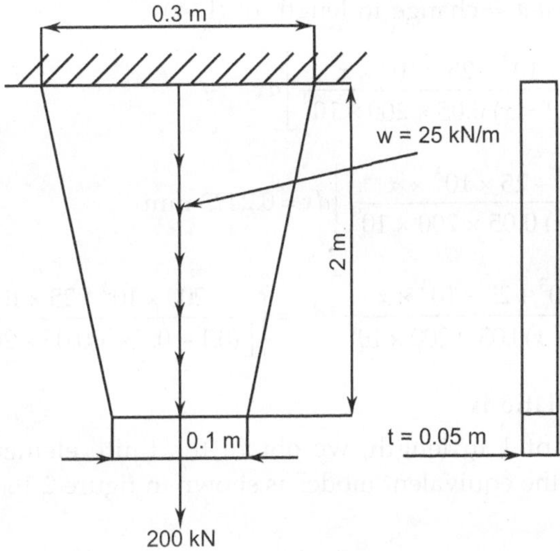

Figure 2.16

SOLUTION. (I) Analytical method. [Refer figure 2.16(*a*)]

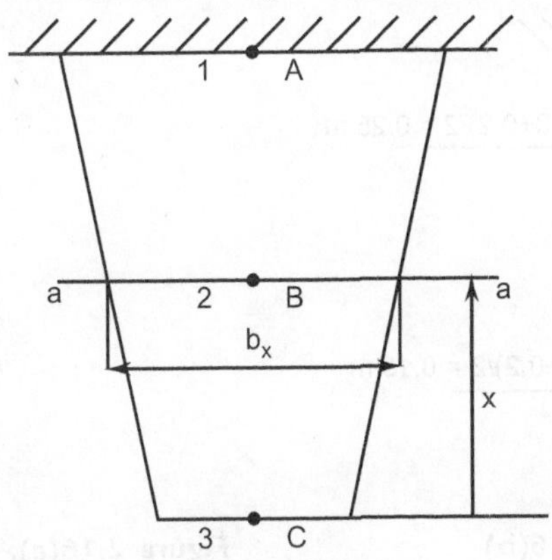

Figure 2.16(a)

L = 2 m.

Stress calculation

$$b_x = 0.1 + (0.3 - 0.1)\frac{x}{2} = 0.1 + 0.1x$$

$$A_x = b_x \times t = (0.1 + 0.1x)0.05$$

$$\sigma_x = \frac{P_x}{A_x} = \frac{200 \times 10^3 + 25 \times 10^3 \times x}{(0.1 + 0.1x)\,0.05}$$

$$\sigma_1 = \sigma_A = \sigma_x\big|_{x=2} = \frac{200 \times 10^3 + 25 \times 10^3 \times 2}{(0.1 + 0.1 \times 2)0.05} = 16.67 \text{ MPa}$$

$$\sigma_2 = \sigma_B = \sigma_x\big|_{x=1} = 22.5 \text{ MPa}$$

$$\sigma_3 = \sigma_C = \sigma_x\big|_{x=0} = 40 \text{ MPa}$$

Displacement calculation

Displacement at section *a-a* = change in length of (L–x)

$$\delta = \int \frac{P_x}{A_x E} = \int_{L-x}^{L} \left[\frac{200 \times 10^3 + 25 \times 10^3 \times x}{(0.1 + 0.1 \times x)\,0.05 \times 200 \times 10^9}\right] dx$$

$$\delta_3 = \delta_C = \int_0^2 \left[\frac{200 \times 10^3 + 25 \times 10^3 \times x}{(0.1 + 0.1 \times x)\,0.05 \times 200 \times 10^9}\right] dx = 0.2423 \text{ mm}$$

$$\delta_2 = \delta_B = \int_{L-1}^{L} \left[\frac{200 \times 10^3 + 25 \times 10^3 \times x}{(0.1 + 0.1 \times x)0.05 \times 200 \times 10^9}\right] dx = \int_1^2 \left[\frac{200 \times 10^3 + 25 \times 10^3 \times x}{(0.1 + 0.1 \times x)0.05 \times 200 \times 10^9}\right] dx = 0.096 \text{ mm}$$

(II) FEM by hand calculations

Using 2 elements each of 1 m length, we obtain the finite element model as shown in figure 2.16(*c*). We can write the equivalent model as shown in figure 2.16(*b*). At the middle of the bar width is,

$$\frac{(0.3 + 0.1)}{2} = 0.2 \text{ m}.$$

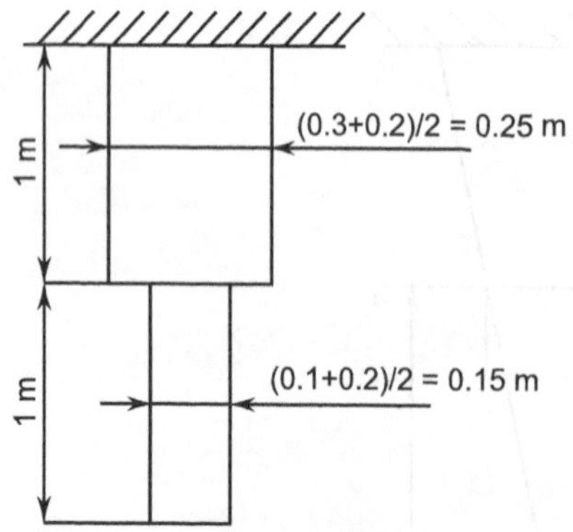

Figure 2.16(b)

Figure 2.16(c). Finite element model

$$A_1 = 0.25 \times 0.05 = 0.0125 \text{ m}^2$$
$$A_2 = 0.15 \times 0.05 = 0.0075 \text{ m}^2$$
$$E_1 = 200 \times 10^9 \text{ N/m}^2 \quad L_1 = 1 \text{ m}$$
$$E_2 = 200 \times 10^9 \text{ N/m}^2 \quad L_2 = 1 \text{ m}$$

Stiffness matrix for element 1 is,

$$[k_1] = \frac{A_1E_1}{L_1}\begin{bmatrix} 1 & -1 \\ -1 & 1 \end{bmatrix} = \frac{0.0125 \times 200 \times 10^9}{1}\begin{bmatrix} 1 & -1 \\ -1 & 1 \end{bmatrix} = 2.5 \times 10^9 \begin{array}{c} \begin{matrix} 1 & 2 \end{matrix} \\ \begin{bmatrix} 1 & -1 \\ -1 & 1 \end{bmatrix} \end{array}\begin{matrix} 1 \\ 2 \end{matrix}$$

Stiffness matrix for element 2 is,

$$[k_2] = \frac{A_2E_2}{L_2}\begin{bmatrix} 1 & -1 \\ -1 & 1 \end{bmatrix} = \frac{0.0075 \times 200 \times 10^9}{1}\begin{bmatrix} 1 & -1 \\ -1 & 1 \end{bmatrix} = 1.5 \times 10^9 \begin{array}{c} \begin{matrix} 2 & 3 \end{matrix} \\ \begin{bmatrix} 1 & -1 \\ -1 & 1 \end{bmatrix} \end{array}\begin{matrix} 2 \\ 3 \end{matrix}$$

Distributed load calculation for elements 1 and 2,

$$W_1 = \begin{Bmatrix} \dfrac{w \times L_1}{2} \\ \dfrac{w \times L_1}{2} \end{Bmatrix} = \begin{Bmatrix} \dfrac{25 \times 1}{2} \\ \dfrac{25 \times 1}{2} \end{Bmatrix} \times 10^3 = \begin{Bmatrix} 12.5 \\ 12.5 \end{Bmatrix}\begin{matrix} 1 \\ 2 \end{matrix} \times 10^3$$

$$W_2 = \begin{Bmatrix} \dfrac{w \times L_2}{2} \\ \dfrac{w \times L_2}{2} \end{Bmatrix} = \begin{Bmatrix} \dfrac{25 \times 1}{2} \\ \dfrac{25 \times 1}{2} \end{Bmatrix} \times 10^3 = \begin{Bmatrix} 12.5 \\ 12.5 \end{Bmatrix}\begin{matrix} 2 \\ 3 \end{matrix} \times 10^3$$

Global equation is,

$$[K]\,\{r\} = \{R\}$$

$$10^9 \begin{array}{c} \begin{matrix} 1 & 2 & 3 \end{matrix} \\ \begin{bmatrix} 2.5 & -2.5 & 0 \\ -2.5 & 2.5+1.5 & -1.5 \\ 0 & -1.5 & 1.5 \end{bmatrix} \end{array}\begin{matrix} 1 \\ 2 \\ 3 \end{matrix}\begin{Bmatrix} u_1 \\ u_2 \\ u_3 \end{Bmatrix} = \begin{Bmatrix} 12.5 + R_1 \\ 25 \\ 12.5 + 200 \end{Bmatrix} \times 10^3 \quad \text{...(i)}$$

Boundary conditions are, at node 1, $u_1 = 0$

By using elimination method the above matrix reduces to,

$$10^9 \begin{bmatrix} 4 & -1.5 \\ -1.5 & 1.5 \end{bmatrix}\begin{Bmatrix} u_2 \\ u_3 \end{Bmatrix} = \begin{Bmatrix} 25 \\ 212.5 \end{Bmatrix} \times 10^3$$

By matrix multiplication, we get

$$10^9(4 \times u_2 - 1.5 \times u_3) = 25 \times 10^3 \quad \text{...(ii)}$$
$$10^9(-1.5 \times u_2 + 1.5 \times u_3) = 212.5 \times 10^3 \quad \text{...(iii)}$$

By solving equations (*ii*) and (*iii*),

we get

$$u_2 = 9.5 \times 10^{-5} \text{ m} = 0.095 \text{ mm}$$
$$u_3 = 2.37 \times 10^{-4} \text{ m} = 0.237 \text{ mm}$$

Stress (σ) calculation: Stress in element 1,

$$\{\sigma_1\} = \frac{E_1}{L_1}[-1 \quad 1]\begin{Bmatrix} u_1 \\ u_2 \end{Bmatrix} = \frac{2 \times 10^5}{1000}[-1 \quad 1]\begin{Bmatrix} 0 \\ 0.095 \end{Bmatrix} = 19 \text{ MPa}$$

Stress in element 2,

$$\{\sigma_2\} = \frac{E_2}{L_2}[-1 \quad 1]\begin{Bmatrix} u_2 \\ u_3 \end{Bmatrix} = \frac{2 \times 10^5}{1000}[-1 \quad 1]\begin{Bmatrix} 0.095 \\ 0.237 \end{Bmatrix} = 28.4 \text{ MPa}$$

(III) Software results

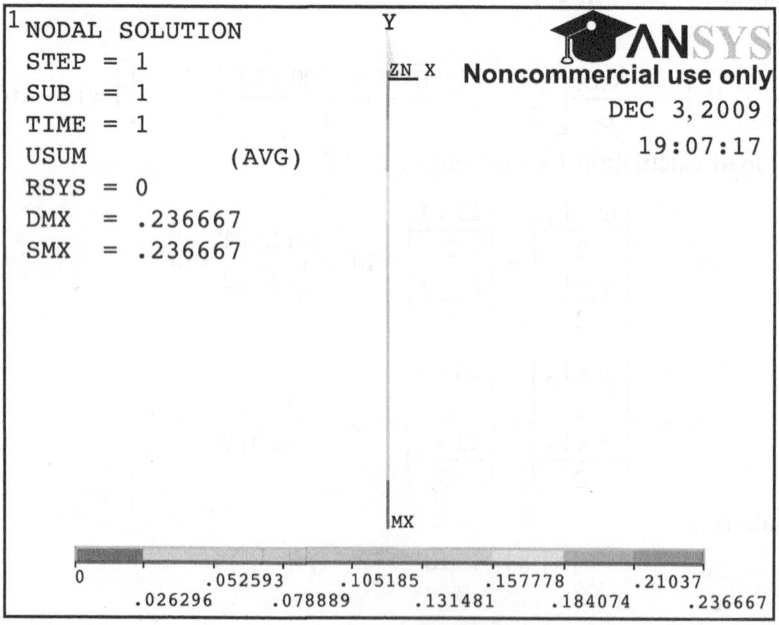

Figure 2.16(d). Deflection pattern for a tapered bar (Refer to appendix for colour figures)

Deflection values at nodes

The following degree of freedom results are in Global coordinates

NODE	UX	UY	UZ	USUM
1	0.0000	0.0000	0.0000	0.0000
2	0.0000	-0.95000E-01	0.0000	0.95000E-01
3	0.0000	-0.23667	0.0000	0.23667

Maximum absolute values

NODE	0	3	0	3
VALUE	0.0000	-0.23667	0.0000	0.23667

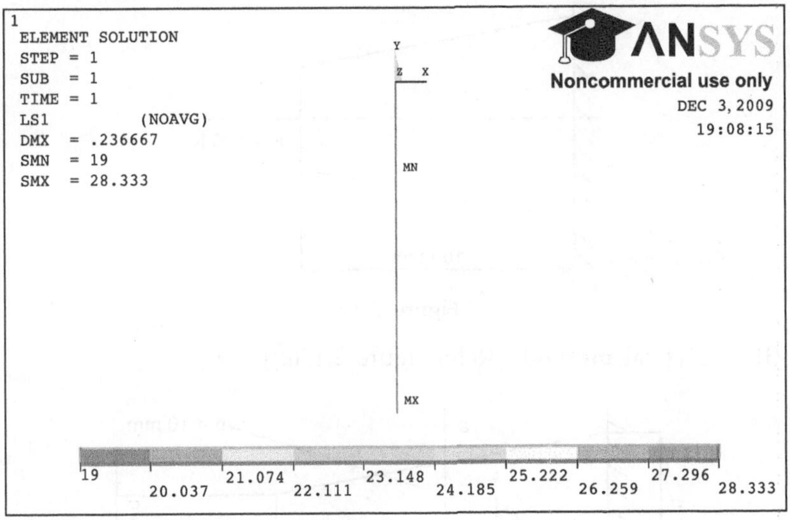

Figure 2.16(e). Stress pattern for a tapered bar (Refer to appendix for colour figures)

Stress values at elements

STAT	CURRENT
ELEM	LS1
3	19.000
4	28.333

Reaction value

The following X,Y,Z solutions are in Global coordinates

NODE	FX	FY
1	0.0000	0.25000E+06

Answers of Problem 2.6

Parameter	Analytical method	FEM-hand calculations	Software results
Displacement at node 2	0.096 mm	0.095 mm	0.095 mm
Displacement at node 3	0.2423 mm	0.237 mm	0.23667 mm
Stress in element 1	16.67 MPa to 22.5 MPa	19 MPa	19 MPa
Stress in element 2	22.5 MPa to 40 MPa	28.4 MPa	28.33 MPa

In the above problem two elements are used for solving the problem by hand calculation and by software. To get the convergence of the solution with the analytical method more number of elements are to be used.

Problem 2.7. *Find the displacement and stress distribution in the tapered bar shown in figure 2.17 using two finite elements under an axial load of P = 100 N.*

Cross-sectional area at fixed end = 200 mm^2

Cross-sectional area at free end = 100 mm^2

Young's modulus E = 200 GPa

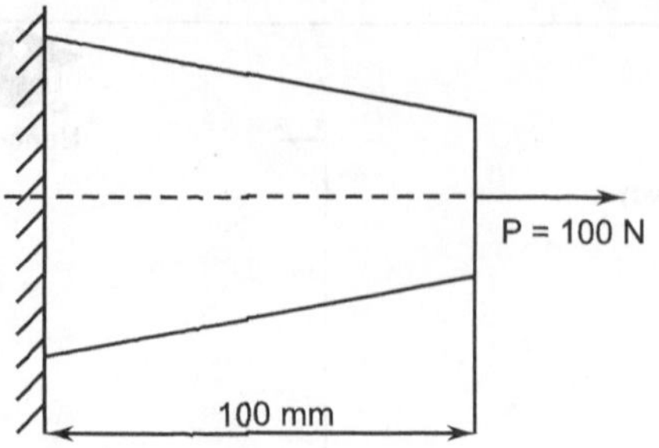

Figure 2.17

SOLUTION. (I) Analytical method. [Refer figure 2.17(*a*)]

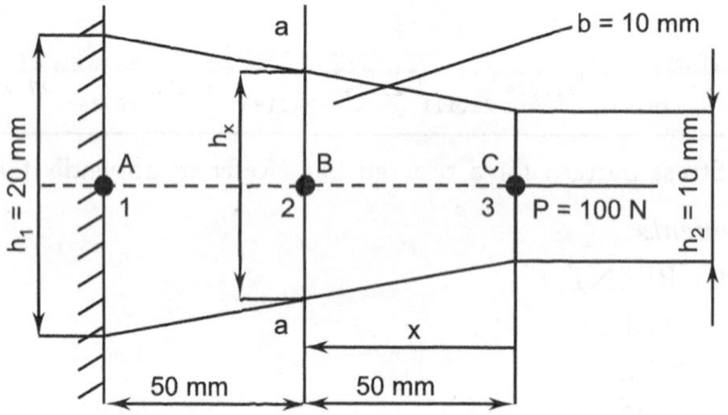

Figure 2.17(a)

Assume, b = thickness = 10 mm

Area at section a–a = $b \times h_x$

$$h_x = \left[h_2 + \left(\frac{h_1 - h_2}{L}\right)x\right]$$

$$A_x = 10\left[10 + \left(\frac{20-10}{100}\right)x\right] = 10(10 + 0.1x)$$

Stress calculation

$$\sigma_x = \frac{P}{A_x} = \frac{100}{10(10 + 0.1x)}$$

$$\sigma_1 = \sigma_{x=100} = \frac{100}{10(10 + 0.1 \times 100)} = 0.5 \text{ MPa}$$

$$\sigma_2 = \sigma_{x=50} = 0.667 \text{ MPa}$$

$$\sigma_3 = \sigma_{x=0} = 1 \text{ MPa}$$

Displacement calculation

$$\delta_3 = \delta_C = \frac{PL}{Eb(h_1 - h_2)} \ln\frac{h_1}{h_2} = \frac{100 \times 100}{2 \times 10^5 \times 10(20-10)} \ln\frac{20}{10} = 3.47 \times 10^{-4} \text{ mm}$$

$$\delta_2 = \delta_B = \frac{100 \times 50}{2 \times 10^5 \times 10(20-15)} \ln\frac{20}{15} = 1.44 \times 10^{-4} \text{ mm}$$

(II) FEM by hand calculations

Using 2 elements each of 50 mm length, we obtain the finite element model as shown in figure 2.17(*c*). We can write the equivalent model as shown in figure 2.17(*b*). At the middle, area of cross-section of bar is $\dfrac{(200+100)}{2} = 150 \text{ mm}^2$.

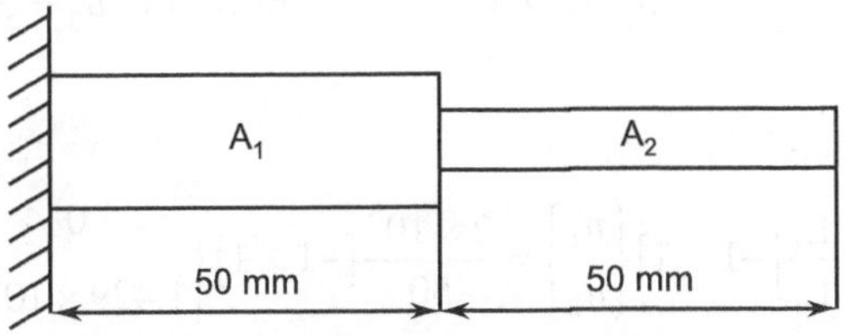

Figure 2.17(b)

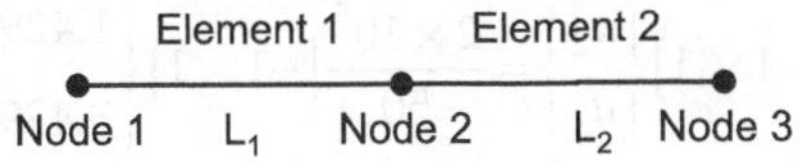

Figure 2.17(c). Finite element model

$$A_1 = \frac{200+150}{2} = 175 \text{ mm}^2$$

$$A_2 = \frac{150+100}{2} = 125 \text{ mm}^2$$

$$L_1 = L_2 = 50 \text{ mm}$$

$$E_1 = E_2 = 2 \times 10^5 \text{ N/mm}^2$$

Stiffness matrix for element 1 is,

$$[k_1] = \frac{A_1E_1}{L_1}\begin{bmatrix} 1 & -1 \\ -1 & 1 \end{bmatrix} = \frac{175 \times 2 \times 10^5}{50}\begin{bmatrix} 1 & -1 \\ -1 & 1 \end{bmatrix} = 7 \times 10^5 \begin{matrix} \quad 1 & \quad 2 & \\ \begin{bmatrix} 1 & -1 \\ -1 & 1 \end{bmatrix} & \begin{matrix} 1 \\ 2 \end{matrix} \end{matrix}$$

Stiffness matrix for element 2 is,

$$[k_2] = \frac{A_2E_2}{L_2}\begin{bmatrix} 1 & -1 \\ -1 & 1 \end{bmatrix} = \frac{125 \times 2 \times 10^5}{50}\begin{bmatrix} 1 & -1 \\ -1 & 1 \end{bmatrix} = 5 \times 10^5 \begin{matrix} \quad 2 & \quad 3 & \\ \begin{bmatrix} 1 & -1 \\ -1 & 1 \end{bmatrix} & \begin{matrix} 2 \\ 3 \end{matrix} \end{matrix}$$

Global equation is,

$$[K]\,\{r\} = \{R\}$$

$$10^5 \begin{matrix} \quad 1 & \quad 2 & \quad 3 & \\ \begin{bmatrix} 7 & -7 & 0 \\ -7 & 7+5 & -5 \\ 0 & -5 & 5 \end{bmatrix} & \begin{matrix} 1 \\ 2 \\ 3 \end{matrix} \end{matrix} \begin{Bmatrix} u_1 \\ u_2 \\ u_3 \end{Bmatrix} = \begin{Bmatrix} R_1 \\ 0 \\ 100 \end{Bmatrix} \qquad ...(i)$$

Boundary conditions are, at node 1, $u_1 = 0$

By using elimination method the above matrix reduces to,

$$10^5 \begin{bmatrix} 12 & -5 \\ -5 & 5 \end{bmatrix} \begin{Bmatrix} u_2 \\ u_3 \end{Bmatrix} = \begin{Bmatrix} 0 \\ 100 \end{Bmatrix}$$

By matrix multiplication, we get

$$10^5(12 \times u_2 - 5 \times u_3) = 0 \quad ...(ii)$$

$$10^5(-5 \times u_2 + 5 \times u_3) = 100 \quad ...(iii)$$

By solving equations (*ii*) and (*iii*), we get

$$u_2 = 1.429 \times 10^{-4} \text{ mm and } u_3 = 3.429 \times 10^{-4} \text{ mm}$$

Stress calculation

Stress in element 1,

$$\{\sigma_1\} = \frac{E_1}{L_1}\begin{bmatrix}-1 & 1\end{bmatrix}\begin{Bmatrix}u_1\\u_2\end{Bmatrix} = \frac{2\times10^5}{50}\begin{bmatrix}-1 & 1\end{bmatrix}\begin{Bmatrix}0\\1.429\times10^{-4}\end{Bmatrix} = 0.5716 \text{ MPa}$$

Stress in element 2,

$$\{\sigma_2\} = \frac{E_2}{L_2}\begin{bmatrix}-1 & 1\end{bmatrix}\begin{Bmatrix}u_2\\u_3\end{Bmatrix} = \frac{2\times10^5}{50}\begin{bmatrix}-1 & 1\end{bmatrix}\begin{Bmatrix}1.429\times10^{-4}\\3.429\times10^{-4}\end{Bmatrix} = 0.8 \text{ MPa}$$

(III) Software results

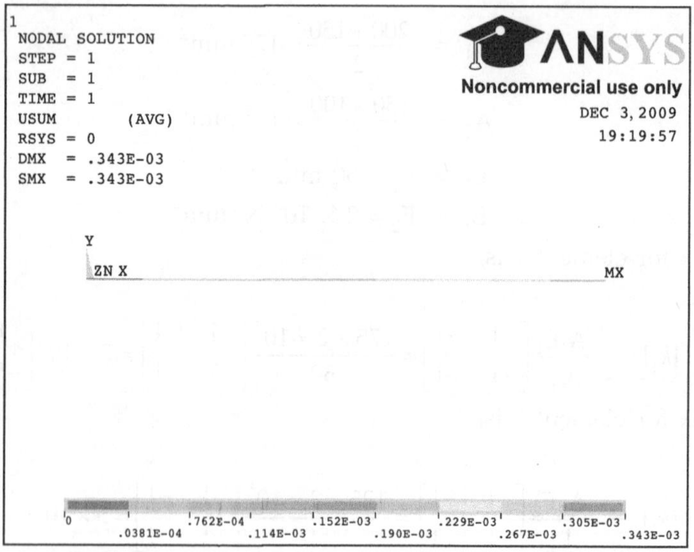

Figure 2.17(d). Deflection pattern for a tapered bar (Refer to appendix for colour figures)

Deflection values at nodes

The following degree of freedom results are in Global coordinates

NODE	UX	UY	UZ	USUM
1	0.0000	0.0000	0.0000	0.0000
2	0.14286E-03	0.0000	0.0000	0.14286E-03
3	0.34286E-03	0.0000	0.0000	0.34286E-03

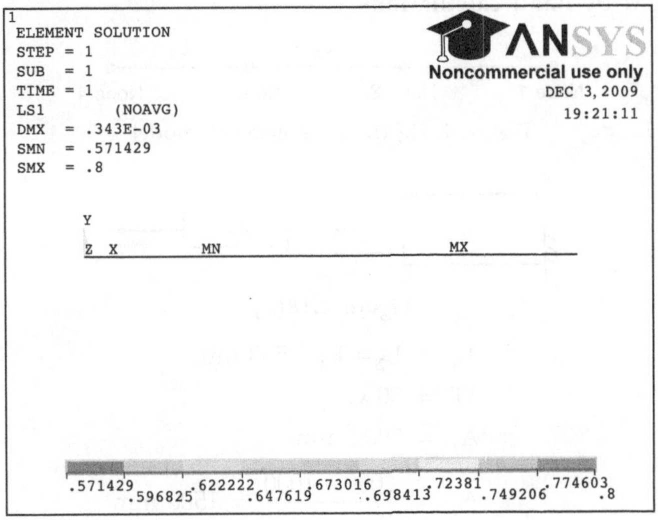

Figure 2.17(e). Stress pattern for a tapered bar (Refer to appendix for colour figures)

Stress values at elements

STAT ELEM	CURRENT LS1
1	0.57143
2	0.80000

ANSWERS OF PROBLEM 2.7

Parameter	Analytical method	FEM-hand calculations	Software results
Displacement at node 2	1.44×10^{-4} mm	1.429×10^{-4} mm	1.4286×10^{-4} mm
Displacement at node 3	3.47×10^{-4} mm	3.429×10^{-4} mm	3.4286×10^{-4} mm
Stress in element 1	0.5 MPa to 0.667 MPa	0.5716 MPa	0.57143 MPa
Stress in element 2	0.667 MPa to 1 MPa	0.8 MPa	0.8 MPa

PROBLEM 2.8. *Find the nodal displacements, element stresses and reactions in the tapered bar subjected to a load of 6000 N as shown in figure 2.18. Further the member experiences a temperature increase of 30°C. Use three equal length elements for finite element model. Take E = 200 GPa, $\nu = 0.3$ and $\alpha = 7 \times 10^{-6}$/°C.*

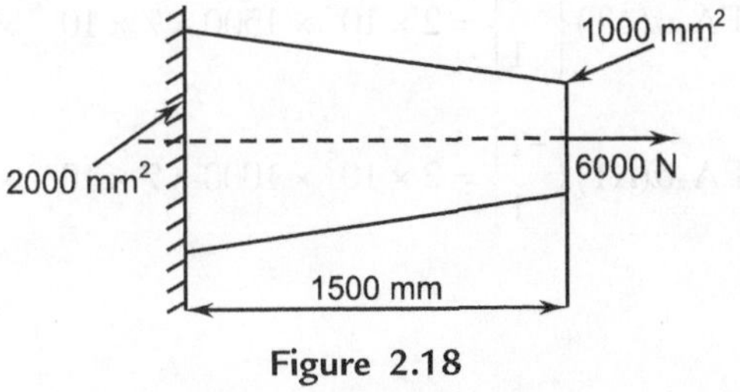

Figure 2.18

SOLUTION. **(I) FEM by hand calculations**

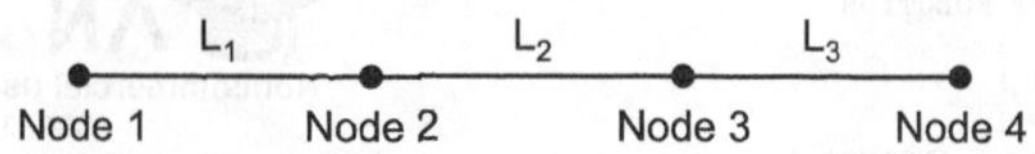

Figure 2.18(a). Finite element model

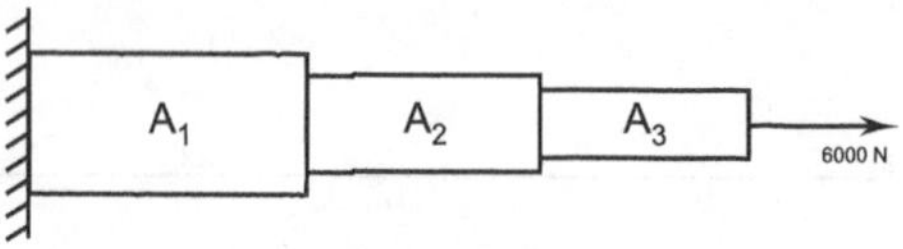

Figure 2.18(b)

$$L_1 = L_2 = L_3 = 500 \text{ mm}$$

$$\Delta T = 30°C$$

$$A_1 = 2000 \text{ mm}^2$$

$$A_2 = \frac{2000 + 1000}{2} = 1500 \text{ mm}^2$$

$$A_3 \approx 1000 \text{ mm}^2$$

Element stiffness matrices are,

$$[k_1] = \frac{A_1E}{L_1}\begin{bmatrix} 1 & -1 \\ -1 & 1 \end{bmatrix} = \frac{2000 \times 2 \times 10^5}{500}\begin{bmatrix} 1 & -1 \\ -1 & 1 \end{bmatrix} = 8 \times 10^5 \begin{array}{c} \begin{matrix} 1 & 2 \end{matrix} \\ \begin{bmatrix} 1 & -1 \\ -1 & 1 \end{bmatrix} \end{array} \begin{matrix} 1 \\ 2 \end{matrix}$$

$$[k_2] = \frac{A_2E}{L_2}\begin{bmatrix} 1 & -1 \\ -1 & 1 \end{bmatrix} = \frac{1500 \times 2 \times 10^5}{500}\begin{bmatrix} 1 & -1 \\ -1 & 1 \end{bmatrix} = 6 \times 10^5 \begin{array}{c} \begin{matrix} 2 & 3 \end{matrix} \\ \begin{bmatrix} 1 & -1 \\ -1 & 1 \end{bmatrix} \end{array} \begin{matrix} 2 \\ 3 \end{matrix}$$

$$[k_3] = \frac{A_3E}{L_3}\begin{bmatrix} 1 & -1 \\ -1 & 1 \end{bmatrix} = \frac{1000 \times 2 \times 10^5}{500}\begin{bmatrix} 1 & -1 \\ -1 & 1 \end{bmatrix} = 4 \times 10^5 \begin{array}{c} \begin{matrix} 3 & 4 \end{matrix} \\ \begin{bmatrix} 1 & -1 \\ -1 & 1 \end{bmatrix} \end{array} \begin{matrix} 3 \\ 4 \end{matrix}$$

Nodal loads due to thermal effect are,

$$\{Q_{1(th)}\} = EA_1\alpha(\Delta T)\begin{Bmatrix} -1 \\ 1 \end{Bmatrix} = 2 \times 10^5 \times 2000 \times 7 \times 10^{-6} \times 30\begin{Bmatrix} -1 \\ 1 \end{Bmatrix} = 84 \times 10^3 \begin{Bmatrix} -1 \\ 1 \end{Bmatrix} \begin{matrix} 1 \\ 2 \end{matrix}$$

$$\{Q_{2(th)}\} = EA_2\alpha(\Delta T)\begin{Bmatrix} -1 \\ 1 \end{Bmatrix} = 2 \times 10^5 \times 1500 \times 7 \times 10^{-6} \times 30\begin{Bmatrix} -1 \\ 1 \end{Bmatrix} = 63 \times 10^3 \begin{Bmatrix} -1 \\ 1 \end{Bmatrix} \begin{matrix} 2 \\ 3 \end{matrix}$$

$$\{Q_{3(th)}\} = EA_3\alpha(\Delta T)\begin{Bmatrix} -1 \\ 1 \end{Bmatrix} = 2 \times 10^5 \times 1000 \times 7 \times 10^{-6} \times 30\begin{Bmatrix} -1 \\ 1 \end{Bmatrix} = 42 \times 10^3 \begin{Bmatrix} -1 \\ 1 \end{Bmatrix} \begin{matrix} 3 \\ 4 \end{matrix}$$

Global forced vector {R}

$$= \begin{Bmatrix} -84 \times 10^3 \\ 84 \times 10^3 - 63 \times 10^3 \\ 63 \times 10^3 - 42 \times 10^3 \\ 42 \times 10^3 \end{Bmatrix} \begin{matrix} 1 \\ 2 \\ 3 \\ 4 \end{matrix} = \begin{Bmatrix} -84 \times 10^3 \\ 21 \times 10^3 \\ 21 \times 10^3 \\ 42 \times 10^3 \end{Bmatrix} \begin{matrix} 1 \\ 2 \\ 3 \\ 4 \end{matrix}$$

Global equation is,

$$10^5 \begin{bmatrix} 8 & -8 & 0 & 0 \\ -8 & 8+6 & -6 & 0 \\ 0 & -6 & 6+4 & -4 \\ 0 & 0 & -4 & 4 \end{bmatrix} \begin{matrix} 1 \\ 2 \\ 3 \\ 4 \end{matrix} \begin{Bmatrix} u_1 \\ u_2 \\ u_3 \\ u_4 \end{Bmatrix} = \begin{Bmatrix} -84 \times 10^3 + R_1 \\ 21 \times 10^3 \\ 21 \times 10^3 \\ 42 \times 10^3 + 6000 \end{Bmatrix} \quad \ldots(i)$$

Using elimination method of applying boundary conditions *i.e.*, $u_1 = 0$.

The equation (*i*) reduces to,

$$10^5 \begin{bmatrix} 14 & -6 & 0 \\ -6 & 10 & -4 \\ 0 & -4 & 4 \end{bmatrix} \begin{Bmatrix} u_2 \\ u_3 \\ u_4 \end{Bmatrix} = \begin{Bmatrix} 21 \times 10^3 \\ 21 \times 10^3 \\ 48 \times 10^3 \end{Bmatrix}$$

Solving the above matrix and equations, we get

$$u_2 = 0.1125 \text{ mm}$$
$$u_3 = 0.2275 \text{ mm}$$
$$u_4 = 0.3475 \text{ mm}$$

Stress calculation

$$\sigma_1 = E \times \frac{1}{L_1}[-1 \quad 1] \begin{Bmatrix} u_1 \\ u_2 \end{Bmatrix} - E\alpha(\Delta T) = \frac{2 \times 10^5}{500}[-1 \quad 1] \begin{Bmatrix} 0 \\ 0.1125 \end{Bmatrix} - 2 \times 10^5 \times 7 \times 10^{-6} \times 30 = 3 \text{ MPa}$$

$$\sigma_2 = E \times \frac{1}{L_2}[-1 \quad 1] \begin{Bmatrix} u_2 \\ u_3 \end{Bmatrix} - E\alpha(\Delta T) = \frac{2 \times 10^5}{500}[-1 \quad 1] \begin{Bmatrix} 0.1125 \\ 0.2275 \end{Bmatrix} - 2 \times 10^5 \times 7 \times 10^{-6} \times 30 = 4 \text{ MPa}$$

$$\sigma_3 = E \times \frac{1}{L_3}[-1 \quad 1] \begin{Bmatrix} u_3 \\ u_4 \end{Bmatrix} - E\alpha(\Delta T) = \frac{2 \times 10^5}{500}[-1 \quad 1] \begin{Bmatrix} 0.2275 \\ 0.3475 \end{Bmatrix} - 2 \times 10^5 \times 7 \times 10^{-6} \times 30 = 6 \text{ MPa}$$

Reaction calculation: From equation (*i*),

$$8 \times 10^5 u_1 - 8 \times 10^5 u_2 = -84 \times 10^3 + R_1$$

$$\therefore \quad R_1 = -6000 \text{ N}$$

(II) Software results

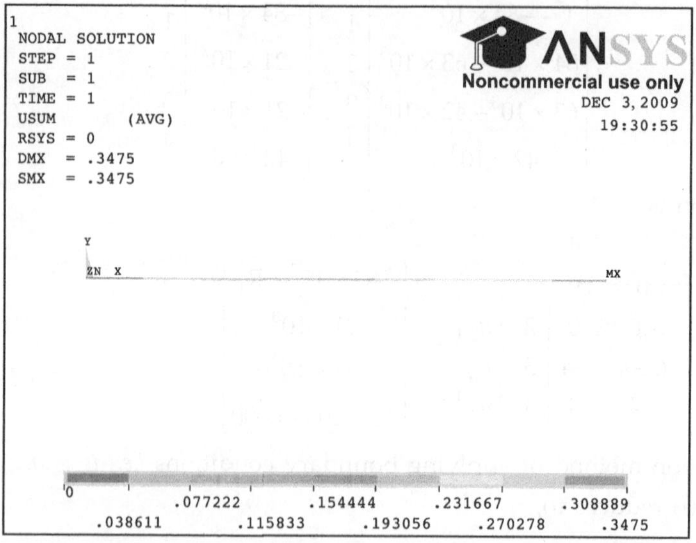

Figure 2.18(c). Deflection pattern for a tapered bar (Refer to appendix for colour figures)

Deflection values at nodes

The following degree of freedom results are in Global coordinates

NODE	UX	UY	UZ	USUM
1	0.0000	0.0000	0.0000	0.0000
2	0.11250	0.0000	0.0000	0.11250
3	0.22750	0.0000	0.0000	0.22750
4	0.34750	0.0000	0.0000	0.34750

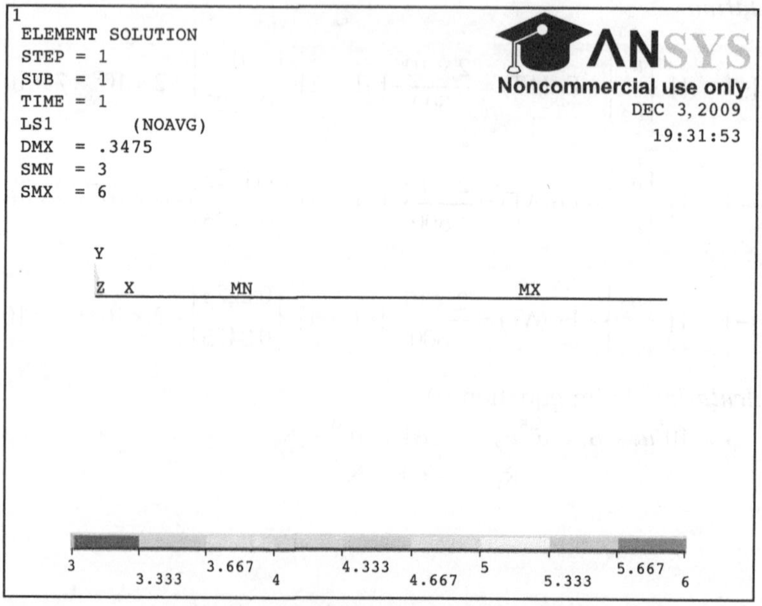

Figure 2.18(d). Stress pattern for a tapered bar (Refer to appendix for colour figures)

Stress values at elements

STAT	CURRENT
ELEM	LS1
1	3.0000
2	4.0000
3	6.0000

Reaction value

The following X,Y,Z solutions are in Global coordinates

NODE	FX	FY
1	-6000.0	0.0000

ANSWERS OF PROBLEM 2.8

Parameter	FEM-hand calculations	Software results
Deflection of node 2	0.1125 mm	0.1125 mm
Deflection of node 3	0.2275 mm	0.2275 mm
Deflection of node 4	0.3475 mm	0.3475 mm
Stress in element 1	3 MPa	3 MPa
Stress in element 2	4 MPa	4 MPa
Stress in element 3	6 MPa	6 MPa
Reaction at fixed end	-6000 N	-6000 N

Procedure for solving the problem using ANSYS® 12.0 academic teaching software

FOR PROBLEM 2.6

PREPROCESSING

1. **Main Menu > Preprocessor > Element Type > Add/Edit/Delete > Add >Structural Link> 2D spar 1 > OK>Close**

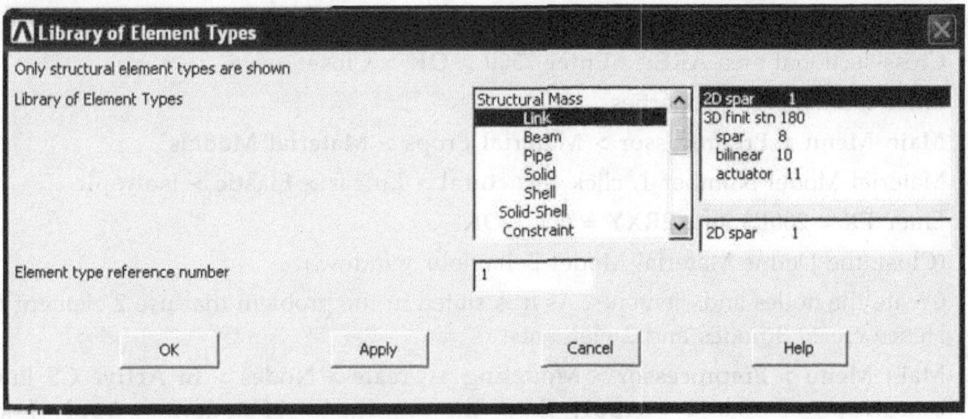

Figure 2.19. Element selection

2. **Main Menu > Preprocessor > Real Constants > Add/Edit/Delete > Add > OK**

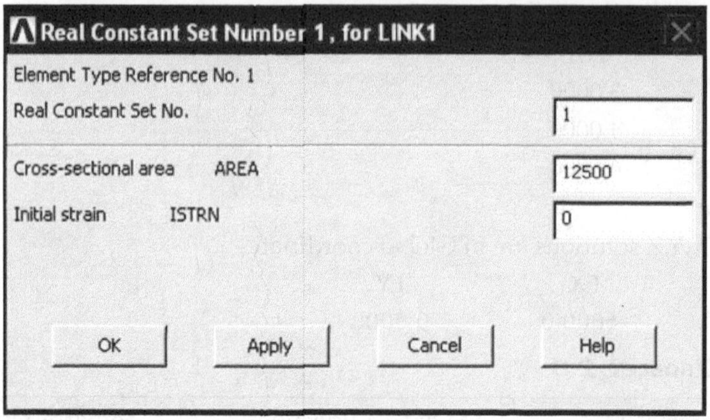

Figure 2.20. Enter the cross-sectional area of 1st element

Cross-sectional area AREA **>Enter 12500 > OK > Add>OK**

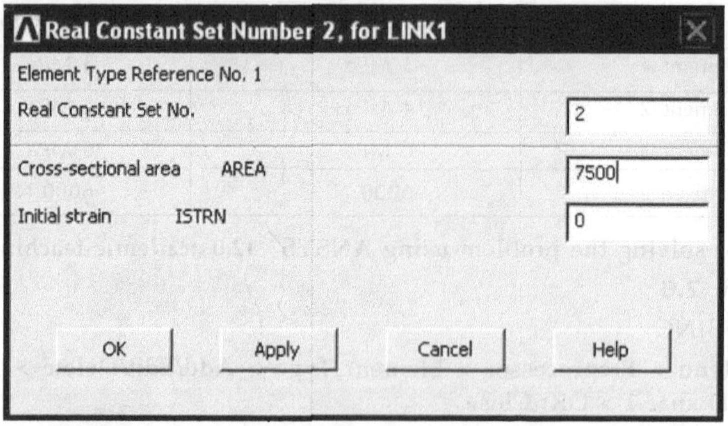

Figure 2.21. Enter the cross-sectional area of 2nd element

Cross-sectional area AREA **>Enter 7500 > OK > Close**

Enter the material properties.

3. **Main Menu > Preprocessor > Material Props > Material Models**

   Material Model Number 1, click **Structural > Linear > Elastic > Isotropic**

   Enter **EX = 200E3 and PRXY = 0.3 > OK**

   (**Close** the Define Material Model Behaviour window.)

   Create the nodes and elements. As it is stated in the problem that use 2 element model. Hence create 3 nodes and 2 elements.

4. **Main Menu > Preprocessor > Modeling > Create > Nodes > In Active CS** Enter the coordinates of node 1 **> Apply** Enter the coordinates of node 2 **> Apply** Enter the coordinates of node 3 **> OK.**

Node locations		
Node number	**X coordinate**	**Y coordinate**
1	0	0
2	0	–1000
3	0	–2000

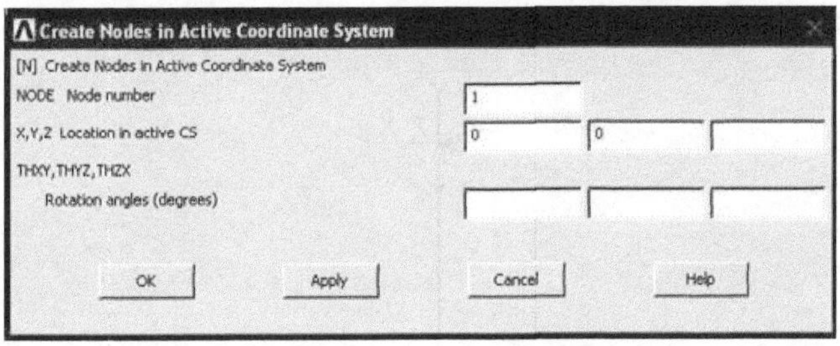

Figure 2.22. Enter the node coordinates

5. **Main Menu > Preprocessor > Modeling > Create > Elements > Elem Attributes > OK > Auto Numbered > Thru nodes** Pick the 1st and 2nd node **> OK**

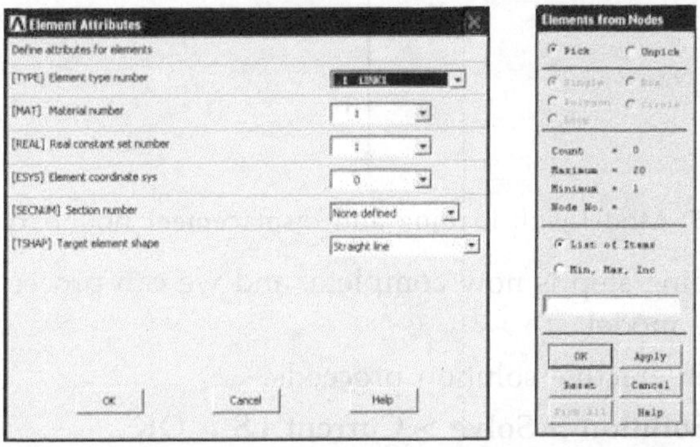

Figure 2.23. Assigning element attributes to element 1 and creating element 1

> Elem Attributes > change the Real constant set number to 2 > OK > Auto Numbered > Thru nodes Pick the 2nd and 3rd node **> OK**

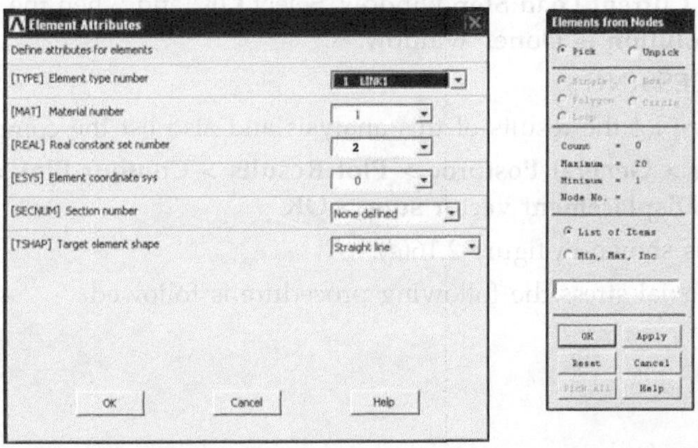

Figure 2.24. Assigning element attributes to element 2 and creating element 2

Apply the displacement boundary conditions and loads.

6. **Main Menu > Preprocessor > Loads > Define Loads > Apply > Structural >Displacement > On Nodes** Pick the 1st node > **Apply > All DOF = 0 > OK**
7. **Main Menu > Preprocessor > Loads > Define Loads > Apply > Structural >Force/ Moment > On Nodes** Pick the 3rd node **> OK > Force/Moment value = –212.5e3 in FY direction> OK > Force/Moment > On Nodes** Pick the 2nd node > **OK > Force/Moment value = –25e3 in FY direction> OK > Force/Moment>On Nodes** Pick the 1st node>**OK>Force/Moment value = –12.5e3 in FY direction>OK**

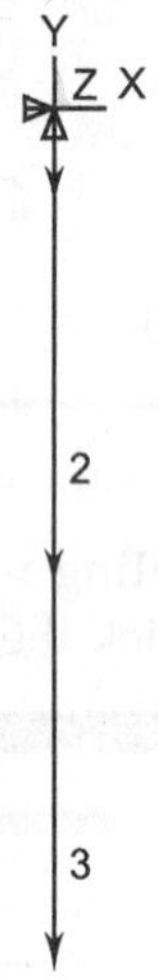

Figure 2.25. Model with loading and displacement boundary conditions

The model-building step is now complete, and we can proceed to the solution. First to be safe, save the model.

SOLUTION. The interactive solution proceeds.

8. **Main Menu > Solution > Solve > Current LS > OK**

The **/STATUS Command** window displays the problem parameters and **the Solve Current Load Step** window is shown. Check the solution options in the **/STATUS** window and if all is OK, select **File > Close**

In the Solve **Current Load Step** window, Select **OK**, and when the solution is complete, **close** the **'Solution is Done!'** window.

POSTPROCESSING

We can now plot the results of this analysis and also list the computed values.

9. **Main Menu > General Postproc > Plot Results > Contour Plot > Nodal Solu >DOF Solution > Displacement vector sum > OK**

This result is shown in figure 2.16(*d*).

To find the axial stress the following procedure is followed.

10. **Main Menu > General Postproc > Element Table > Define Table > Add**

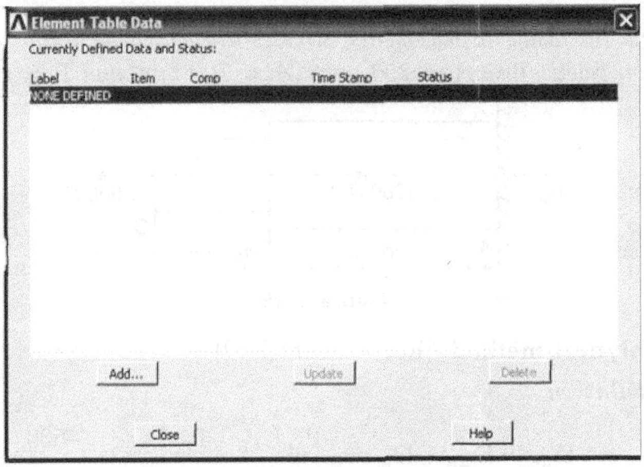

Figure 2.26. Defining the element table

Select **By sequence num and LS** and type **1 after LS** as shown in figure 2.27.

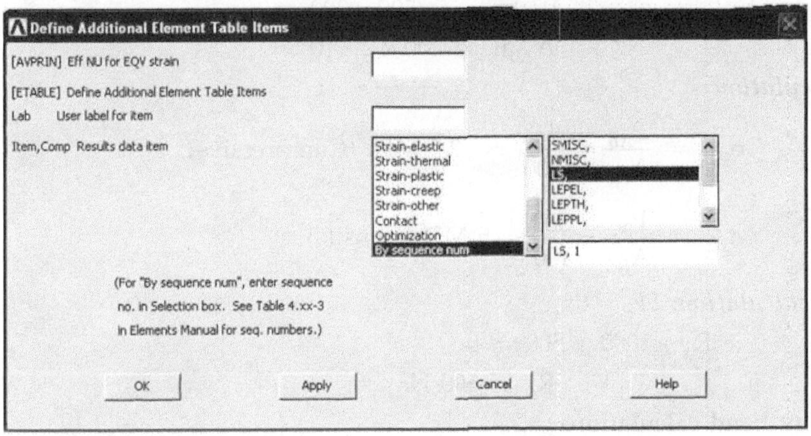

Figure 2.27. Selecting options in element table

> OK

11. **Main Menu > General Postproc > Plot Results > Contour Plot > Elem Table >** Select **LS1 > OK**

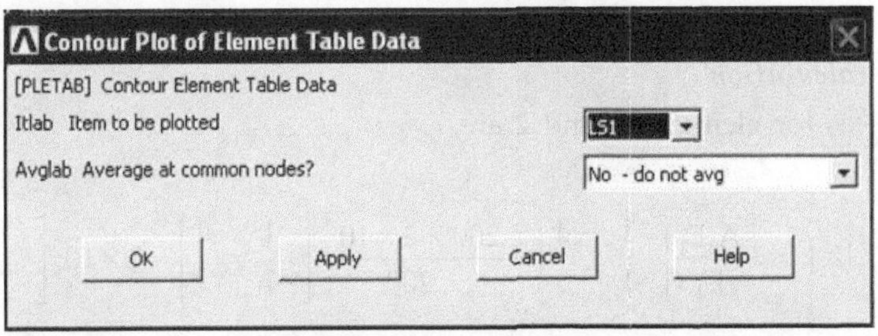

Figure 2.28. Selecting options for finding out axial stress

This result is shown in figure 2.16(*e*).

2.4 STEPPED BAR

PROBLEM 2.9. *Find the nodal displacements, stresses in each element and reaction at the fixed end for the figure 2.29 shown below. Take $A_1 = 200\ mm^2$, $A_2 = 100\ mm^2$ and $E_1 = E_2 = 200$ GPa.*

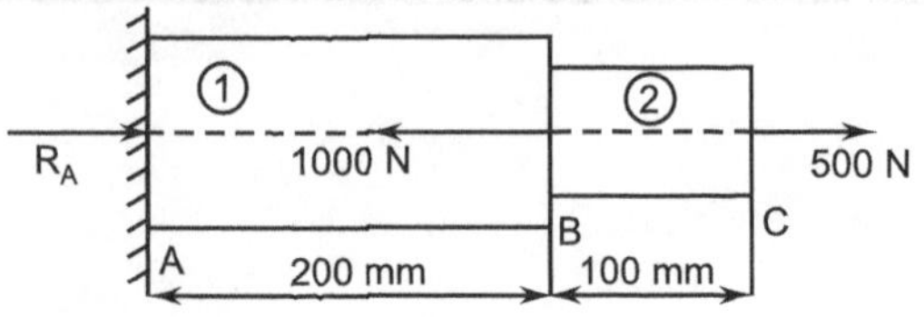

Figure 2.29

SOLUTION. (I) Analytical method. [Refer figure 2.29]

Displacement calculation

$$d_C = D_{AC} = D_{AB} + D_{BC} = \frac{-P_{AB}L_{AB}}{A_{AB}E} + \frac{P_{BC}L_{BC}}{A_{BC}E} = \frac{-500 \times 200}{200 \times 2 \times 10^5} + \frac{500 \times 100}{100 \times 2 \times 10^5}$$

$$= -2.5 \times 10^{-3} + 2.5 \times 10^{-3} = 0$$

$$d_B = D_{AB} = \frac{-P_{AB}L_{AB}}{A_{AB}E} = \frac{-500 \times 200}{200 \times 2 \times 10^5} = -2.5 \times 10^{-3} \text{ mm}$$

Stress calculation

$$\sigma_{AB} = \frac{P_{AB}}{A_{AB}} = \frac{-500}{200} = -2.5 \text{ MPa (Compressive)}$$

$$\sigma_{BC} = \frac{P_{BC}}{A_{BC}} = \frac{500}{100} = 5 \text{ MPa (Tensile)}$$

Reaction calculation: $\Sigma F_x = 0$

$$R_A - 1000 + 500 = 0$$

$$R_A = 500 \text{ N}$$

(II) FEM by hand calculations

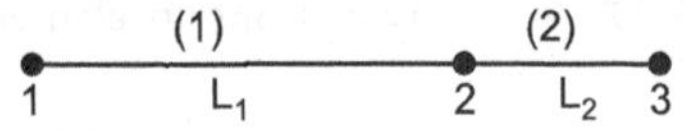

Figure 2.29(a). Finite element model

$$L_1 = 200 \text{ mm}$$

$$L_2 = 100 \text{ mm}$$

Displacement calculation

Stiffness matrices for elements 1 and 2 are,

$$[k_1] = \frac{A_1E_1}{L_1}\begin{bmatrix} 1 & -1 \\ -1 & 1 \end{bmatrix} = \frac{200 \times 2 \times 10^5}{200}\begin{bmatrix} 1 & -1 \\ -1 & 1 \end{bmatrix} = 2 \times 10^5 \begin{array}{c} \begin{array}{cc} 1 & 2 \end{array} \\ \begin{bmatrix} 1 & -1 \\ -1 & 1 \end{bmatrix} \end{array}\begin{array}{c} \\ 1 \\ 2 \end{array}$$

$$[k_2] = \frac{A_2E_2}{L_2}\begin{bmatrix} 1 & -1 \\ -1 & 1 \end{bmatrix} = \frac{100 \times 2 \times 10^5}{100}\begin{bmatrix} 1 & -1 \\ -1 & 1 \end{bmatrix} = 2 \times 10^5 \overset{\quad 2 \qquad 3}{\begin{bmatrix} 1 & -1 \\ -1 & 1 \end{bmatrix}}\begin{matrix} 2 \\ 3 \end{matrix}$$

Global equation is,

$$2 \times 10^5 \overset{\quad 1 \qquad 2 \qquad 3}{\begin{bmatrix} 1 & -1 & 0 \\ -1 & 1+1 & -1 \\ 0 & -1 & 1 \end{bmatrix}}\begin{matrix} 1 \\ 2 \\ 3 \end{matrix}\begin{Bmatrix} u_1 \\ u_2 \\ u_3 \end{Bmatrix} = \begin{Bmatrix} R_1 \\ -1000 \\ 500 \end{Bmatrix} \qquad ...(i)$$

Using elimination method and applying boundary conditions at node 1, u_1=0

The equation (*i*) reduces to

$$2 \times 10^5 \begin{bmatrix} 2 & -1 \\ -1 & 1 \end{bmatrix}\begin{Bmatrix} u_2 \\ u_3 \end{Bmatrix} = \begin{Bmatrix} -1000 \\ 500 \end{Bmatrix}$$

By solving the above matrix and equations,

we get,

$$u_2 = -2.5 \times 10^{-3} \text{ mm}$$
$$u_3 = 0$$

Stress calculation

$$\sigma_1 = E_1 \times \frac{1}{L_1}[-1 \quad 1]\begin{Bmatrix} u_1 \\ u_2 \end{Bmatrix} = \frac{2 \times 10^5}{200}[-1 \quad 1]\begin{Bmatrix} 0 \\ -2.5 \times 10^{-3} \end{Bmatrix} = -2.5 \text{ MPa (Compressive)}$$

$$\sigma_2 = E_2 \times \frac{1}{L_2}[-1 \quad 1]\begin{Bmatrix} u_2 \\ u_3 \end{Bmatrix} = \frac{2 \times 10^5}{100}[-1 \quad 1]\begin{Bmatrix} -2.5 \times 10^{-3} \\ 0 \end{Bmatrix} = 5 \text{ MPa (Tensile)}$$

Reaction calculation: From equation (*i*)

$$2 \times 10^5(u_1 - u_2) = R_1$$
$$2 \times 10^5[0 - (-2.5 \times 10^{-3})] = R_1$$
$$R_1 = 500 \text{ N}$$

(III) Software results

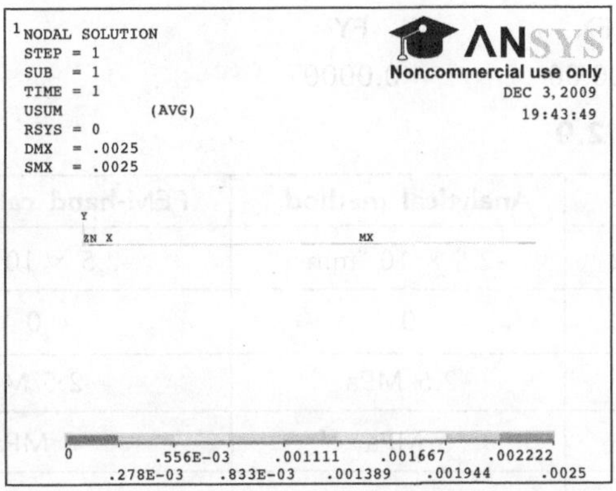

Figure 2.29(b). Deflection pattern for a stepped bar (Refer to appendix for colour figures)

Deflection values at nodes

The following Degree of Freedom results are in Global coordinates

NODE	UX	UY	UZ	USUM
1	0.0000	0.0000	0.0000	0.0000
2	-0.25000E-02	0.0000	0.0000	0.25000E-02
3	0.0000	0.0000	0.0000	0.0000

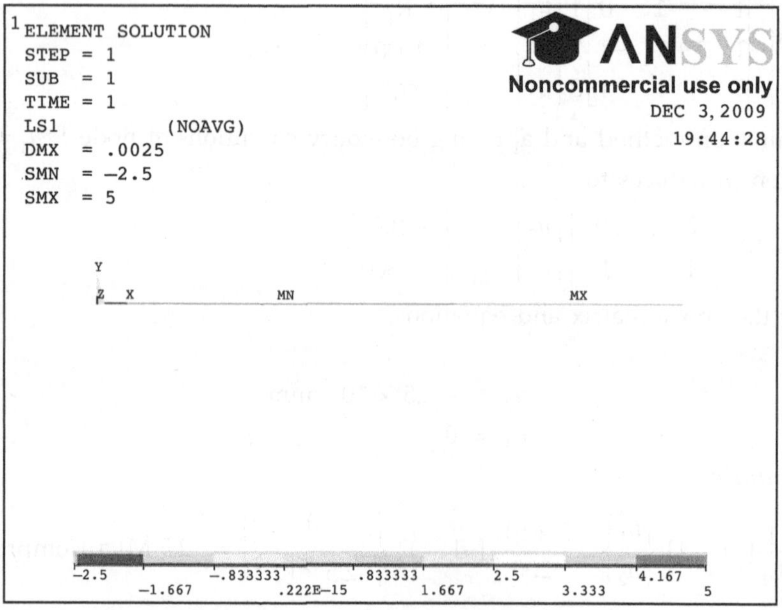

Figure 2.29(c). Stress pattern for a stepped bar (Refer to appendix for colour figures)

Stress values at elements

STAT	CURRENT
ELEM	LS1
1	−2.5000
3	5.0000

Reaction value

The following X,Y,Z solutions are in Global coordinates

NODE	FX	FY
1	500.00	0.0000

Answers of Problem 2.9

Parameter	Analytical method	FEM-hand calculations	Software results
Displacement at node 2	-2.5×10^{-3}mm	-2.5×10^{-3}mm	-0.25×10^{-2} mm
Displacement at node 3	0	0	0
Stress in element 1	-2.5 MPa	-2.5 MPa	-2.5 MPa
Stress in element 2	5 MPa	5 MPa	5 MPa
Reaction at fixed end	500 N	500 N	500 N

PROBLEM 2.10. *Find the nodal displacements, stress in each element and reaction at the fixed end for the figure 2.30 shown below. Take $E_1 = 2 \times 10^5$ N/mm^2 and $E_2 = 1 \times 10^5$ N/mm^2.*

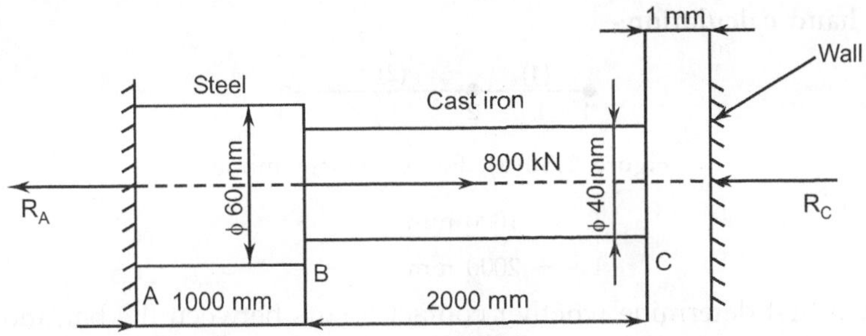

Figure 2.30

SOLUTION: (I) Analytical method. [Refer figure 2.30]

$$A_{AB} = \frac{\pi}{4} d_1^2 = \frac{\pi}{4}(60)^2 = 2827.43 \text{ mm}^2$$

$$A_{BC} = \frac{\pi}{4} d_2^2 = \frac{\pi}{4}(40)^2 = 1256.64 \text{ mm}^2$$

In the absence of the right wall,

$$\Delta_L = \Delta_{AB} = \frac{P_{AB} \times L_{AB}}{A_{AB} \times E_{AB}} = \frac{800 \times 10^3 \times 1000}{2827.43 \times 2 \times 10^5} = 1.415 \text{ mm}$$

Hence the contact does occur with the right will since $u_3 = 1.415$ mm

Let R_A and R_C be the reactions developed due to constraint.

$$R_A + R_C = 800 \times 10^3 \quad ...(i)$$

$$\frac{R_A L_{AB}}{A_{AB} E_{AB}} + \frac{(-R_C) L_{BC}}{A_{BC} E_{BC}} = 1$$

$$\frac{R_A \times (1000)}{2827.43 \times 2 \times 10^5} + \frac{(-R_C) \times 2000}{1256.64 \times 1 \times 10^5} = 1$$

$$1.7684 \times 10^{-6} \times R_A - 1.5915 \times 10^{-5} \times R_C = 1 \quad ...(ii)$$

By solving equations (*i*) and (*ii*),

we get

$$R_A = 776547.49 \text{ N}$$

$$R_C = 23452.51 \text{ N}$$

Displacement calculation

$$\delta_B = \Delta_{AB} = \frac{R_A L_{AB}}{A_{AB} E_{AB}} = \frac{776547.49 \times 1000}{2827.43 \times 2 \times 10^5} = 1.373 \text{ mm}$$

$$\delta_C = 1 \text{ mm}$$

Stress calculation

$$\sigma_{AB} = \frac{R_A}{A_{AB}} = \frac{776547.49}{2827.43} = 274.65 \text{ MPa}$$

$$\sigma_{BC} = \frac{(-R_C)}{A_{BC}} = \frac{-23452.51}{1256.64} = -18.66 \text{ MPa}$$

(II) FEM by hand calculations

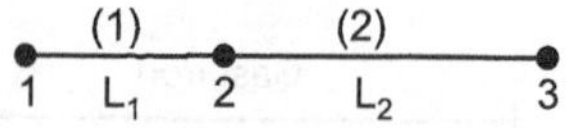

Figure 2.30(a). Finite element model

$$L_1 = 1000 \text{ mm}$$
$$L_2 = 2000 \text{ mm}$$

In this problem first determine whether contact occurs between the bar and the wall. To do this, assume that wall does not exist. Then the solution to the problem is (consider 2 element model),

Stiffness matrix for element 1 is,

$$[k_1] = \frac{A_1E_1}{L_1}\begin{bmatrix} 1 & -1 \\ -1 & 1 \end{bmatrix} = \frac{2827.43 \times 2 \times 10^5}{1000}\begin{bmatrix} 1 & -1 \\ -1 & 1 \end{bmatrix} = 5.655 \times 10^5 \begin{bmatrix} 1 & -1 \\ -1 & 1 \end{bmatrix}\begin{matrix} 1 \\ 2 \end{matrix}$$

(columns 1, 2)

Stiffness matrix for element 2 is,

$$[k_2] = \frac{A_2E_2}{L_2}\begin{bmatrix} 1 & -1 \\ -1 & 1 \end{bmatrix} = \frac{1256.64 \times 1 \times 10^5}{2000}\begin{bmatrix} 1 & -1 \\ -1 & 1 \end{bmatrix} = 0.628 \times 10^5 \begin{bmatrix} 1 & -1 \\ -1 & 1 \end{bmatrix}\begin{matrix} 2 \\ 3 \end{matrix}$$

(columns 2, 3)

Global equation is,

$$[K]\,\{r\} = \{R\}$$

$$10^5\begin{bmatrix} 5.655 & -5.655 & 0 \\ -5.655 & 5.655 + 0.628 & -0.628 \\ 0 & -0.628 & 0.628 \end{bmatrix}\begin{matrix} 1 \\ 2 \\ 3 \end{matrix}\begin{Bmatrix} u_1 \\ u_2 \\ u_3 \end{Bmatrix} = \begin{Bmatrix} R_1 \\ 800 \times 10^3 \\ 0 \end{Bmatrix} \quad \text{...}(i)$$

(columns 1, 2, 3; row 1 and column 1 struck out)

Boundary conditions are at node 1, $u_1 = 0$

By using elimination method the above matrix reduces to,

$$10^5\begin{bmatrix} 6.283 & -0.628 \\ -0.628 & 0.628 \end{bmatrix}\begin{Bmatrix} u_2 \\ u_3 \end{Bmatrix} = \begin{Bmatrix} 800 \times 10^3 \\ 0 \end{Bmatrix}$$

By matrix multiplication, we get

$$10^5(6.283 \times u_2 - 0.628 \times u_3) = 800 \times 10^3 \quad \text{...}(ii)$$
$$10^5(-0.628 \times u_2 + 0.628 \times u_3) = 0 \quad \text{...}(iii)$$

By solving equations (*ii*) and (*iii*)

we get $u_2 = 1.415$ mm and $u_3 = 1.415$ mm

Since the displacement of node 3 is 1.415 mm we can say that contact does occur. The problem has to be resolved. Since the boundary conditions are now different. The displacement at node 3 is given as 1 mm.

Global equation is,

$$10^5 \begin{bmatrix} 5.655 & -5.655 & 0 \\ -5.655 & 5.655+0.628 & -0.628 \\ 0 & -0.628 & 0.628 \end{bmatrix} \begin{Bmatrix} u_1 \\ u_2 \\ u_3 \end{Bmatrix} = \begin{Bmatrix} R_1 \\ 800 \times 10^3 \\ 0 \end{Bmatrix} \quad ...(i)$$

Boundary conditions at node 1, $u_1 = 0$ and at node 3 $u_3 = 1$ mm

By using elimination method the above matrix reduces to,

$$10^5\,[6.283]\,\{u_2\} = [800 \times 10^3] - 1[10^5 \times (-0.628)]$$
$$10^5\,[6.283]\{u_2\} = 800 \times 10^3 + 0.628 \times 10^5$$
$$u_2 = 1.373 \text{ mm}$$

Stress calculation

$$\sigma_1 = E_1 \times \frac{1}{L_1}[-1 \;\; 1]\begin{Bmatrix} u_1 \\ u_2 \end{Bmatrix} = \frac{2\times10^5}{1000}[-1 \;\; 1]\begin{Bmatrix} 0 \\ 1.373 \end{Bmatrix} = 274.6 \text{ MPa}$$

$$\sigma_2 = E_2 \times \frac{1}{L_2}[-1 \;\; 1]\begin{Bmatrix} u_2 \\ u_3 \end{Bmatrix} = \frac{1\times10^5}{2000}[-1 \;\; 1]\begin{Bmatrix} 1.373 \\ 1 \end{Bmatrix} = 18.65 \text{ MPa}$$

Reaction calculation

From equation (*i*)

$$5.655 \times 10^5 \times u_1 - 5.655 \times 10^5 \times u_2 = R_1$$
$$0 - 5.655 \times 10^5 \times 1.373 = R_1$$
$$R_1 = -776431.5 \text{ N (Direction is leftwards)}$$

We know that,

$$R_1 + P + R_3 = 0$$
$$-776431.5 + 800 \times 10^3 + R_3 = 0$$
$$R_3 = -23568.5 \text{ N (Direction is leftwards)}$$

(III) Software results

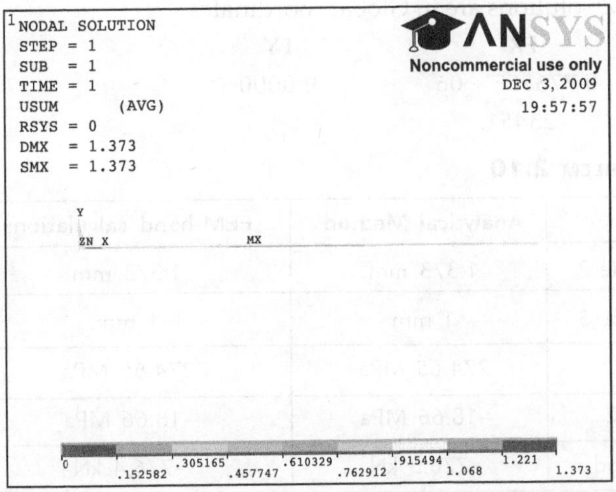

Figure 2.30(b). Deflection pattern for a stepped bar (Refer to appendix for colour figures)

Deflection values at nodes

The following degree of freedom results are in Global coordinates

NODE	UX	UY	UZ	USUM
1	0.0000	0.0000	0.0000	0.0000
2	1.3732	0.0000	0.0000	1.3732
3	1.0000	0.0000	0.0000	1.0000

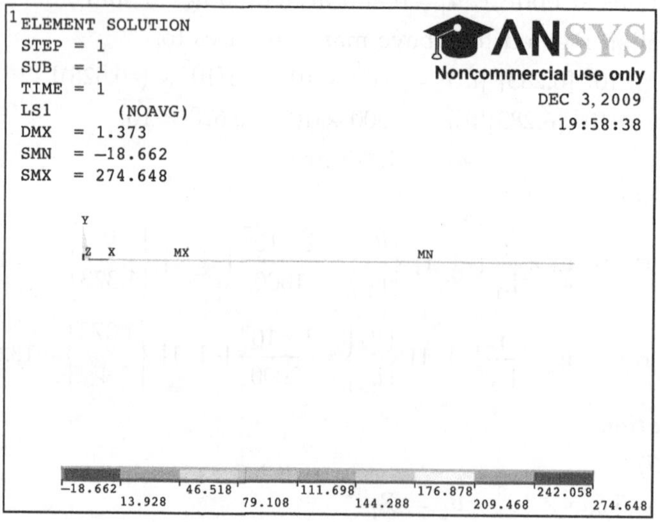

Figure 2.30(c). Stress pattern for a stepped bar (Refer to appendix for colour figures)

Stress values at elements

STAT ELEM	CURRENT LS1
1	274.65
2	–18.662

Reaction values

The following X,Y,Z solutions are in Global coordinates

NODE	FX	FY
1	-0.77655E + 06	0.0000
3	-23451.	

Answers of Problem 2.10

Parameter	Analytical Method	FEM-hand calculations	Software results
Displacement at node 2	1.373 mm	1.373 mm	1.3732 mm
Displacement at node 3	1 mm	1 mm	1 mm
Stress in element 1	274.65 MPa	274.65 MPa	274.65 MPa
Stress in element 2	-18.66 MPa	-18.66 MPa	-18.662 MPa
Reaction at fixed end	-776.5 kN	-776.4 kN	-776.55 kN
Reaction at wall	-23.45 kN	-23.57 kN	-23.451 kN

PROBLEM 2.11. *Find the nodal displacement, stress in each element and reaction at fixed ends for the figure 2.31 shown below. If the structure is subjected to an increase in temperature, $\Delta T = 75°C$, $P_1 = 50$ kN, $P_2 = 75$ kN.*

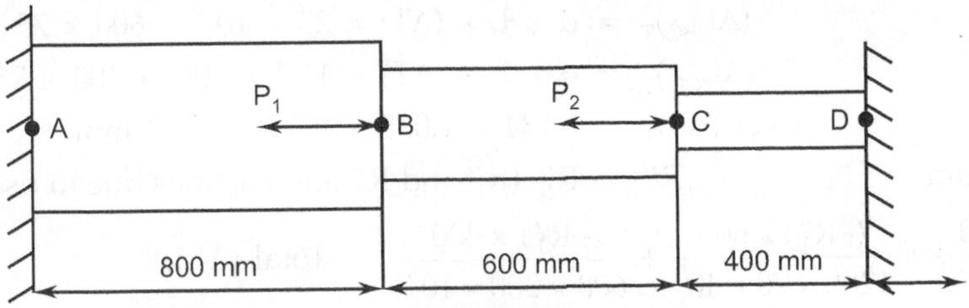

Figure 2.31

	Bronze	**Aluminum**	**Steel**
A =	2400 mm^2	1200 mm^2	600 mm^2
E =	83 GPa	70 GPa	200 GPa
α =	$18.9 \times 10^{-6}/°C$	$23 \times 10^{-6}/°C$	$11.7 \times 10^{-6}/°C$

SOLUTION. (I) Analytical method. [Refer figure 2.31]

Problem can be solved by method of superposition by considering load and temperature separately.

STEP 1. Consider only the loads P_1 and P_2 and neglect rise in temperature.

$$R_1' + R_2' = 125 \times 10^3 \quad ...(i) \quad (R_1' \text{ and } R_2' \text{ are reactions due to } P_1 \text{ and } P_2)$$

$$\frac{(-P_{AB}) \times 800}{2400 \times 83 \times 10^3} + \frac{(-P_{BC}) \times 600}{1200 \times 70 \times 10^3} + \frac{P_{CD} \times 400}{600 \times 200 \times 10^3} = 0$$

But $\quad P_{AB} = R_1',\ P_{CD} = R_2'$ and $P_{BC} = R_1' - 50 \times 10^3$

$$\frac{(-R_1') \times 800}{2400 \times 83 \times 10^3} - \frac{(R_1' - 50 \times 10^3) \times 600}{1200 \times 70 \times 10^3} + \frac{R_2' \times 400}{600 \times 200 \times 10^3} = 0 \quad ...(ii)$$

Solving equations (*i*) and (*ii*)

$$R_1' = 53.39 \text{ kN}$$

$$R_2' = 71.61 \text{ kN}$$

$$\therefore \quad \sigma'_{AB} = \frac{(-53.39 \times 10^3)}{2400} = -22.25 \text{ MPa}$$

$$\therefore \quad \sigma'_{BC} = \frac{-(53.39 \times 10^3 - 50 \times 10^3)}{1200} = -2.825 \text{ MPa}$$

$$\therefore \quad \sigma'_{CD} = \frac{71.61 \times 10^3}{600} = 119.35 \text{ MPa}$$

$$\Delta'_{AB} = \frac{-22.25 \times 800}{83 \times 10^3} = -0.2144 \text{ mm}$$

$$\Delta'_{BC} = \frac{-2.825 \times 600}{70 \times 10^3} = -0.0242 \text{ mm}$$

$$\Delta'_{CD} = \frac{119.35 \times 400}{200 \times 10^3} = 0.2395 \text{ mm}$$

Step 2. Consider only rise in temperature and neglect P_1 and P_2.

Free expansions due to $\Delta T = 75°C$ are

$$(\Delta L_{AB})_T = \alpha \times L \times (\Delta T) = 18.9 \times 10^{-6} \times 800 \times 75 = 1.134 \text{ mm}$$
$$(\Delta L_{BC})_T = \alpha \times L \times (\Delta T) = 23 \times 10^{-6} \times 600 \times 75 = 1.035 \text{ mm}$$
$$(\Delta L_{CD})_T = \alpha \times L \times (\Delta T) = 11.7 \times 10^{-6} \times 400 \times 75 = 0.351 \text{ mm}$$
$$\text{Total } (\Delta L)_T = 1.134 + 1.035 + 0.351 = 2.52 \text{ mm}$$

For equilibrium $R_1'' = R_2''$ (R_1'' and R_2'' are reactions due to rise in temperature)

$$\frac{(-R_1'') \times 800}{2400 \times 83 \times 10^3} + \frac{(-R_1'') \times 600}{1200 \times 70 \times 10^3} + \frac{(-R_1'') \times 400}{600 \times 200 \times 10^3} = -\text{Total } (\Delta L)_T$$

$$\frac{(-R_1'') \times 800}{2400 \times 83 \times 10^3} + \frac{(-R_1'') \times 600}{1200 \times 70 \times 10^3} + \frac{(-R_1'') \times 400}{600 \times 200 \times 10^3} = -2.52$$

Solving, $R_1'' = 173.89$ kN

$$\sigma''_{AB} = \frac{-173890}{2400} = -72.45 \text{ MPa}$$

$$\sigma''_{BC} = \frac{-173890}{1200} = -144.91 \text{ MPa}$$

$$\sigma''_{CD} = \frac{-173890}{600} = -289.82 \text{ MPa}$$

$$(\Delta L_{AB})_{Load} = \frac{-72.45 \times 800}{83 \times 10^3} = -0.698 \text{ mm}$$

$$(\Delta L_{BC})_{Load} = \frac{-144.91 \times 600}{70 \times 10^3} = -1.242 \text{ mm}$$

$$(\Delta L_{CD})_{Load} = \frac{-289.82 \times 400}{200 \times 10^3} = -0.5796 \text{ mm}$$

$$\Delta''_{AB} = 1.134 - 0.698 = 0.436 \text{ mm}$$
$$\Delta''_{BC} = 1.035 - 1.242 = -0.207 \text{ mm}$$
$$\Delta''_{CD} = 0.351 - 0.5796 = -0.2286 \text{ mm}$$

Step 3. Use method of superposition and combine steps (1) and (2)

Stresses are, $\sigma_{AB} = \sigma'_{AB} + \sigma''_{AB} = -22.45 - 72.45 = -94.7$ MPa.

Similarly, $\sigma_{BC} = \sigma'_{BC} + \sigma''_{BC} = -147.74$ MPa

$$\sigma_{CD} = \sigma'_{CD} + \sigma''_{CD} = -170.47 \text{ MPa}$$

Change in lengths are,

$$\Delta_{AB} = \Delta'_{AB} + \Delta''_{AB} = -0.2144 + 0.436 = 0.2216 \text{ mm}$$
$$\Delta_{BC} = \Delta'_{BC} + \Delta''_{BC} = -0.0242 - 0.207 = -0.2312 \text{ mm}$$
$$\Delta_{CD} = \Delta'_{CD} + \Delta''_{CD} = 0.2395 - 0.2286 = 0.0109 \text{ mm}$$

$$\therefore \quad u_2 = \Delta_{AB} = 0.2216 \text{ mm}$$
$$u_3 = \Delta_{CD} = -0.0109 \text{ mm}$$

Reactions are,

$$R_1 = R_1' + R_1'' = 53.39 + 173.89 = 227.28 \text{ kN}$$

$$R_2 = R_2' + R_2'' = 71.61 - 173.89 = -102.28 \text{ kN}$$

(II) FEM by hand calculations

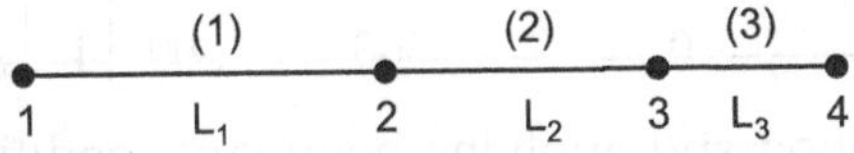

Figure 2.31(a). Finite element model

$L_1 = 800$ mm,	$L_2 = 600$ mm,	$L_3 = 400$ mm
$A_1 = 2400$ mm^2,	$A_2 = 1200$ mm^2,	$A_3 = 600$ mm^2
$E_1 = 83 \times 10^3$ N/mm^2,	$E_2 = 70 \times 10^3$ N/mm^2,	$E_3 = 200 \times 10^3$ N/mm^2
$\alpha_1 = 18.9 \times 10^{-6}$/°C,	$\alpha_2 = 23 \times 10^{-6}$/°C,	$\alpha_3 = 11.7 \times 10^{-6}$/°C

Element stiffness matrices are,

$$[k_1] = \frac{A_1 E_1}{L_1}\begin{bmatrix} 1 & -1 \\ -1 & 1 \end{bmatrix} = \frac{2400 \times 83 \times 10^3}{800}\begin{bmatrix} 1 & -1 \\ -1 & 1 \end{bmatrix} = 249 \times 10^3 \begin{bmatrix} 1 & -1 \\ -1 & 1 \end{bmatrix}\begin{matrix} 1 \\ 2 \end{matrix} \quad (\text{columns } 1, 2)$$

$$[k_2] = \frac{A_2 E_2}{L_2}\begin{bmatrix} 1 & -1 \\ -1 & 1 \end{bmatrix} = \frac{1200 \times 70 \times 10^3}{600}\begin{bmatrix} 1 & -1 \\ -1 & 1 \end{bmatrix} = 140 \times 10^3 \begin{bmatrix} 1 & -1 \\ -1 & 1 \end{bmatrix}\begin{matrix} 2 \\ 3 \end{matrix} \quad (\text{columns } 2, 3)$$

$$[k_3] = \frac{A_3 E_3}{L_3}\begin{bmatrix} 1 & -1 \\ -1 & 1 \end{bmatrix} = \frac{600 \times 200 \times 10^3}{400}\begin{bmatrix} 1 & -1 \\ -1 & 1 \end{bmatrix} = 300 \times 10^3 \begin{bmatrix} 1 & -1 \\ -1 & 1 \end{bmatrix}\begin{matrix} 3 \\ 4 \end{matrix} \quad (\text{columns } 3, 4)$$

Effect of temperature and thermal loads are,

$$\{Q_{1(th)}\} = E_1 A_1 \alpha_1(\Delta T)\begin{Bmatrix} -1 \\ 1 \end{Bmatrix} = 83 \times 10^3 \times 2400 \times 18.9 \times 10^{-6} \times 75 \begin{Bmatrix} -1 \\ 1 \end{Bmatrix} = 282.37 \times 10^3 \begin{Bmatrix} -1 \\ 1 \end{Bmatrix}\begin{matrix} 1 \\ 2 \end{matrix}$$

$$\{Q_{2(th)}\} = E_2 A_2 \alpha_2(\Delta T)\begin{Bmatrix} -1 \\ 1 \end{Bmatrix} = 70 \times 10^3 \times 1200 \times 23 \times 10^{-6} \times 75 \begin{Bmatrix} -1 \\ 1 \end{Bmatrix} = 144.9 \times 10^3 \begin{Bmatrix} -1 \\ 1 \end{Bmatrix}\begin{matrix} 2 \\ 3 \end{matrix}$$

$$\{Q_{3(th)}\} = E_3 A_3 \alpha_3(\Delta T)\begin{Bmatrix} -1 \\ 1 \end{Bmatrix} = 200 \times 10^3 \times 600 \times 11.7 \times 10^{-6} \times 75 \begin{Bmatrix} -1 \\ 1 \end{Bmatrix} = 105.3 \times 10^3 \begin{Bmatrix} -1 \\ 1 \end{Bmatrix}\begin{matrix} 3 \\ 4 \end{matrix}$$

Global Force vector {R}

$$= \begin{Bmatrix} -282.37 \times 10^3 \\ 282.37 \times 10^3 - 144.9 \times 10^3 \\ 144.9 \times 10^3 - 105.3 \times 10^3 \\ 105.3 \times 10^3 \end{Bmatrix}\begin{matrix} 1 \\ 2 \\ 3 \\ 4 \end{matrix} = \begin{Bmatrix} -282.37 \times 10^3 \\ 137.47 \times 10^3 \\ 39.6 \times 10^3 \\ 105.3 \times 10^3 \end{Bmatrix}\begin{matrix} 1 \\ 2 \\ 3 \\ 4 \end{matrix}$$

$$10^3 \begin{bmatrix} 249 & -249 & 0 & 0 \\ -249 & 249+140 & -140 & 0 \\ 0 & -140 & 140+300 & -300 \\ 0 & 0 & -300 & 300 \end{bmatrix} \begin{Bmatrix} u_1 \\ u_2 \\ u_3 \\ u_4 \end{Bmatrix} = 10^3 \begin{Bmatrix} -282.37 + R_1 \\ 137.47 - 50 \\ 39.6 - 75 \\ 105.3 + R_4 \end{Bmatrix} \quad ...(i)$$

Using the elimination method and applying boundary conditions,

i.e., $u_1 = u_4 = 0$

The equation (*i*) reduces to,

$$\cancel{10^3} \begin{bmatrix} 389 & -140 \\ -140 & 440 \end{bmatrix} \begin{Bmatrix} u_2 \\ u_3 \end{Bmatrix} = \cancel{10^3} \begin{Bmatrix} 87.47 \\ -35.4 \end{Bmatrix}$$

By solving the above matrix and equation,

we get $u_2 = 0.2212$ mm and $u_3 = -0.0101$ mm

Stress calculation

$$\sigma_1 = \frac{E_1}{L_1}[-1 \ \ 1] \begin{Bmatrix} u_1 \\ u_2 \end{Bmatrix} - E_1 \alpha_1(\Delta T)$$

$$= \frac{83 \times 10^3}{800}[-1 \ \ 1] \begin{Bmatrix} 0 \\ 0.2212 \end{Bmatrix} - 83 \times 10^3 \times 18.9 \times 10^{-6} \times 75 = -94.7 \text{ MPa}$$

$$\sigma_2 = \frac{E_2}{L_2}[-1 \ \ 1] \begin{Bmatrix} u_2 \\ u_3 \end{Bmatrix} - E_2 \alpha_2(\Delta T)$$

$$= \frac{70 \times 10^3}{600}[-1 \ \ 1] \begin{Bmatrix} 0.2212 \\ -0.0101 \end{Bmatrix} - 70 \times 10^3 \times 23 \times 10^{-6} \times 75 = -145.38 \text{ MPa}$$

$$\sigma_3 = \frac{E_3}{L_3}[-1 \ \ 1] \begin{Bmatrix} u_3 \\ u_4 \end{Bmatrix} - E_3 \alpha_3(\Delta T)$$

$$= \frac{200 \times 10^3}{400}[-1 \ \ 1] \begin{Bmatrix} -0.0101 \\ 0 \end{Bmatrix} - 200 \times 10^3 \times 11.7 \times 10^{-6} \times 75 = -170.45 \text{ MPa}$$

Reaction calculation

$$249 \times 10^3 \times u_1 - 249 \times 10^3 \times u_2 = -282.37 \times 10^3 + R_1$$

$$0 - 249 \times 10^3 \times 0.2212 = -282.37 \times 10^3 + R_1$$

$$R_1 = 227.29 \text{ kN}$$

$$-300 \times 10^3 \times u_3 + 300 \times 10^3 \times u_4 = 105.3 \times 10^3 + R_4$$

$$-300 \times 10^3 \times (-0.0101) + 0 = 105.3 \times 10^3 + R_4$$

$$R_4 = -102.27 \text{ kN}$$

(III) Software results

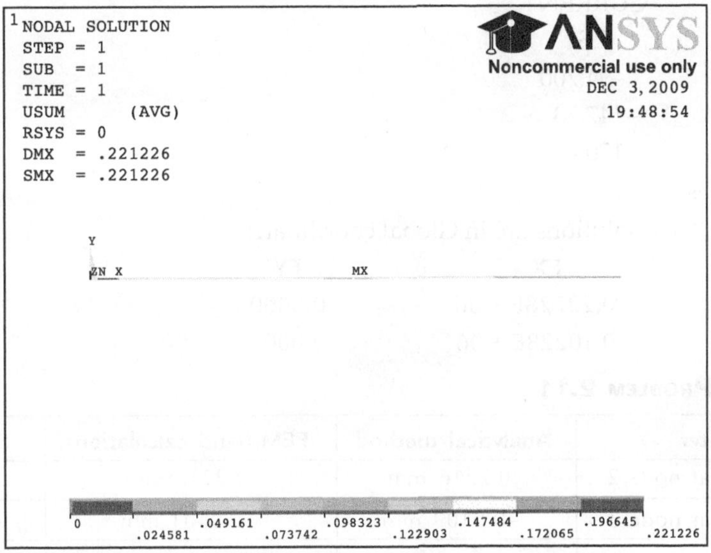

Figure 2.31(b). Deflection pattern for a stepped bar (Refer to appendix for colour figures)

Deflection values at nodes

The following degree of freedom results are in Global coordinates

NODE	UX	UY	UZ	USUM
1	0.0000	0.0000	0.0000	0.0000
2	0.22123	0.0000	0.0000	0.22123
3	-0.10064E-01	0.0000	0.0000	0.10064E-01
4	0.0000	0.0000	0.0000	0.0000

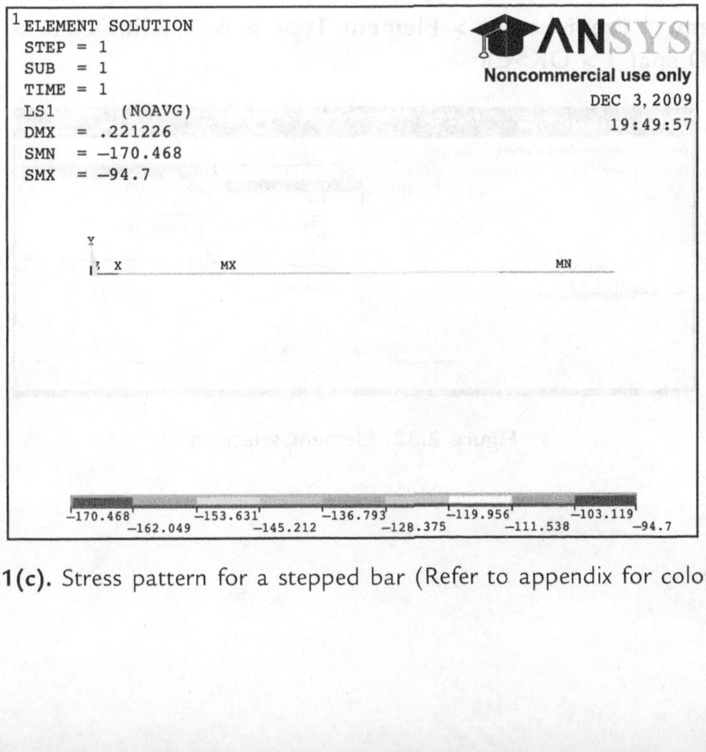

Figure 2.31(c). Stress pattern for a stepped bar (Refer to appendix for colour figures)

Stress values at elements

STAT ELEM	CURRENT LS1
1	–94.700
2	–147.73
3	–170.47

Reaction values

The following X,Y,Z solutions are in Global coordinates

NODE	FX	FY
1	0.22728E + 06	0.0000
4	-0.10228E + 06	0.0000

Answers of Problem 2.11

Parameter	Analytical method	FEM-hand calculations	Software results
Displacement at node 2	0.2216 mm	0.2212 mm	0.22123 mm
Displacement at node 3	-0.0109 mm	-0.0101 mm	-0.010064 mm
Stress in element 1	-94.7 MPa	-94.7 MPa	-94.7 MPa
Stress in element 2	-147.74 MPa	-145.38 MPa	-147.73 MPa
Stress in element 3	-170.47 MPa	-170.45 MPa	-170.47 MPa
Reaction at fixed end	227.28 kN	227.2912 kN	227.28 kN
Reaction at wall	-102.28 kN	-102.27 kN	-102.28 kN

Procedure for solving the problem using ANSYS® 12.0 academic teaching software

For Problem 2.11

PREPROCESSING

1. **Main Menu > Preprocessor > Element Type > Add/Edit/Delete > Add > Structural Link > 2D spar 1 > OK>Close**

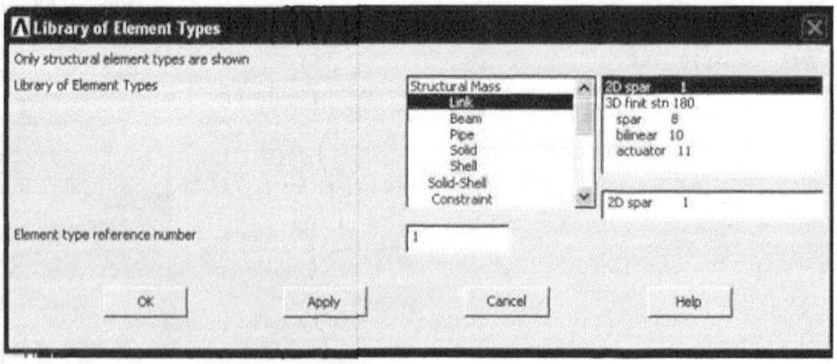

Figure 2.32. Element selection

2. **Main Menu > Preprocessor > Real Constants > Add/Edit/Delete > Add > OK**

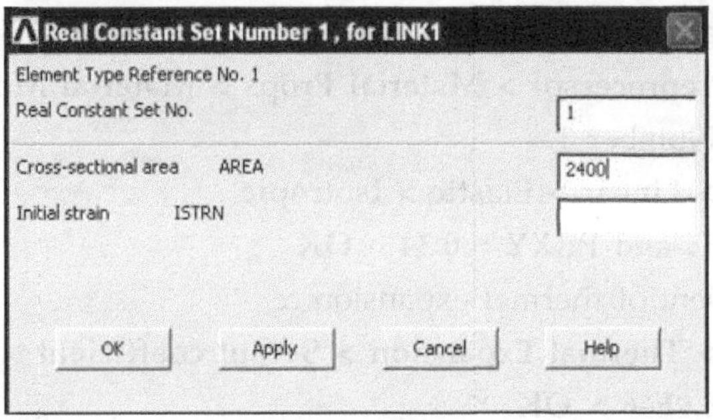

Figure 2.33. Enter the cross-sectional area of 1st element

Cross-sectional area AREA > **Enter 2400 > OK > Add > OK**

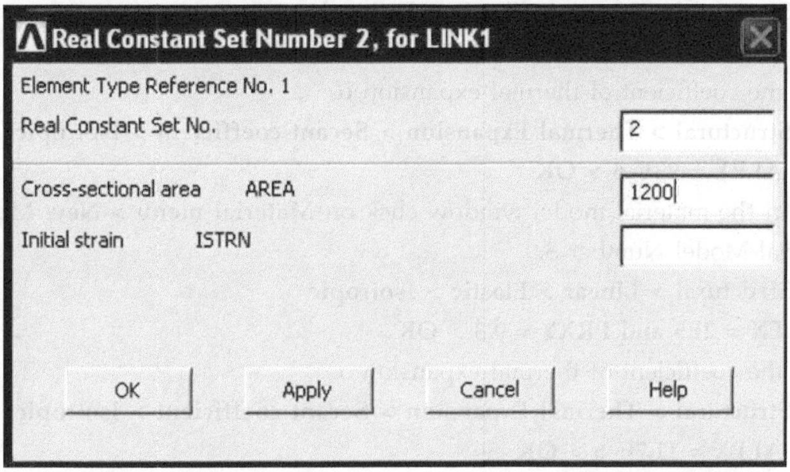

Figure 2.34. Enter the cross-sectional area of 2nd element

Cross-sectional area AREA > **Enter 1200 > OK > Add > OK**

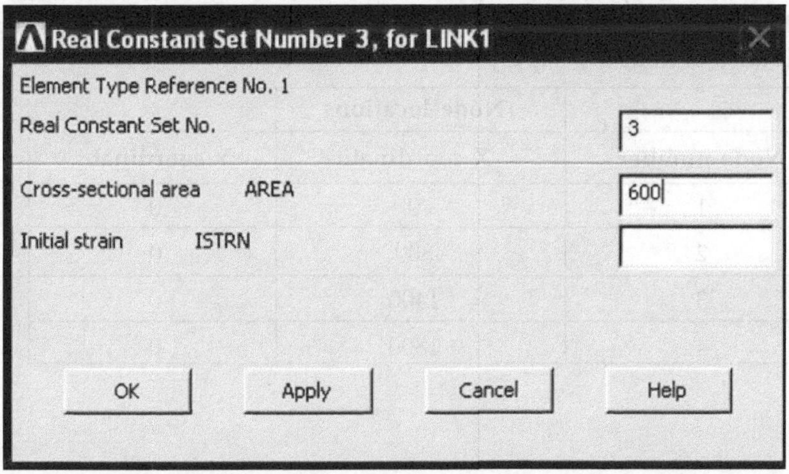

Figure 2.35. Enter the cross-sectional area of 3rd element

Cross-sectional area AREA > **Enter 600 > OK > Add > OK > Close**

Enter the material properties.

3. **Main Menu > Preprocessor > Material Props > Material Models**

Material Model Number 1,

click **Structural > Linear > Elastic > Isotropic**

Enter **EX = 0.83E5 and PRXY = 0.34** > **OK**

Enter the coefficient of thermal expansion α

click **Structural > Thermal Expansion > Secant coefficient > Isotropic**

Enter **ALPX = 18.9E–6 > OK**

Then in the material model window click on **Material menu > New Model > OK**

Material Model Number 2,

click **Structural > Linear > Elastic > Isotropic**

Enter **EX = 0.7E5 and PRXY = 0.35** > **OK**

Enter the coefficient of thermal expansion α

click **Structural > Thermal Expansion > Secant coefficient > Isotropic**

Enter **ALPX = 23E–6 > OK**

Then in the material model window click on **Material menu > New Model > OK**

Material Model Number 3,

click **Structural > Linear > Elastic > Isotropic**

Enter **EX = 2E5 and PRXY = 0.3** > **OK**

Enter the coefficient of thermal expansion α

click **Structural > Thermal Expansion > Secant coefficient > Isotropic**

Enter **ALPX = 11.7E–6 > OK**

(**Close** the Define Material Model Behaviour window.)

Create the nodes and elements. Use 3 element model. Hence create 4 nodes and 3 elements.

4. **Main Menu > Preprocessor > Modeling > Create > Nodes > In Active CS** Enter the coordinates of node 1 > **Apply** Enter the coordinates of node2 > **Apply** Enter the coordinates of node 3 > **Apply** >Enter the coordinates of node 4 > **OK.**

	Node locations	
Node number	**X coordinate**	**Y coordinate**
1	0	0
2	800	0
3	1400	0
4	1800	0

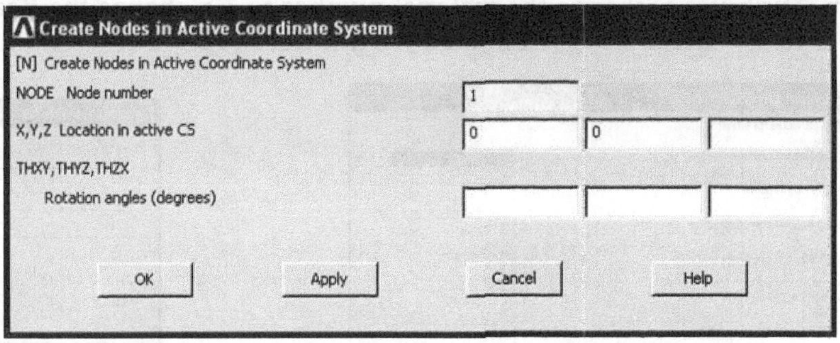

Figure 2.36. Enter the node coordinates.

5. **Main Menu > Preprocessor > Modeling > Create > Elements > Elem Attributes > OK > Auto Numbered > Thru nodes** Pick the 1st and 2nd node **> OK**

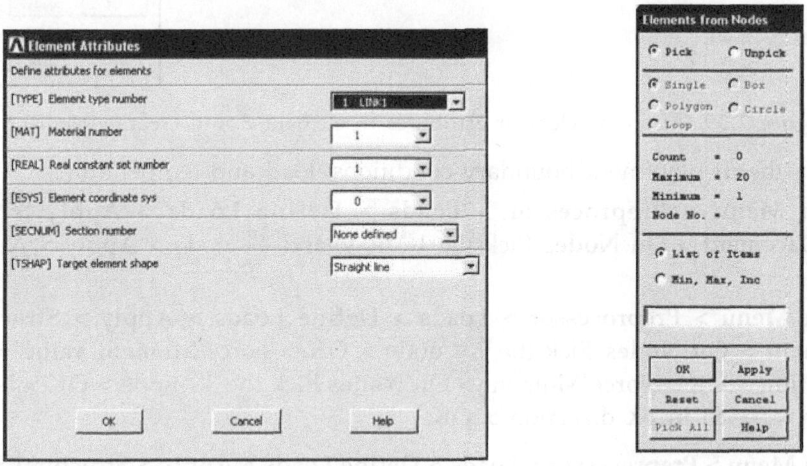

Figure 2.37. Assigning element attributes to element 1 and creating element 1

> Elem Attributes > change the material number to 2 > change the Real constant set number to 2 > OK > Auto Numbered > Thru nodes Pick the 2nd and 3rd node **> OK**

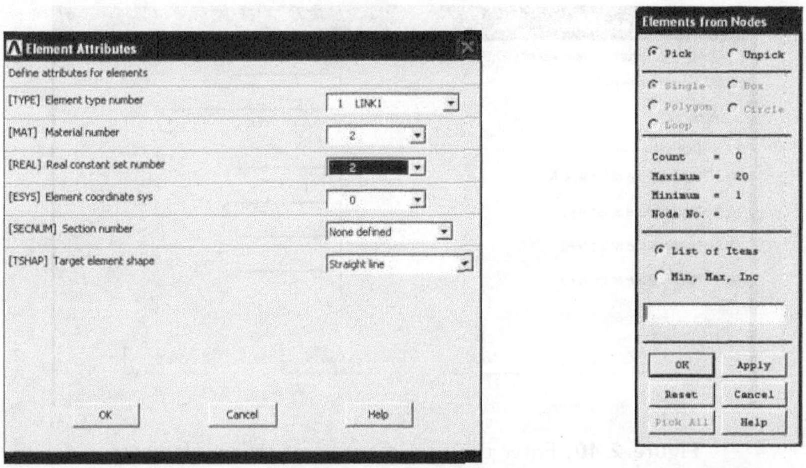

Figure 2.38. Assigning element attributes to element 2 and creating element 2

> Elem Attributes > change the material number to 3 > change the Real constant set number to 3 > OK > Auto Numbered > Thru nodes Pick the 3rd and 4th node **> OK**

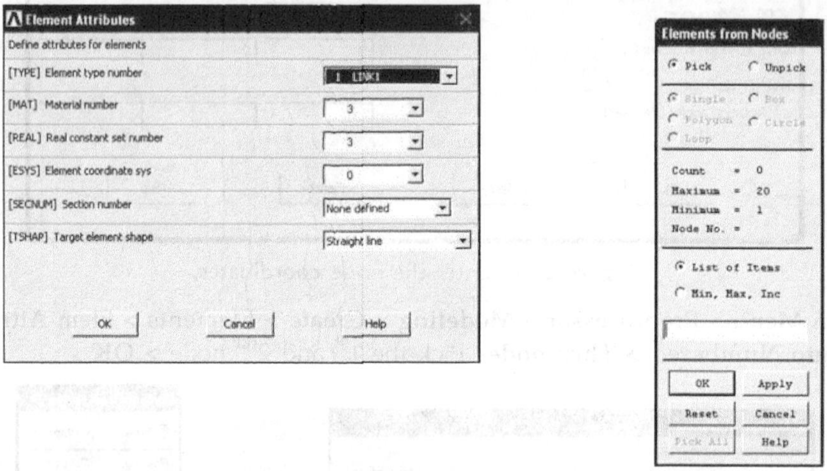

Figure 2.39. Assigning element attributes to element 3 and creating element 3

Apply the displacement boundary conditions, load and temperature.

6. **Main Menu > Preprocessor > Loads > Define Loads > Apply > Structural > Displacement > On Nodes** Pick the 1st node and 4th node **> Apply > All DOF = 0. > OK**
7. **Main Menu > Preprocessor > Loads > Define Loads > Apply > Structural >Force/ Moment > On Nodes** Pick the 2nd node **> OK > Force/Moment value = –50e3 in FX direction > OK > Force/Moment > On Nodes** Pick the 3rd node > **OK > Force/Moment value = –75e3 in FX direction > OK**
8. **Main Menu > Preprocessor > Loads > Define Loads > Apply > Structural >Temperature > On Elements** Pick the 1st element, 2nd element and 3rd element **> OK**

Enter **Temperature at location N = 75** as shown in figure 2.38

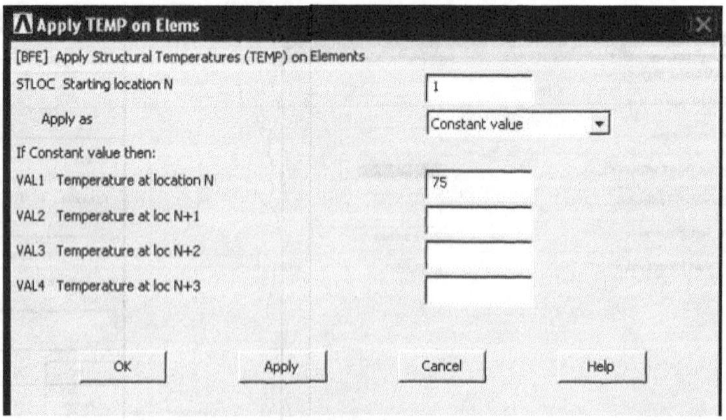

Figure 2.40. Enter the rise in temperature on elements

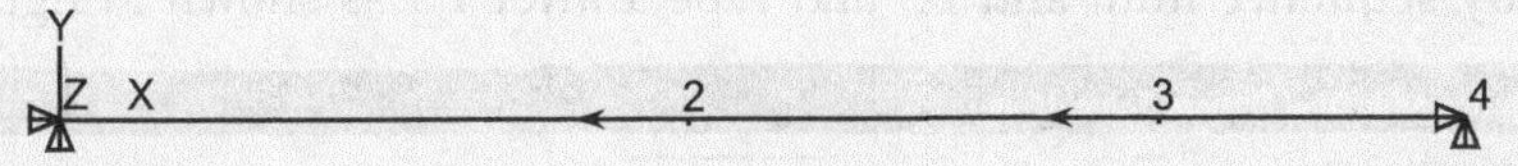

Figure 2.41. Model with loading and displacement boundary conditions

The model-building step is now complete, and we can proceed to the solution. First to be safe, save the model.

SOLUTION. The interactive solution proceeds.

9. **Main Menu > Solution > Solve > Current LS > OK**

   The **/STATUS Command** window displays the problem parameters and **the Solve Current Load Step** window is shown. Check the solution options in the **/STATUS** window and if all is OK, select **File > Close**

   In the Solve **Current Load Step** window, Select **OK**, and when the solution is complete, **close** the **'Solution is Done!'** window.

POSTPROCESSING

We can now plot the results of this analysis and also list the computed values.

10. **Main Menu > General Postproc > Plot Results > Contour Plot > Nodal Solu >DOF Solution > Displacement vector sum > OK**

    This result is shown in figure 2.31(*b*).

    To find the axial stress the following procedure is followed.

11. **Main Menu > General Postproc > Element Table > Define Table > Add**

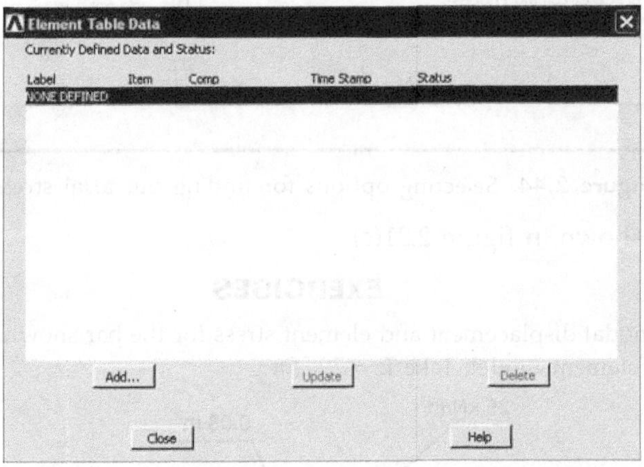

Figure 2.42. Defining the element table

Select **By sequence num and LS** and type **1 after LS** as shown in figure 2.43.

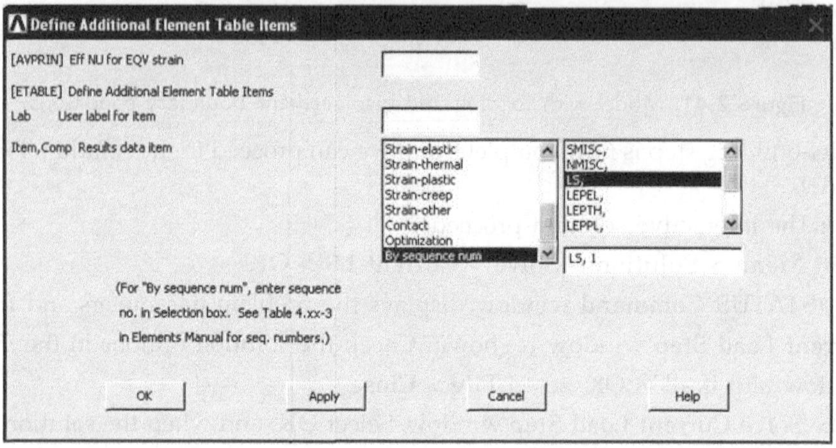

Figure 2.43. Selecting options in element table

> **OK**

12. **Main Menu > General Postproc > Plot Results > Contour Plot > Elem Table >** Select **LS1 > OK**

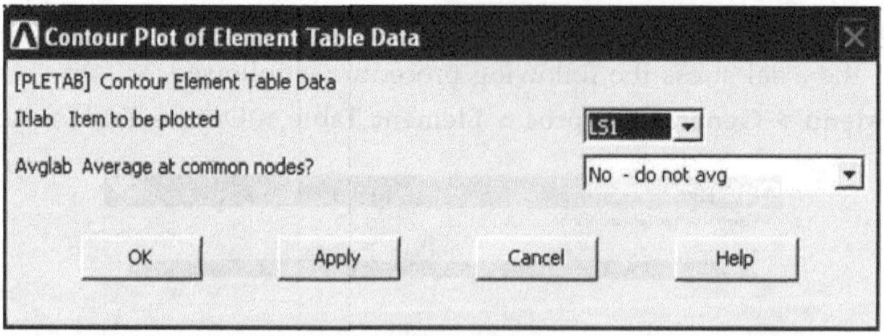

Figure 2.44. Selecting options for finding out axial stress

This result is shown in figure 2.31(*c*).

EXERCISES

1. Determine the nodal displacement and element stress for the bar shown in figure 2.45. Take three elements finite element model. Take E = 70GPa

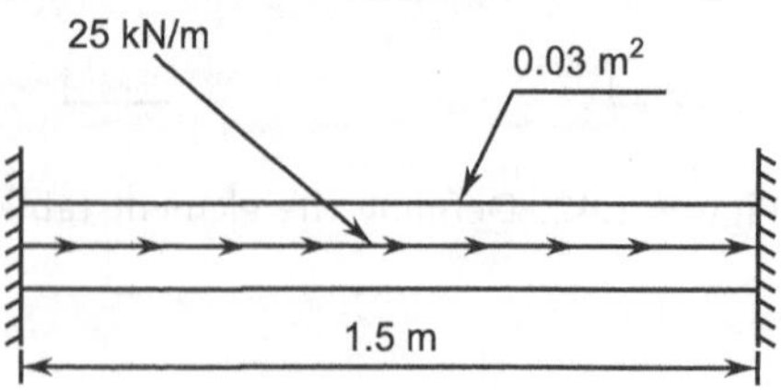

Figure 2.45

2. Determine the nodal displacements and stresses in the element for the axial distributed loading shown in figure 2.46. Take one element model. Take E = 200 GPa, A = 5 × 10^{-4} m^2.

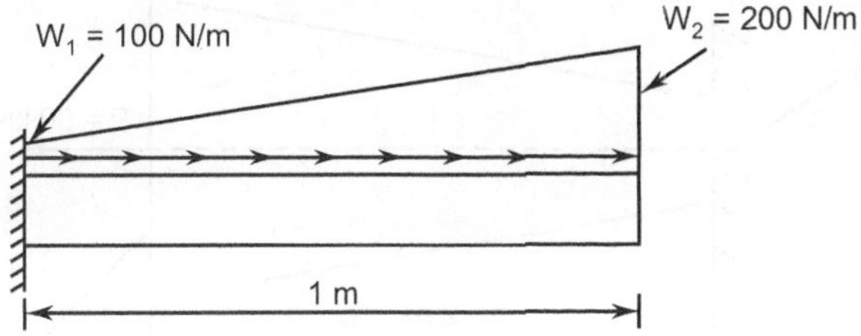

Figure 2.46

3. For the bar assembly shown in figure 2.47, determine the nodal displacements, stresses in each element and reactions. Take E = 210 GPa, $A = 5 \times 10^{-4}$ m^2.

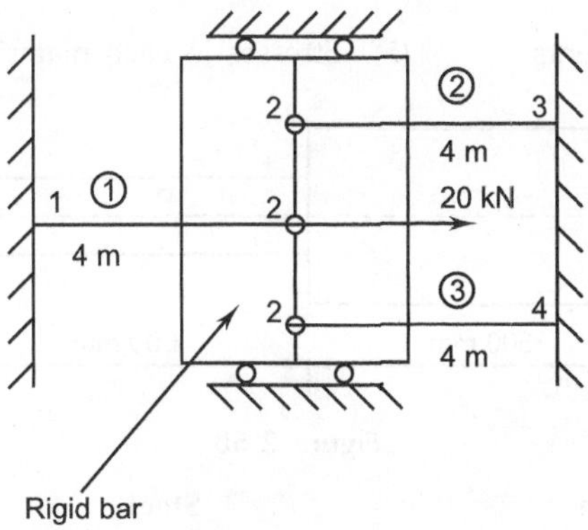

Figure 2.47

4. Find the deflection at the free end under its own weight for a tapered bar shown in figure 2.48. Use 2 element model. Take E = 200 GPa, weight density ρ = 7800 kg/m^3

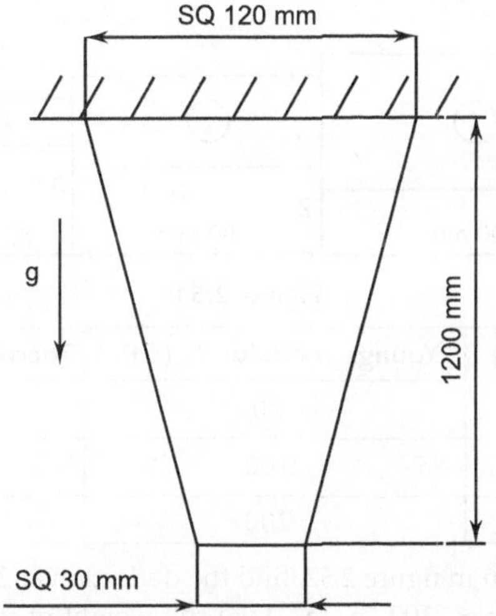

Figure 2.48

5. Determine the displacement, element stresses and reactions for the tapered bar shown in figure 2.49. Use 2 elements finite element model. Take E = 200 GPa, A_1 = 2000 mm^2, A_2 = 4000 mm^2.

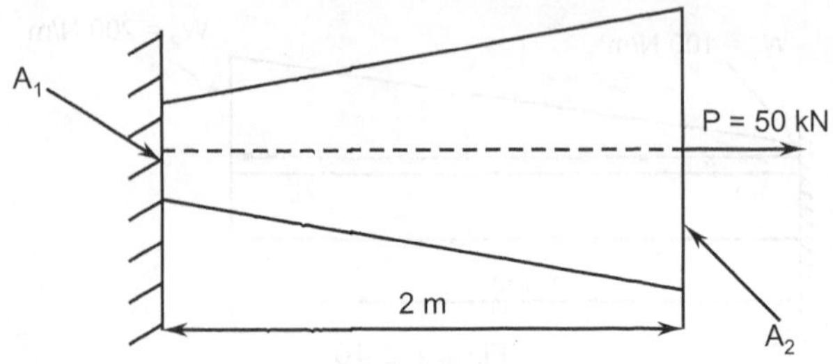

Figure 2.49

6. Consider the bar shown in figure 2.50. An axial load P = 500 kN is applied as shown. Determine the

   (*a*) Nodal displacements (*b*) Stresses in each material (*c*) Reaction forces.

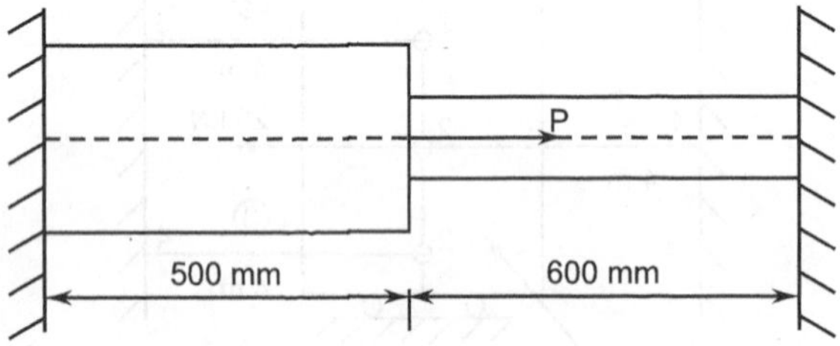

Figure 2.50

Aluminum	**Steel**
$A_1 = 3000\ mm^2$	$A_2 = 1000\ mm^2$
$E_1 = 70$ GPa	$E_2 = 200$ GPa

7. In figure 2.51, determine displacements at 2 and 3, stresses in the members and reactions if the temperature is increased by 60°.

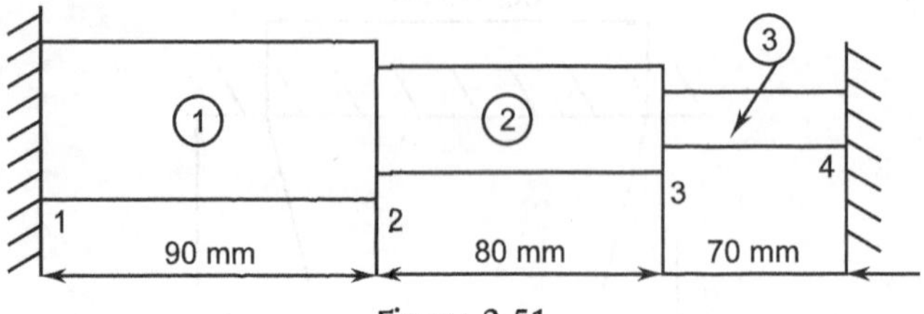

Figure 2.51

Member	Area A (mm^2)	Youngs modulus E (GPa)	Thermal expansion coefficient α (/°C)
1	1000	70	23×10^{-6}
2	500	100	19×10^{-6}
3	300	200	12×10^{-6}

8. For the vertical bar shown in figure 2.52, find the deflection at 2 and 3 and stress distribution. Take E = 25 GPa and density ρ = 2100 kg/m^3. Take self weight of the bar into consideration and solve the problem using 2 elements.

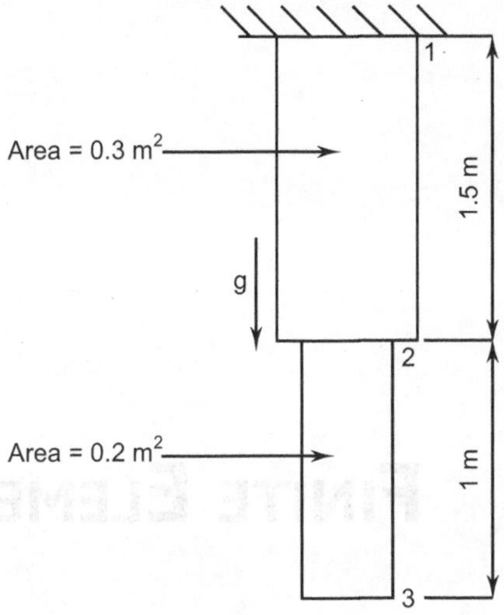

Figure 2.52

9. Find displacement and stresses shown in figure 2.53. Take E = 200 GPa.

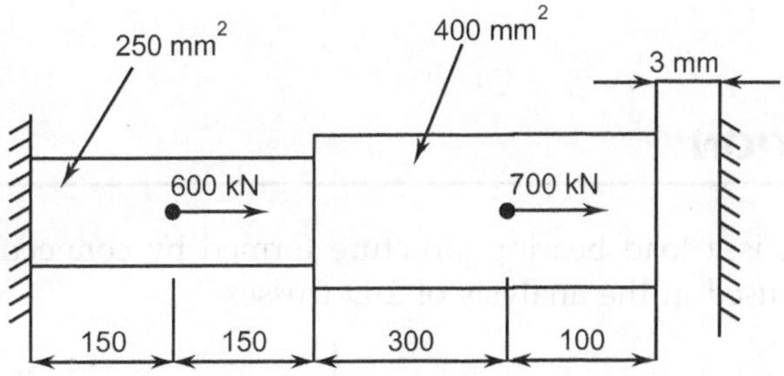

Figure 2.53. All dimensions are in mm

Chapter 3

FINITE ELEMENT ANALYSIS OF TRUSSES

3.1 INTRODUCTION

Truss by definition, is a load bearing structure formed by connecting members using pin joints. Truss element is used in the analysis of 2-D trusses.

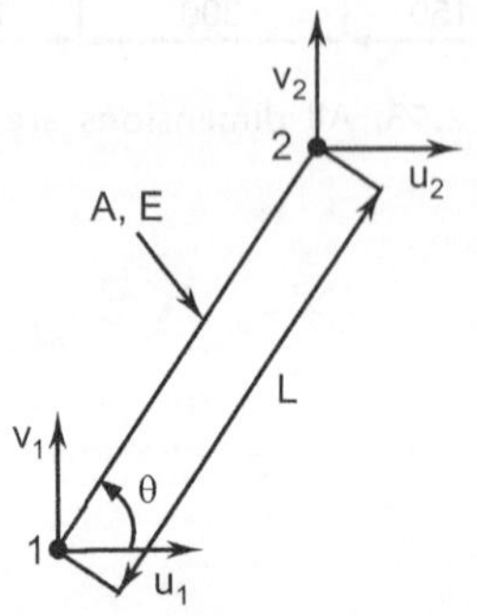

Figure 3.1

This element has two nodes each having two degrees of freedom namely translations along x and y-axes.

The element stiffness matrix and element stress for a truss element are given by,

$$[k] = \frac{AE}{L}\begin{bmatrix} \cos^2\theta & \cos\theta.\sin\theta & -\cos^2\theta & -\cos\theta.\sin\theta \\ \cos\theta.\sin\theta & \sin^2\theta & -\cos\theta.\sin\theta & -\sin^2\theta \\ -\cos^2\theta & -\cos\theta.\sin\theta & \cos^2\theta & \cos\theta.\sin\theta \\ -\cos\theta.\sin\theta & -\sin^2\theta & \cos\theta.\sin\theta & \sin^2\theta \end{bmatrix} = \frac{AE}{L}\begin{bmatrix} c^2 & cs & -c^2 & -cs \\ cs & s^2 & -cs & -s^2 \\ -c^2 & -cs & c^2 & cs \\ -cs & -s^2 & cs & s^2 \end{bmatrix}$$

$$\{\sigma\} = \frac{E}{L}[-\cos\theta \quad -\sin\theta \quad \cos\theta \quad \sin\theta]\{q\}, \quad \text{where } q = \begin{Bmatrix} u_1 \\ v_1 \\ u_2 \\ v_2 \end{Bmatrix}$$

θ = Angle of truss element at node 1 with positive x-axis (in degrees)

PROBLEM 3.1. *Determine the nodal displacements, element stresses and support reactions for the three member truss shown in figure 3.2. Take A = 800 mm² and E = 200 GPa for all members.*

(M.Tech. VTU, Dec. 06/Jan. 07)

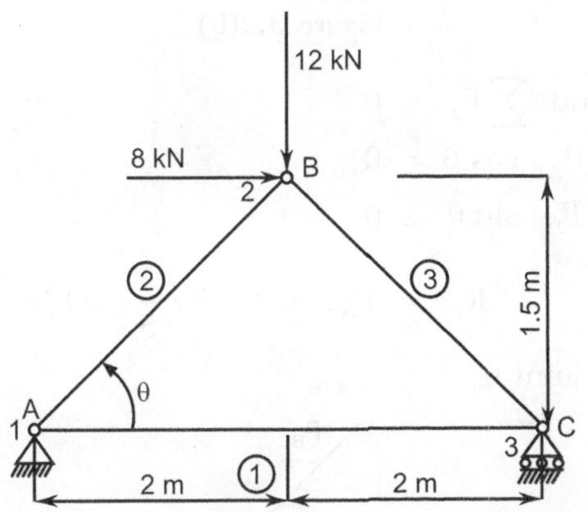

Figure 3.2

SOLUTION. (I) Analytical method. [Refer figure 3.2]

$$AB = AC = \sqrt{(2)^2 + (1.5)^2} = 2.5 \text{ m}$$

$$\sin\theta = \frac{3}{5}, \cos\theta = \frac{4}{5}$$

Consider equilibrium of point B,

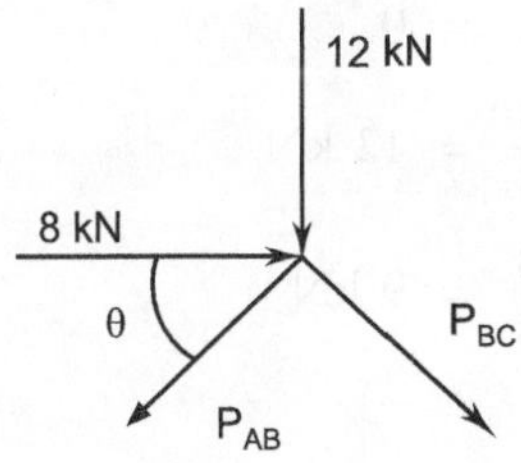

Figure 3.2(a)

$$\sum F_x = 0 \text{ and} \sum F_y = 0$$

$$8 - P_{AB}\cos\theta + P_{BC}\cos\theta = 0 \qquad ...(i)$$

$$-12 - P_{AB}\sin\theta - P_{BC}\sin\theta = 0 \qquad ...(ii)$$

Solving equations (*i*) and (*ii*)

$$P_{AB} = -5 \text{ kN and } P_{BC} = -15 \text{ kN} \qquad (P_{AB} \text{ and } P_{BC} \text{ are compressive})$$

Consider equilibrium of joint A,

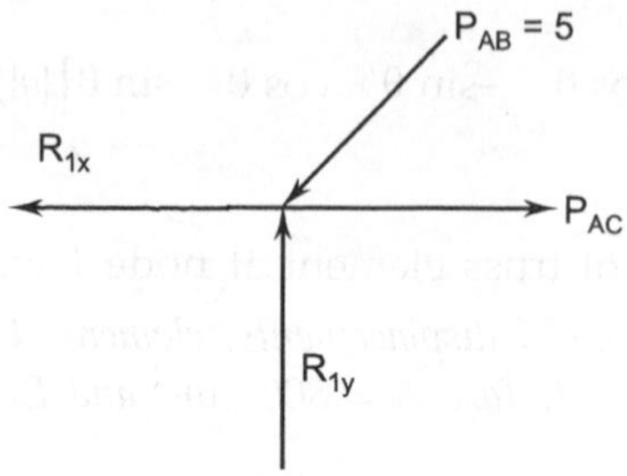

Figure 3.2(b)

$$\sum F_x = 0 \text{ and } \sum F_y = 0$$

$$-R_{1x} + P_{AC} - P_{AB}\cos\theta = 0 \qquad ...(iii)$$

$$R_{1y} - P_{AB}\sin\theta = 0$$

$$R_{1y} = P_{AB}\sin\theta = 5 \times \frac{3}{5} = 3 \text{ kN}$$

Consider equilibrium of joint C,

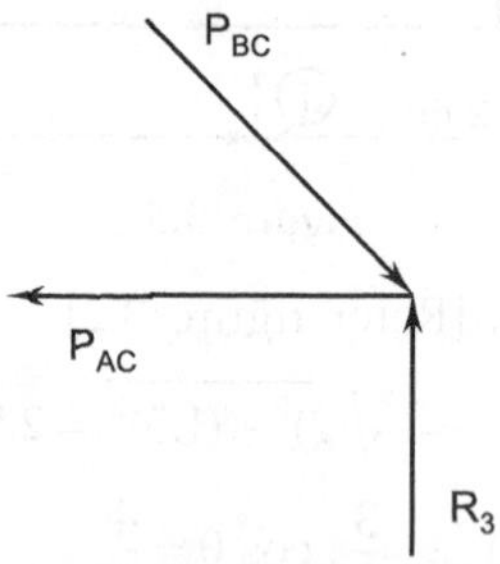

Figure 3.2(c)

$$\sum F_x = 0 \text{ and } \sum F_y = 0$$

$$P_{AC} = P_{BC}\cos\theta = 15 \times \frac{4}{5} = 12 \text{ kN}$$

$$R_3 = P_{BC}\sin\theta = 15 \times \frac{3}{5} = 9 \text{ kN}$$

For equation (*iii*)

$$R_{1x} = P_{AC} - P_{AB}\cos\theta = 12 - 5 \times \frac{4}{5} = 8 \text{ kN}$$

$$\sigma_{AB} = \sigma_2 = \frac{P_{AB}}{A_{AB}} = \frac{-5 \times 10^3}{800} = -6.25 \text{ MPa (Compressive)}$$

$$\sigma_{BC} = \sigma_3 = \frac{P_{BC}}{A_{BC}} = \frac{-15 \times 10^3}{800} = -18.75 \text{ MPa (Compressive)}$$

$$\sigma_{AC} = \sigma_1 = \frac{P_{AC}}{A_{AC}} = \frac{12 \times 10^3}{800} = 15 \text{ MPa (Tensile)}$$

$$\Delta_{AB} = \frac{P_{AB} L_{AB}}{A_{AB} E_{AB}} = \frac{-5 \times 10^3 \times 2500}{800 \times 2 \times 10^5} = -0.078125 \text{ mm}$$

$$\Delta_{BC} = \frac{P_{BC} L_{BC}}{A_{BC} E_{BC}} = \frac{-15 \times 10^3 \times 2500}{800 \times 2 \times 10^5} = -0.234375 \text{ mm}$$

$$\Delta_{AC} = \frac{P_{AC} L_{AC}}{A_{AC} E_{AC}} = \frac{12 \times 10^3 \times 4000}{800 \times 2 \times 10^5} = 0.3 \text{ mm}$$

Calculation of nodal displacements u_2, v_2 and u_3

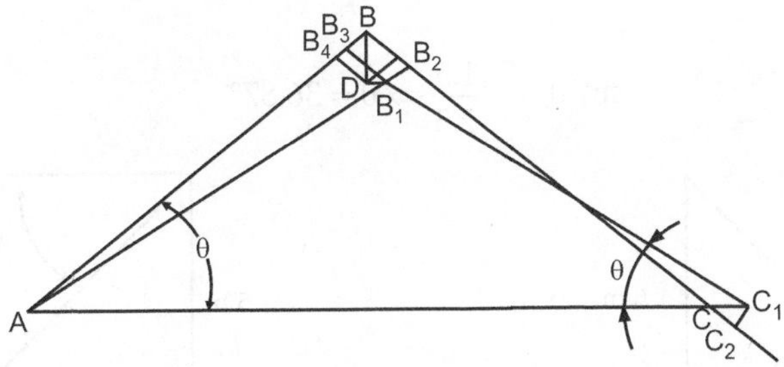

Figure 3.2(d)

$$DB_1 = u_2, \; BD = v_2 \text{ and } CC_1 = u_3$$

$$BB_3 = \Delta_{AB}$$

$$BB_2 = \Delta_{BC}$$

$$CC_1 = u_3 = \Delta_{AC} = 0.3 \text{ mm}$$

$$CC_2 = CC_1 \cos\theta = \Delta_{AC} \cos\theta = 0.3 \times \frac{4}{5} = 0.24 \text{ mm}$$

From geometry [Refer figure 3.2(*d*)]

$$BB_3 = \Delta_{AB} = BD \sin\theta - DB_1 \cos\theta = v_2 \sin\theta - u_2 \cos\theta \qquad \text{...}(iv)$$

$$BB_2 = BC - B_2C = BC - [B_2C_2 - CC_2] = (BC - B_2C_2) + CC_2$$

$$BB_2 = \Delta_{BC} + CC_2$$

$$BB_2 = \Delta_{BC} + CC_2 = BD \sin\theta + DB_1 \cos\theta$$

$$\Delta_{BC} + CC_2 = v_2 \sin\theta + u_2 \cos. \qquad \text{...}(v)$$

Substituting in (*iv*) and (*v*)

$$0.078125 = v_2 \times \frac{3}{5} - u_2 \times \frac{4}{5} \qquad \text{...}(vi)$$

$$0.234375 + 0.24 = v_2 \times \frac{3}{5} + u_2 \times \frac{4}{5}$$

$$0.474375 = v_2 \times \frac{3}{5} + u_2 \times \frac{4}{5} \qquad \text{...}(vii)$$

Solving equations (*vi*) and (*vii*), we get,

$$u_2 = 0.2477 \text{ mm}$$

$v_2 = 0.4604$ mm (since point B moves downwards) Hence $v_2 = -0.4604$ mm

(II) FEM by hand calculation (Refer to figure 3.2)

Elements	Node numbers		θ	cos θ	sin θ	L (mm)
	Local 1	Local 2				
1	1	3	0	1	0	4000
2	1	2	36.87	0.8	0.6	2500
3	2	3	-36.87	0.8	-0.6	2500

Angle calculation

For element 2

$$\sin\theta = \frac{1.5}{2.5} \Rightarrow \theta = 36.87°$$

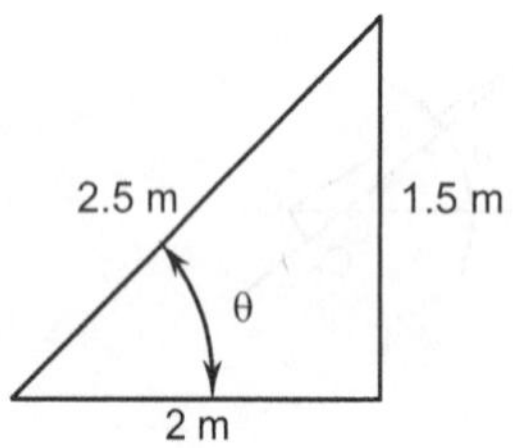

Figure 3.2(e)

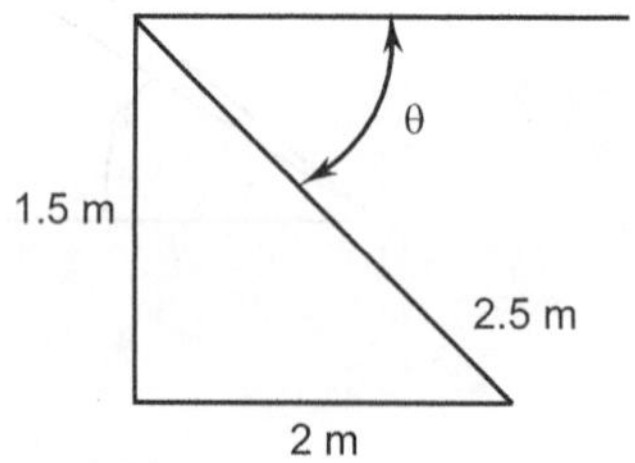

Figure 3.2(f)

For element 3

$$\sin\theta = \frac{1.5}{2.5} \Rightarrow \theta = -36.87°$$

Element stiffness matrix for element 1 is,

$$[k_1] = \frac{AE}{L}\begin{bmatrix} \cos^2\theta & \cos\theta.\sin\theta & -\cos^2\theta & -\cos\theta.\sin\theta \\ \cos\theta.\sin\theta & \sin^2\theta & -\cos\theta.\sin\theta & -\sin^2\theta \\ -\cos^2\theta & -\cos\theta.\sin\theta & \cos^2\theta & \cos\theta.\sin\theta \\ -\cos\theta.\sin\theta & -\sin^2\theta & \cos\theta.\sin\theta & \sin^2\theta \end{bmatrix}$$

$$[k_1] = \frac{800\times 200\times 10^3}{4000}\begin{bmatrix} \cos^2 0 & \cos 0.\sin 0 & -\cos^2 0 & -\cos 0.\sin 0 \\ \cos 0.\sin 0 & \sin^2 0 & -\cos 0.\sin 0 & -\sin^2 0 \\ -\cos^2 0 & -\cos 0.\sin 0 & \cos^2 0 & \cos 0.\sin 0 \\ -\cos 0.\sin 0 & -\sin^2 0 & \cos 0.\sin 0 & \sin^2 0 \end{bmatrix} = 40\times 10^3 \begin{matrix} & u_1 & v_1 & u_3 & v_3 & \\ \end{matrix}\begin{bmatrix} 1 & 0 & -1 & 0 \\ 0 & 0 & 0 & 0 \\ -1 & 0 & 1 & 0 \\ 0 & 0 & 0 & 0 \end{bmatrix}\begin{matrix} u_1 \\ v_1 \\ u_3 \\ v_3 \end{matrix}$$

Element stiffness matrix for element 2 is,

$$[k_2] = \frac{AE}{L}\begin{bmatrix} c^2 & cs & -c^2 & -cs \\ cs & s^2 & -cs & -s^2 \\ -c^2 & -cs & c^2 & cs \\ -cs & -s^2 & cs & s^2 \end{bmatrix} = \frac{800\times 200\times 10^3}{2500}\begin{bmatrix} (0.8)^2 & 0.8\times 0.6 & -(0.8)^2 & -0.8\times 0.6 \\ 0.8\times 0.6 & (0.6)^2 & -0.8\times 0.6 & -(0.6)^2 \\ -(0.8)^2 & -0.8\times 0.6 & (0.8)^2 & 0.8\times 0.6 \\ -0.8\times 0.6 & -(0.6)^2 & 0.8\times 0.6 & (0.6)^2 \end{bmatrix}$$

$$= 64 \times 10^3 \begin{bmatrix} 0.64 & 0.48 & -0.64 & -0.48 \\ 0.48 & 0.36 & -0.48 & -0.36 \\ -0.64 & -0.48 & 0.64 & 0.48 \\ -0.48 & -0.36 & 0.48 & 0.36 \end{bmatrix} \begin{matrix} u_1 \\ v_1 \\ u_2 \\ v_2 \end{matrix}$$

(columns: $u_1 \quad v_1 \quad u_2 \quad v_2$)

Element stiffness matrix for element 3 is,

$$[k_3] = 64 \times 10^3 \begin{bmatrix} 0.64 & -0.48 & -0.64 & 0.48 \\ -0.48 & 0.36 & 0.48 & -0.36 \\ -0.64 & 0.48 & 0.64 & -0.48 \\ 0.48 & -0.36 & -0.48 & 0.36 \end{bmatrix} \begin{matrix} u_2 \\ v_2 \\ u_3 \\ v_3 \end{matrix}$$

(columns: $u_2 \quad v_2 \quad u_3 \quad v_3$)

Global Stiffness matrix is,

$$[K] = 10^3 \begin{bmatrix} 40+40.96 & 30.72 & -40.96 & -30.72 & -40 & 0 \\ 30.72 & 23.04 & -30.72 & -23.04 & 0 & 0 \\ -40.96 & -30.72 & 40.96+40.96 & 30.72-30.72 & -40.96 & 30.72 \\ -30.72 & -23.04 & 30.72+30.72 & 23.04+23.04 & 30.72 & -23.04 \\ -40 & 0 & -40.96 & 30.72 & 40+40.96 & -30.72 \\ 0 & 0 & 30.72 & -23.04 & -30.72 & 23.04 \end{bmatrix} \begin{matrix} u_1 \\ v_1 \\ u_2 \\ v_2 \\ u_3 \\ v_3 \end{matrix}$$

(columns: $u_1 \quad v_1 \quad u_2 \quad v_2 \quad u_3 \quad v_3$)

$$[K] = 10^3 \begin{bmatrix} 80.96 & 30.72 & -40.96 & -30.72 & -40 & 0 \\ 30.72 & 23.04 & -30.72 & -23.04 & 0 & 0 \\ -40.96 & -30.72 & 81.92 & 0 & -40.96 & 30.72 \\ -30.72 & -23.04 & 0 & 46.08 & 30.72 & -23.04 \\ -40 & 0 & -40.96 & 30.72 & 80.96 & -30.72 \\ 0 & 0 & 30.72 & -23.04 & -30.72 & 23.04 \end{bmatrix} \begin{matrix} u_1 \\ v_1 \\ u_2 \\ v_2 \\ u_3 \\ v_3 \end{matrix}$$

(columns: $u_1 \quad v_1 \quad u_2 \quad v_2 \quad u_3 \quad v_3$)

Global equation is,

$$10^3 \begin{bmatrix} 80.96 & 30.72 & -40.96 & -30.72 & -40 & 0 \\ 30.72 & 23.04 & -30.72 & -23.04 & 0 & 0 \\ -40.96 & -30.72 & 81.92 & 0 & -40.96 & 30.72 \\ -30.72 & -23.04 & 0 & 46.08 & 30.72 & -23.04 \\ -40 & 0 & -40.96 & 30.72 & 80.96 & -30.72 \\ 0 & 0 & 30.72 & -23.04 & -30.72 & 23.04 \end{bmatrix} \begin{matrix} u_1 \\ v_1 \\ u_2 \\ v_2 \\ u_3 \\ v_3 \end{matrix} \begin{Bmatrix} u_1 \\ v_1 \\ u_2 \\ v_2 \\ u_3 \\ v_3 \end{Bmatrix} = \begin{Bmatrix} R_{1x} \\ R_{1y} \\ 8 \\ -12 \\ 0 \\ R_{3y} \end{Bmatrix} \times 10^3$$

(columns: $u_1 \quad v_1 \quad u_2 \quad v_2 \quad u_3 \quad v_3$; rows 1, 2 and 6 and columns 1, 2 and 6 are struck out)

Using elimination method for applying boundary conditions,

$$u_1 = v_1 = v_3 = 0$$

Then above matrix reduces to,

$$\cancel{10^3}\begin{bmatrix} 81.92 & 0 & -40.96 \\ 0 & 46.08 & 30.72 \\ -40.96 & 30.72 & 80.96 \end{bmatrix}\begin{bmatrix} u_2 \\ v_2 \\ u_3 \end{bmatrix} = \begin{bmatrix} 8 \\ -12 \\ 0 \end{bmatrix} \times \cancel{10^3}$$

Solving above matrix and equations,

we get $u_2 = 0.2477$ mm, $v_2 = -0.4604$ mm and $v_3 = 0.3$ mm

Stress Calculation

Stress in element 1 is,

$$\sigma_1 = \frac{E}{L_1}[-\cos\theta \quad -\sin\theta \quad \cos\theta \quad \sin\theta]\begin{bmatrix} u_1 \\ v_1 \\ u_3 \\ v_3 \end{bmatrix} = \frac{200 \times 10^3}{4000}[-c \quad -s \quad c \quad s]\begin{bmatrix} u_1 \\ v_1 \\ u_3 \\ v_3 \end{bmatrix}$$

$$= \frac{200 \times 10^3}{4000}[-1 \quad 0 \quad 1 \quad 0]\begin{bmatrix} 0 \\ 0 \\ 0.3 \\ 0 \end{bmatrix} = 15 \text{ MPa}$$

Stress in element 2 is,

$$\sigma_2 = \frac{E}{L_2}[-c \quad -s \quad c \quad s]\begin{bmatrix} u_1 \\ v_1 \\ u_2 \\ v_2 \end{bmatrix} = \frac{200 \times 10^3}{2500}[-0.8 \quad -0.6 \quad 0.8 \quad 0.6]\begin{bmatrix} 0 \\ 0 \\ 0.2477 \\ -0.4604 \end{bmatrix} = -6.249 \text{ MPa}$$

Stress in element 3 is,

$$\sigma_3 = \frac{E}{L_3}[-c \quad -s \quad c \quad s]\begin{bmatrix} u_2 \\ v_2 \\ u_3 \\ v_3 \end{bmatrix} = \frac{200 \times 10^3}{2500}[-0.8 \quad 0.6 \quad 0.8 \quad -0.6]\begin{bmatrix} 0.2477 \\ -0.4604 \\ 0.3 \\ 0 \end{bmatrix} = -18.752 \text{ MPa}$$

Reaction Calculation

From Global equation,

$$-40.96 \times u_2 - 30.72 \times v_2 - 40 \times u_3 = R_{1x}$$

$$-40.96 \times 0.2477 - 30.72 \times (-0.4604) - 40 \times 0.3 = R_{1x}$$

$$R_{1x} = -8 \text{ kN}$$

$$-30.72 \times u_2 - 23.04 \times v_2 = R_{1y}$$

$$-30.72 \times 0.2477 - 23.04 \times (-0.4604) = R_{1y}$$

$$R_{1y} = 3 \text{ kN}$$

$$30.72 \times u_2 - 23.04 \times v_2 - 30.72 \times u_3 = R_{3y}$$

$$30.72 \times 0.2477 - 23.04 \times (-0.4604) - 30.72 \times (0.3) = R_{3y}$$

$$R_{3y} = 9 \text{ kN}$$

(III) Software results

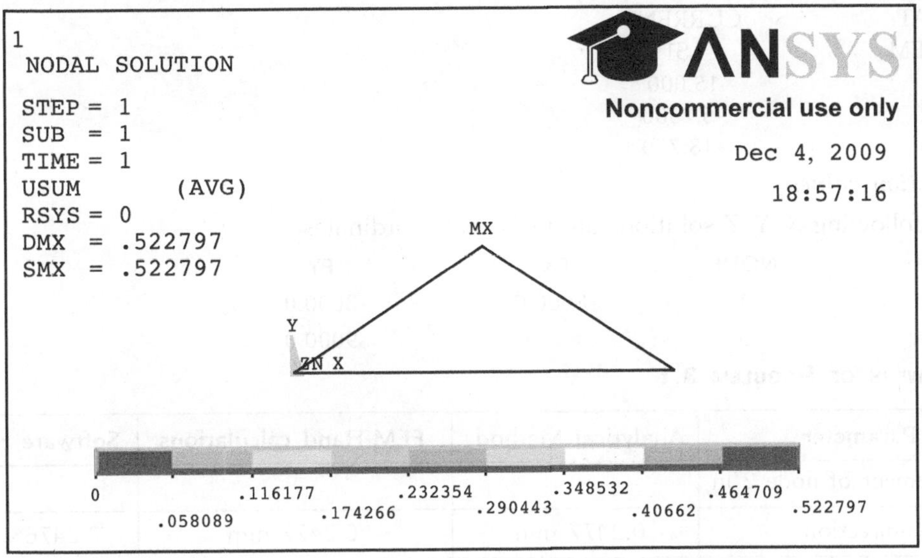

Figure 3.2(g). Deflection pattern for a truss (Refer to appendix for colour figures)

Deflection values at nodes

The following degree of freedom results are in Global coordinates

NODE	UX	UY	UZ	USUM
1	0.0000	0.0000	0.0000	0.0000
2	0.24766	– 0.46042	0.0000	0.52280
3	0.30000	0.0000	0.0000	0.30000

Maximum absolute values

NODE	3	2	0	2
VALUE	0.30000	-0.46042	0.0000	0.52280

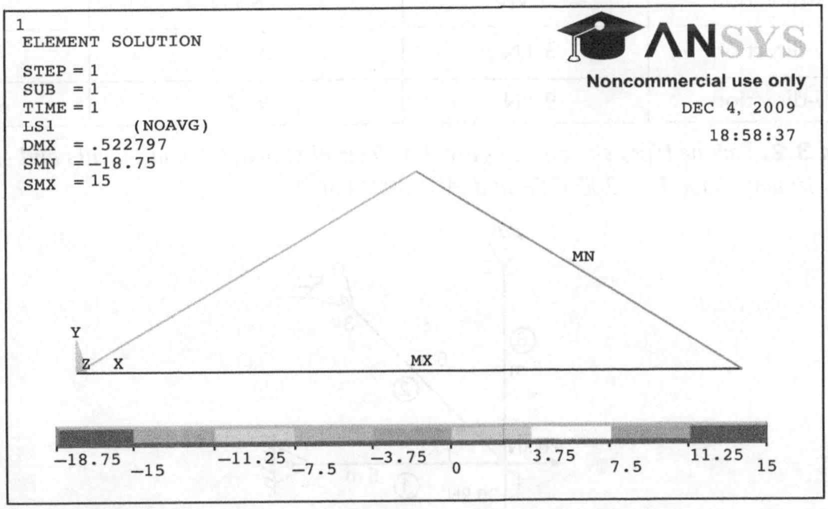

Figure 3.2(h). Stress pattern for a truss (Refer to appendix for colour figures)

Stress values of elements

```
STAT        CURRENT
ELEM          LS1
  1          15.000
  2          -6.2500
  3          -18.750
```

Reaction values

The following X, Y, Z solutions are in Global coordinates

NODE	FX	FY
1	-8000.0	3000.0
3		9000.0

Answers of Problem 3.1

Parameter	Analytical Method	FEM-Hand calculations	Software results
Displacement of node 2 in			
x-direction	0.2477 mm	0.2477 mm	0.24766 mm
y-direction	-0.4604 mm	-0.4604 mm	- 0.46042 mm
Displacement of node 3 in			
x-direction	0.3 mm	0.3 mm	0.3 mm
Stress in			
Element 1	15 MPa	15 MPa	15 MPa
Element 2	-6.25 MPa	-6.248 MPa	-6.25 MPa
Element 3	-18.75 MPa	-18.752 MPa	-18.75 MPa
Reaction			
at 1 in *x*-direction	-8 kN	-8 kN	-8 kN
at 1 in *y*-direction	3 kN	3 kN	3 kN
at 3 in *y*-direction	9 kN	9 kN	9 kN

Problem 3.2. *For the truss shown in figure 3.3, determine nodal displacements and stresses in each member. All elements have E = 200 GPa and A = 500 mm^2.*

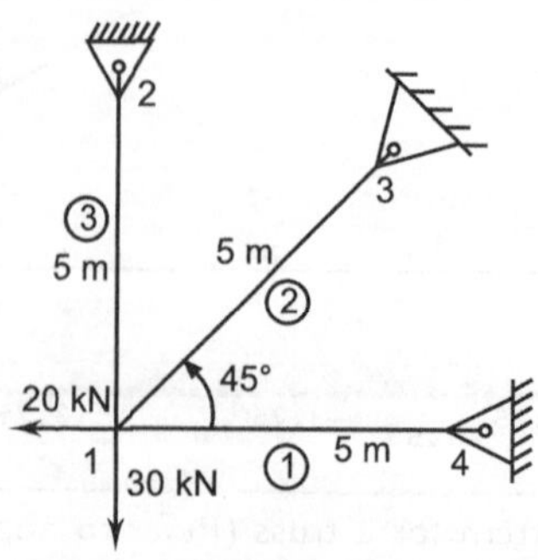

Figure 3.3

SOLUTION.

$$E = 2 \times 10^5 \text{ N/mm}^2$$
$$A = 500 \text{ mm}^2$$

(I) FEM by hand calculation

Elements	Node numbers		θ	cos θ	sin θ	L (mm)
	Local 1	Local 2				
1	1	4	0	1	0	5000
2	1	3	45	0.707	0.707	5000
3	1	2	90	0	1	5000

Stiffness matrices for element 1, 2 and 3 are,

$$[k_1] = \frac{AE}{L}\begin{bmatrix} c^2 & cs & -c^2 & -cs \\ cs & s^2 & -cs & -s^2 \\ -c^2 & -cs & c^2 & cs \\ -cs & -s^2 & cs & s^2 \end{bmatrix} = \frac{500 \times 2 \times 10^5}{5000}\begin{bmatrix} 1 & 0 & -1 & 0 \\ 0 & 0 & 0 & 0 \\ -1 & 0 & 1 & 0 \\ 0 & 0 & 0 & 0 \end{bmatrix}$$

$$= 20 \times 10^3 \begin{array}{c} \begin{matrix} u_1 & v_1 & u_4 & v_4 \end{matrix} \\ \begin{bmatrix} 1 & 0 & -1 & 0 \\ 0 & 0 & 0 & 0 \\ -1 & 0 & 1 & 0 \\ 0 & 0 & 0 & 0 \end{bmatrix} \end{array} \begin{matrix} u_1 \\ v_1 \\ u_4 \\ v_4 \end{matrix}$$

$$[k_2] = 20 \times 10^3 \begin{array}{c} \begin{matrix} u_1 & v_1 & u_3 & v_3 \end{matrix} \\ \begin{bmatrix} 0.5 & 0.5 & -0.5 & -0.5 \\ 0.5 & 0.5 & -0.5 & -0.5 \\ -0.5 & -0.5 & 0.5 & 0.5 \\ -0.5 & -0.5 & 0.5 & 0.5 \end{bmatrix} \end{array} \begin{matrix} u_1 \\ v_1 \\ u_3 \\ v_3 \end{matrix} \quad [k_3] = 20 \times 10^3 \begin{array}{c} \begin{matrix} u_1 & v_1 & u_2 & v_2 \end{matrix} \\ \begin{bmatrix} 0 & 0 & 0 & 0 \\ 0 & 1 & 0 & -1 \\ 0 & 0 & 0 & 0 \\ 0 & -1 & 0 & 1 \end{bmatrix} \end{array} \begin{matrix} u_1 \\ v_1 \\ u_2 \\ v_2 \end{matrix}$$

Global stiffness matrix is,

$$[K] = 20 \times 10^3 \begin{array}{c} \begin{matrix} u_1 & v_1 & u_2 & v_2 & u_3 & v_3 & u_4 & v_4 \end{matrix} \\ \begin{bmatrix} 1+0.5 & 0.5 & 0 & 0 & -0.5 & -0.5 & -1 & 0 \\ 0.5 & 0.5+1 & 0 & -1 & -0.5 & -0.5 & 0 & 0 \\ 0 & 0 & 0 & 0 & 0 & 0 & 0 & 0 \\ 0 & -1 & 0 & 1 & 0 & 0 & 0 & 0 \\ -0.5 & -0.5 & 0 & 0 & 0.5 & 0.5 & 0 & 0 \\ -0.5 & -0.5 & 0 & 0 & 0.5 & 0.5 & 0 & 0 \\ -1 & 0 & 0 & 0 & 0 & 0 & 1 & 0 \\ 0 & 0 & 0 & 0 & 0 & 0 & 0 & 0 \end{bmatrix} \end{array} \begin{matrix} u_1 \\ v_1 \\ u_2 \\ v_2 \\ u_3 \\ v_3 \\ u_4 \\ v_4 \end{matrix}$$

$$[K] = 20 \times 10^3 \begin{array}{c} \begin{matrix} u_1 & v_1 & u_2 & v_2 & u_3 & v_3 & u_4 & v_4 \end{matrix} \\ \begin{bmatrix} 1.5 & 0.5 & 0 & 0 & -0.5 & -0.5 & -1 & 0 \\ 0.5 & 1.5 & 0 & -1 & -0.5 & -0.5 & 0 & 0 \\ 0 & 0 & 0 & 0 & 0 & 0 & 0 & 0 \\ 0 & -1 & 0 & 1 & 0 & 0 & 0 & 0 \\ -0.5 & -0.5 & 0 & 0 & 0.5 & 0.5 & 0 & 0 \\ -0.5 & -0.5 & 0 & 0 & 0.5 & 0.5 & 0 & 0 \\ -1 & 0 & 0 & 0 & 0 & 0 & 1 & 0 \\ 0 & 0 & 0 & 0 & 0 & 0 & 0 & 0 \end{bmatrix} \end{array} \begin{matrix} u_1 \\ v_1 \\ u_2 \\ v_2 \\ u_3 \\ v_3 \\ u_4 \\ v_4 \end{matrix}$$

Global equation is,

$$20 \times 10^3 \begin{array}{c} \begin{matrix} u_1 & v_1 & u_2 & v_2 & u_3 & v_3 & u_4 & v_4 \end{matrix} \\ \begin{bmatrix} 1.5 & 0.5 & 0 & 0 & -0.5 & -0.5 & -1 & 0 \\ 0.5 & 1.5 & 0 & -1 & -0.5 & -0.5 & 0 & 0 \\ 0 & 0 & 0 & 0 & 0 & 0 & 0 & 0 \\ 0 & -1 & 0 & 1 & 0 & 0 & 0 & 0 \\ -0.5 & -0.5 & 0 & 0 & 0.5 & 0.5 & 0 & 0 \\ -0.5 & -0.5 & 0 & 0 & 0.5 & 0.5 & 0 & 0 \\ -1 & 0 & 0 & 0 & 0 & 0 & 1 & 0 \\ 0 & 0 & 0 & 0 & 0 & 0 & 0 & 0 \end{bmatrix} \end{array} \begin{bmatrix} u_1 \\ v_1 \\ u_2 \\ v_2 \\ u_3 \\ v_3 \\ u_4 \\ v_4 \end{bmatrix} = \begin{bmatrix} -20 \\ -30 \\ R_{2x} \\ R_{2y} \\ R_{3x} \\ R_{3y} \\ R_{4x} \\ R_{4y} \end{bmatrix} \times 10^3$$

Using elimination method for applying boundary conditions,

$$u_2 = v_2 = u_3 = v_3 = u_4 = v_4 = 0$$

Then above matrix reduces to,

$$20 \begin{bmatrix} 1.5 & 0.5 \\ 0.5 & 1.5 \end{bmatrix} \begin{bmatrix} u_1 \\ v_1 \end{bmatrix} = \begin{bmatrix} -20 \\ -30 \end{bmatrix}$$

Solving above matrix and equations,

we get,

$$u_1 = -0.375 \text{ mm}$$
$$v_1 = -0.875 \text{ mm}$$

Stress calculation

Stress in element 1 is

$$\sigma_1 = \frac{E}{L_1}[-c \quad -s \quad c \quad s] \begin{bmatrix} u_1 \\ v_1 \\ u_4 \\ v_4 \end{bmatrix} = \frac{2 \times 10^5}{5000}[-1 \quad 0 \quad 1 \quad 0] \begin{bmatrix} -0.375 \\ -0.875 \\ 0 \\ 0 \end{bmatrix} = 15 \text{ MPa}$$

Stress in element 2 is,

$$\sigma_2 = \frac{E}{L_2}\begin{bmatrix}-c & -s & c & s\end{bmatrix}\begin{bmatrix}u_1\\v_1\\u_3\\v_3\end{bmatrix} = \frac{2\times10^5}{5000}\begin{bmatrix}-0.707 & -0.707 & 0.707 & 0.707\end{bmatrix}\begin{bmatrix}-0.375\\-0.875\\0\\0\end{bmatrix} = 35.352 \text{ MPa}$$

Stress in element 3 is,

$$\sigma_3 = \frac{E}{L_3}\begin{bmatrix}-c & -s & c & s\end{bmatrix}\begin{bmatrix}u_1\\v_1\\u_2\\v_2\end{bmatrix} = \frac{2\times10^5}{5000}\begin{bmatrix}0 & -1 & 0 & 1\end{bmatrix}\begin{bmatrix}-0.375\\-0.875\\0\\0\end{bmatrix} = 35 \text{ MPa}$$

(II) Software results

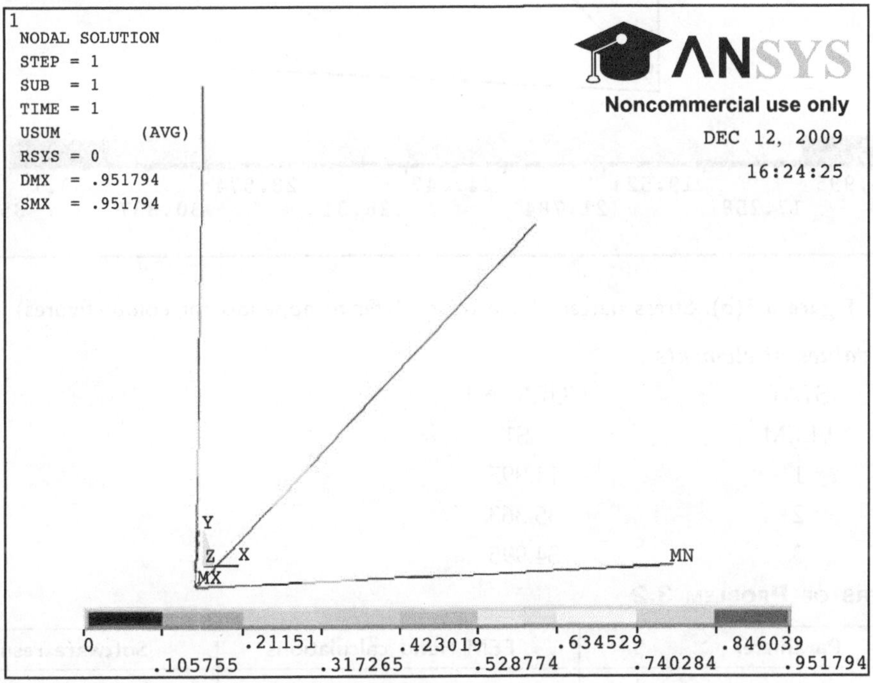

Figure 3.3(a). Deflection pattern for a truss (Refer to appendix for colour figures)

Deflection values at nodes

The following degree of freedom results are in the global coordinate system

NODE	UX	UY	UZ	USUM
1	-0.37486	-0.87486	0.0000	0.95179
2	0.0000	0.0000	0.0000	0.0000
3	0.0000	0.0000	0.0000	0.0000
4	0.0000	0.0000	0.0000	0.0000

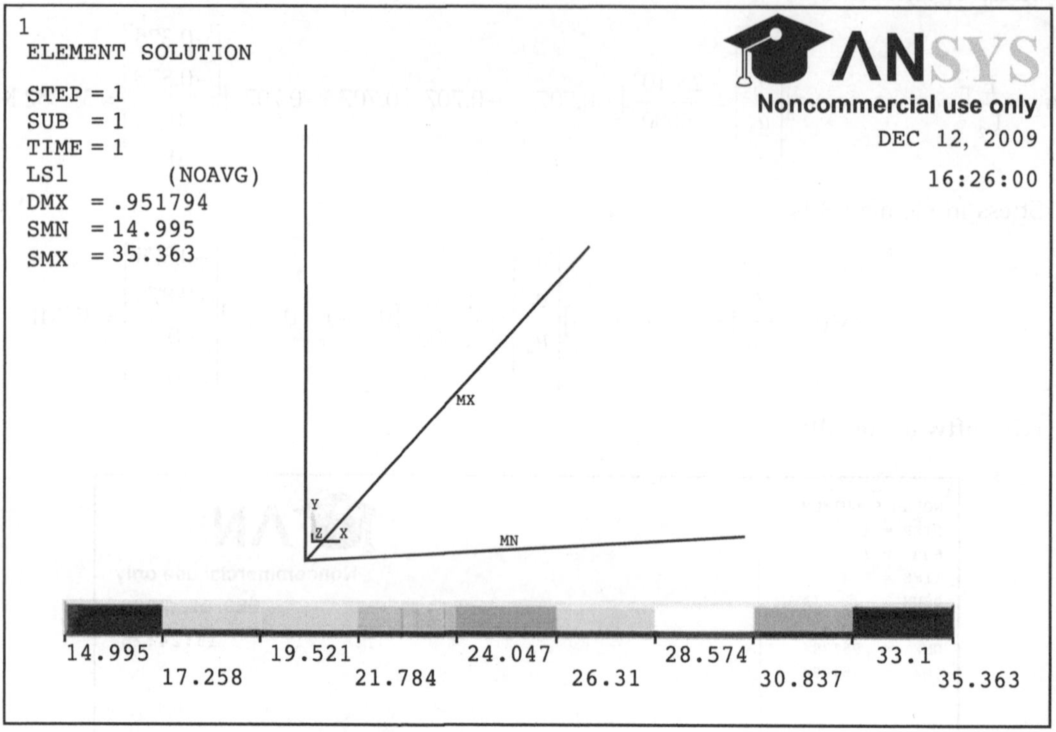

Figure 3.3(b). Stress pattern for a truss (Refer to appendix for colour figures)

Stress values at elements

STAT	CURRENT
ELEM	LS1
1	14.995
2	35.363
3	34.995

Answers of Problem 3.2

Parameter	FEM-Hand calculations	Software results
Displacement of node 1 in		
x-direction	-0.375 mm	-0.37486 mm
y-direction	-0.875 mm	-0.87486 mm
Stress in element 1	15 MPa	14.995 MPa
Stress in element 2	35.352 MPa	35.363 MPa
Stress in element 3	35 MPa	34.995 MPa

Problem 3.3. *For the two bar truss shown in figure 3.4, determine the displacement of node 1 and the axial stress in each member. Take E = 210 GPa and A = 600 mm^2. Solve the problem if node 1 settles an amount of δ = 25 mm in the negative y-direction.*

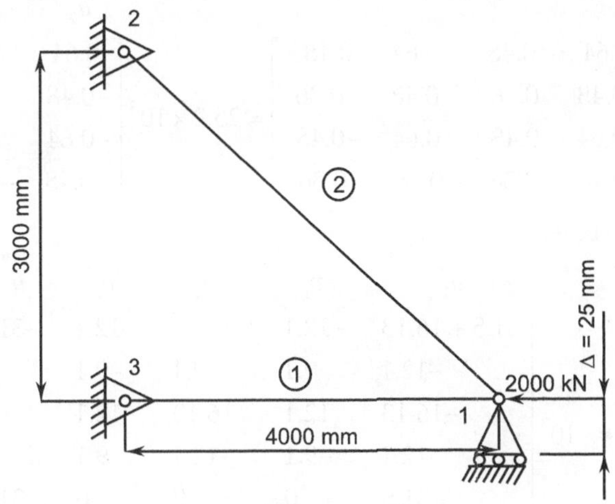

Figure 3.4

SOLUTION. (I) FEM by hand calculation

Elements	Node numbers		θ	cos θ	sin θ	L (mm)
	Local 1	Local 2				
1	3	1	0	1	0	4000
2	2	1	-36.87	0.8	-0.6	5000

Angle calculation

For 2nd element,

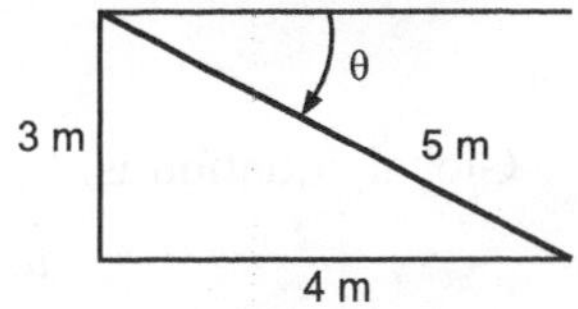

Figure 3.4(a)

$$\sin\theta = \frac{3}{5} = 0.6 \Rightarrow \theta = -36.87°$$

Stiffness matrices for element 1 and 2 are,

$$[k_1] = \frac{AE}{L}\begin{bmatrix} c^2 & cs & -c^2 & -cs \\ cs & s^2 & -cs & -s^2 \\ -c^2 & -cs & c^2 & cs \\ -cs & -s^2 & cs & s^2 \end{bmatrix} = \frac{600\times 210\times 10^3}{4000}\begin{bmatrix} 1 & 0 & -1 & 0 \\ 0 & 0 & 0 & 0 \\ -1 & 0 & 1 & 0 \\ 0 & 0 & 0 & 0 \end{bmatrix}$$

$$= 31.5\times 10^3 \begin{bmatrix} 1 & 0 & -1 & 0 \\ 0 & 0 & 0 & 0 \\ -1 & 0 & 1 & 0 \\ 0 & 0 & 0 & 0 \end{bmatrix} \begin{matrix} u_3 \\ v_3 \\ u_1 \\ v_1 \end{matrix} \quad (\text{columns: } u_3\ v_3\ u_1\ v_1)$$

$$[k_2] = \frac{600 \times 210 \times 10^3}{5000}\begin{bmatrix} 0.64 & -0.48 & -0.64 & 0.48 \\ -0.48 & 0.36 & 0.48 & -0.36 \\ -0.64 & 0.48 & 0.64 & -0.48 \\ 0.48 & -0.36 & -0.48 & 0.36 \end{bmatrix} = 25.2 \times 10^3 \begin{array}{c} \begin{matrix} u_2 & \quad v_2 & \quad u_1 & \quad v_1 \end{matrix} \\ \begin{bmatrix} 0.64 & -0.48 & -0.64 & 0.48 \\ -0.48 & 0.36 & 0.48 & -0.36 \\ -0.64 & 0.48 & 0.64 & -0.48 \\ 0.48 & -0.36 & -0.48 & 0.36 \end{bmatrix} \end{array} \begin{matrix} u_2 \\ v_2 \\ u_1 \\ v_1 \end{matrix}$$

Global Stiffness matrix is,

$$[\mathrm{K}] = 10^3 \begin{array}{c} \begin{matrix} u_1 & \quad v_1 & \quad u_2 & \quad v_2 & \quad u_3 & \quad v_3 \end{matrix} \\ \begin{bmatrix} 31.5+16.13 & -12.1 & -16.13 & 12.1 & -31.5 & 0 \\ -12.1 & 9.1 & 12.1 & -9.1 & 0 & 0 \\ -16.13 & 12.1 & 16.13 & -12.1 & 0 & 0 \\ 12.1 & -9.1 & -12.1 & 9.1 & 0 & 0 \\ -31.5 & 0 & 0 & 0 & 31.5 & 0 \\ 0 & 0 & 0 & 0 & 0 & 0 \end{bmatrix} \end{array} \begin{matrix} u_1 \\ v_1 \\ u_2 \\ v_2 \\ u_3 \\ v_3 \end{matrix}$$

$$[\mathrm{K}] = 10^3 \begin{array}{c} \begin{matrix} u_1 & \quad v_1 & \quad u_2 & \quad v_2 & \quad u_3 & \quad v_3 \end{matrix} \\ \begin{bmatrix} 47.63 & -12.1 & -16.13 & 12.1 & -31.5 & 0 \\ -12.1 & 9.1 & 12.1 & -9.1 & 0 & 0 \\ -16.13 & 12.1 & 16.13 & -12.1 & 0 & 0 \\ 12.1 & -9.1 & -12.1 & 9.1 & 0 & 0 \\ -31.5 & 0 & 0 & 0 & 31.5 & 0 \\ 0 & 0 & 0 & 0 & 0 & 0 \end{bmatrix} \end{array} \begin{matrix} u_1 \\ v_1 \\ u_2 \\ v_2 \\ u_3 \\ v_3 \end{matrix}$$

Global equation is,

$$10^3 \begin{array}{c} \begin{matrix} u_1 & \quad v_1 & \quad u_2 & \quad v_2 & \quad u_3 & \quad v_3 \end{matrix} \\ \begin{bmatrix} 47.63 & -12.1 & -16.13 & 12.1 & -31.5 & 0 \\ -12.1 & 9.1 & 12.1 & -9.1 & 0 & 0 \\ -16.13 & 12.1 & 16.13 & -12.1 & 0 & 0 \\ 12.1 & -9.1 & -12.1 & 9.1 & 0 & 0 \\ -31.5 & 0 & 0 & 0 & 31.5 & 0 \\ 0 & 0 & 0 & 0 & 0 & 0 \end{bmatrix} \end{array} \begin{bmatrix} u_1 \\ v_1 \\ u_2 \\ v_2 \\ u_3 \\ v_3 \end{bmatrix} = 10^3 \begin{bmatrix} -2000 \\ R_{1y} \\ R_{2x} \\ R_{2y} \\ R_{3x} \\ R_{3y} \end{bmatrix}$$

Using elimination method for applying boundary conditions,

i.e., $u_2 = v_2 = u_3 = v_3 = 0$

Then above matrix reduces to,

$$\begin{bmatrix} 47.63 & -12.1 \\ -12.1 & 9.1 \end{bmatrix}\begin{bmatrix} u_1 \\ v_1 \end{bmatrix} = \begin{bmatrix} -2000 \\ R_{1y} \end{bmatrix}$$

We know that $v_1 = -25$ mm substitute this in the above matrix,

Then,

$$\begin{bmatrix} 47.63 & -12.1 \\ -12.1 & 9.1 \end{bmatrix}\begin{bmatrix} u_1 \\ -25 \end{bmatrix} = \begin{bmatrix} -2000 \\ R_{1y} \end{bmatrix}$$

Solving the above matrix and equations we get,

$$u_1 = -48.34 \text{ mm}$$

Stress calculation

Stress in element 1 is,

$$\sigma_1 = \frac{E}{L_1}\begin{bmatrix}-c & -s & c & s\end{bmatrix}\begin{bmatrix}u_3\\v_3\\u_1\\v_1\end{bmatrix} = \frac{210\times10^3}{4000}\begin{bmatrix}-1 & 0 & 1 & 0\end{bmatrix}\begin{bmatrix}0\\0\\-48.34\\-25\end{bmatrix} = -2537.85 \text{ MPa}$$

Stress in element 2 is,

$$\sigma_2 = \frac{E}{L_2}\begin{bmatrix}-c & -s & c & s\end{bmatrix}\begin{bmatrix}u_2\\v_2\\u_1\\v_1\end{bmatrix} = \frac{210\times10^3}{5000}\begin{bmatrix}-0.8 & 0.6 & 0.8 & -0.6\end{bmatrix}\begin{bmatrix}0\\0\\-48.34\\-25\end{bmatrix} = -994.22 \text{ MPa}$$

(II) Software results

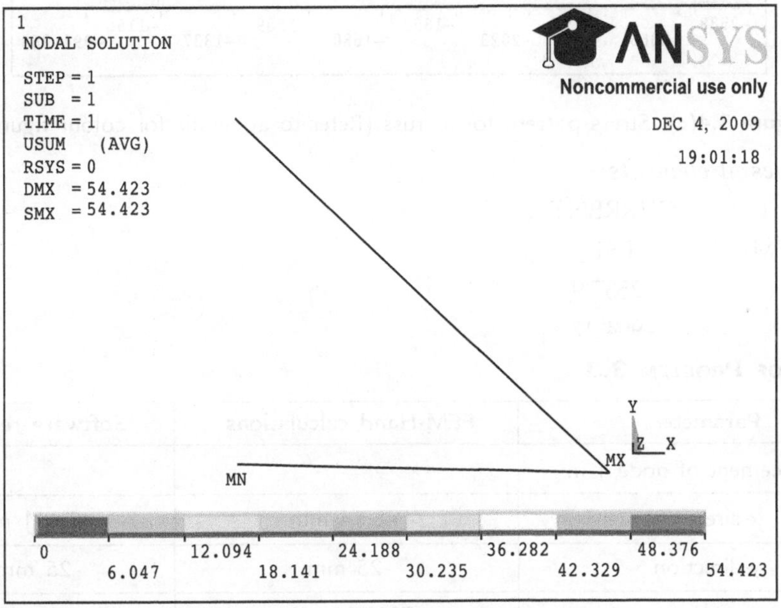

Figure 3.4(b). Deflection pattern for a truss (Refer to appendix for colour figures)

Deflection values at nodes

The following degree of freedom results are in global coordinates

NODE	UX	UY	UZ	USUM
1	-48.341	-25.000	0.0000	54.423
2	0.0000	0.0000	0.0000	0.0000
3	0.0000	0.0000	0.0000	0.0000

Maximum absolute values

NODE	1	1	0	1
VALUE	-48.341	-25.000	0.0000	54.423

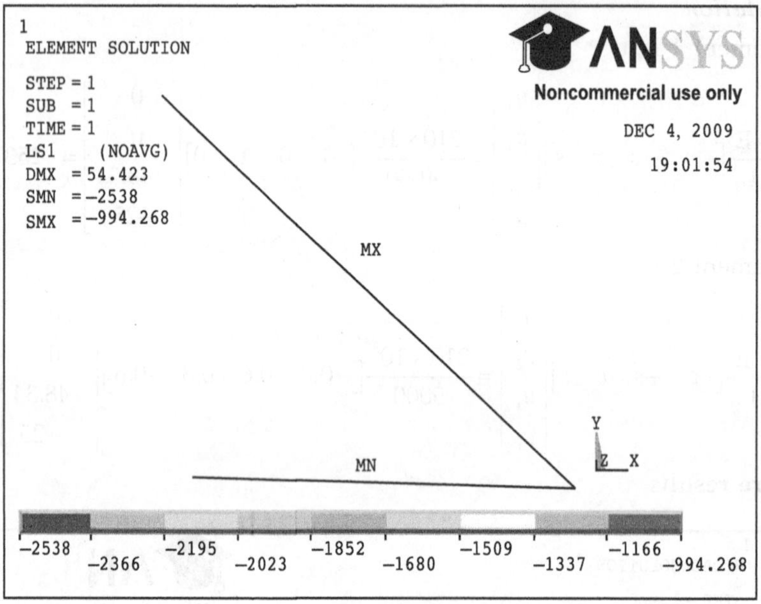

Figure 3.4(c). Stress pattern for a truss (Refer to appendix for colour figures)

Stress values at elements

STAT	CURRENT
ELEM	LS1
1	–2537.9
2	–994.27

ANSWERS OF PROBLEM 3.3

Parameter	FEM-Hand calculations	Software results
Displacement of node 1 in		
x-direction	-48.34 mm	-48.341 mm
y-direction	-25 mm	-25 mm
Stress in element 1	-2537.85 MPa	-2537.9 MPa
Stress in element 2	-994.22 MPa	-994.27 MPa

Procedure for solving the problems using ANSYS®12.0 academic teaching software

FOR PROBLEM 3.3

PREPROCESSING

1. **Main Menu > Preprocessor > Element Type > Add/Edit/Delete > Add >Structural Link > 2D spar 1 > OK>Close**

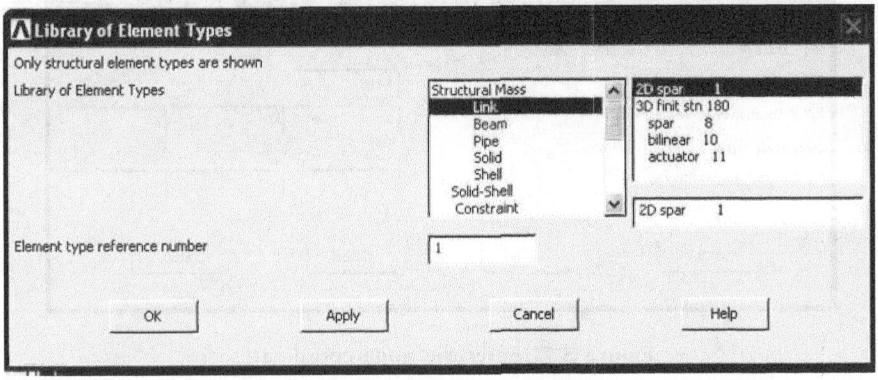

Figure 3.5. Element selection

2. **Main Menu > Preprocessor > Real Constants > Add/Edit/Delete > Add > OK**

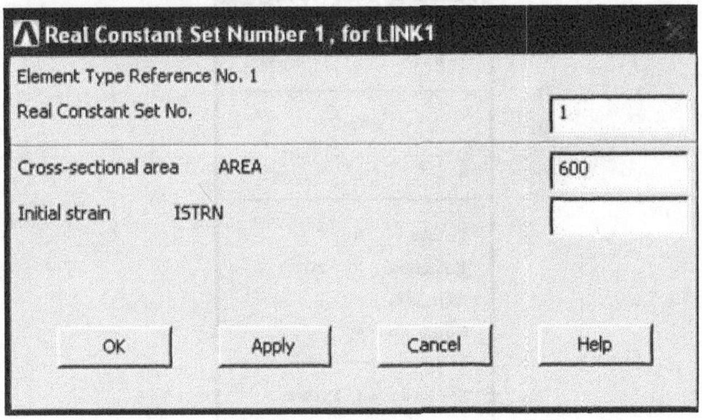

Figure 3.6. Enter the cross-sectional area

Cross-sectional area AREA > **Enter 600 > OK > Close**

Enter the material properties.

3. **Main Menu > Preprocessor > Material Props > Material Models**

   Material Model Number 1, click **Structural > Linear > Elastic > Isotropic**

   Enter **EX = 2.1E5 and PRXY = 0.3 > OK**

   (**Close** the Define Material Model Behaviour window.)

   Create the nodes and elements as shown in the figure.

4. **Main Menu > Preprocessor > Modeling > Create > Nodes > In Active CS** Enter the coordinates of node 1 > **Apply** Enter the coordinates of node 2 > **Apply** Enter the coordinates of node 3 > **OK**

	Node locations	
Node number	**X-coordinate**	**Y-coordinate**
1	0	0
2	-4000	3000
3	-4000	0

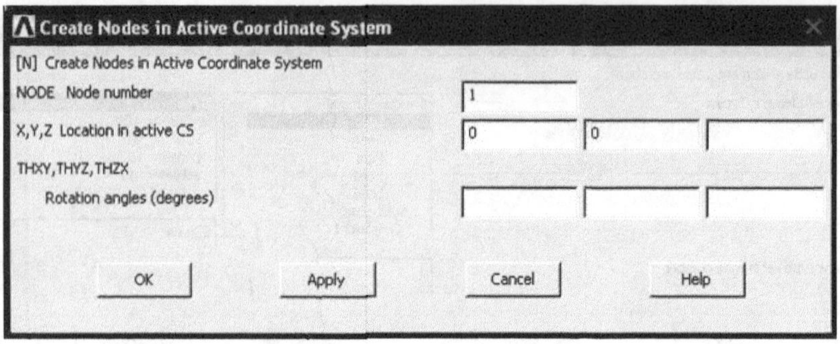

Figure 3.7. Enter the node coordinates

5. **Main Menu > Preprocessor > Modeling > Create > Elements > Auto Numbered > Thru nodes** Pick the 1st and 2nd node > **Apply** Pick the 1st and 3rd node > **OK**

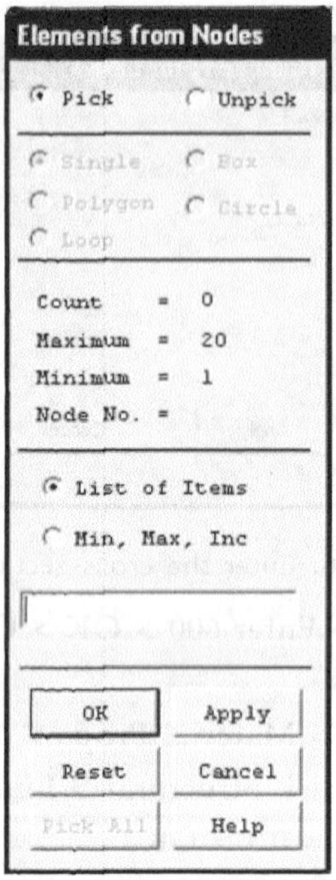

Figure 3.8. Pick the nodes to create elements

Apply the displacement boundary conditions and loads.

6. **Main Menu > Preprocessor > Loads > Define Loads > Apply > Structural >Displacement > On Nodes** Pick the 2nd and 3rd node > **Apply > All DOF = 0 > OK**
7. **Main Menu > Preprocessor > Loads > Define Loads > Apply > Structural >Displacement > On Nodes** Pick the 1st node > **Apply > UY = –25 > OK**
8. **Main Menu > Preprocessor > Loads > Define Loads > Apply > Structural >Force/ Moment > On Nodes** Pick the 1st node > **OK > Force/Moment value = –2000e3 > OK**

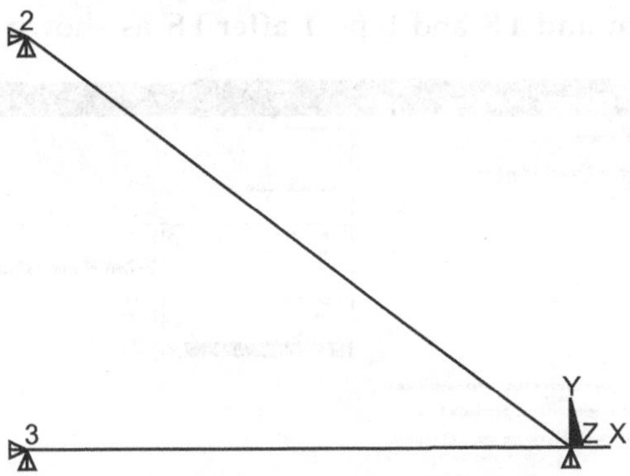

Figure 3.9. Model with loading and displacement boundary conditions

The model-building step is now complete, and we can proceed to the solution. First to be safe, save the model.

SOLUTION. The interactive solution proceeds.

9. **Main Menu > Solution > Solve > Current LS > OK**

The **/STATUS Command** window displays the problem parameters and **the Solve Current Load Step** window is shown. Check the solution options in the **/STATUS** window and if all is OK, select **File > Close**

In the Solve **Current Load Step** window, Select **OK**, and when the solution is complete, **close** the **'Solution is Done!'** window.

POSTPROCESSING

We can now plot the results of this analysis and also list the computed values.

10. **Main Menu > General Postproc > Plot Results > Contour Plot > Nodal Solu >DOF Solution > Displacement vector sum > OK**

This result is shown in figure 3.4(*b*)

To find the axial stress the following procedure is followed.

11. **Main Menu > General Postproc > Element Table > Define Table > Add**

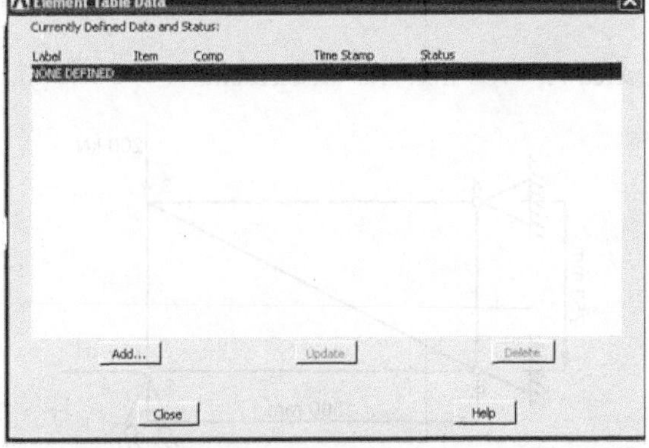

Figure 3.10. Defining the element table

Select **By sequence num and LS** and type **1 after LS** as shown in figure 3.11

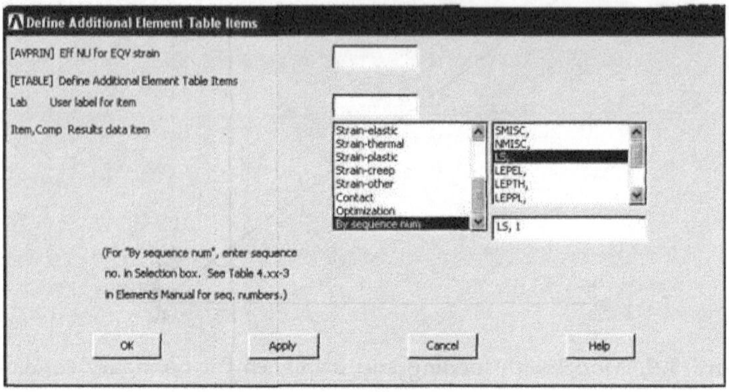

Figure 3.11. Selecting options in element table

> OK

12. **Main Menu > General Postproc > Plot Results > Contour Plot > Elem Table > Select LS1 > OK**

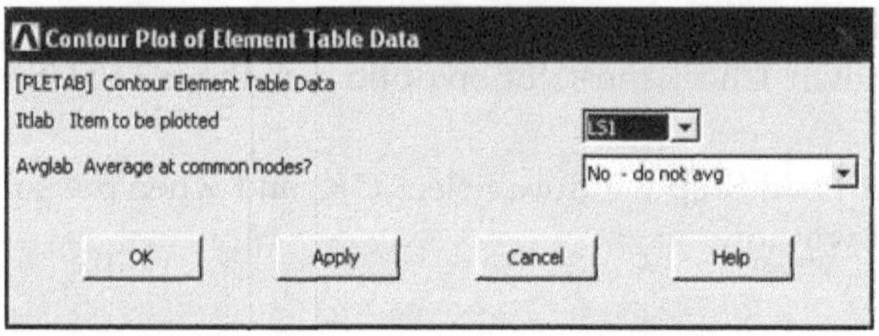

Figure 3.12. Selecting options for finding out axial stress

This result is shown in figure 3.4(*c*)

EXERCISES

1. For a five bar truss shown in figure 3.13, determine the following,
   (*a*) nodal displacements
   (*b*) stresses in each element
   (*c*) reaction forces.

Take E = 200 GPa and Area A = 750 mm^2 for all elements

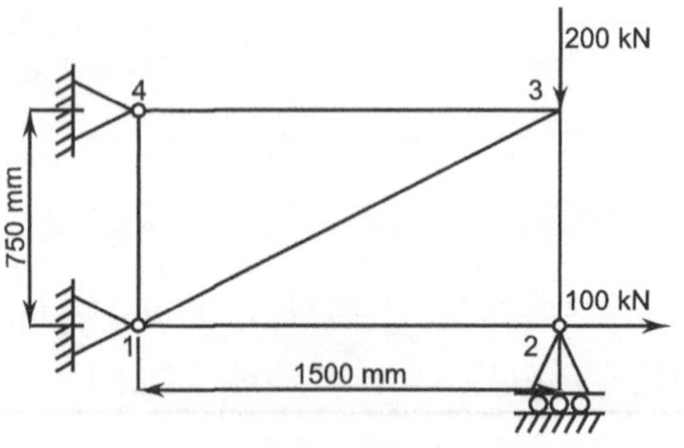

Figure 3.13

2. For the three-bar truss shown in figure 3.14, determine the displacement of node 1 and the stresses in elements. Take A = 300 mm^2 and E = 210 GPa.

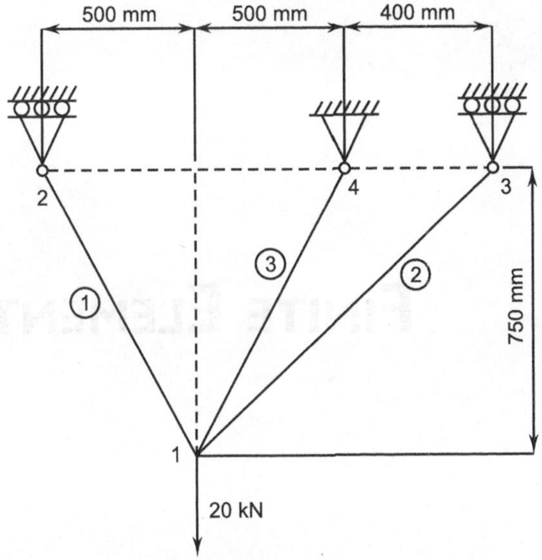

Figure 3.14

3. Consider the truss shown in figure 3.15, determine the nodal displacements, element stresses and reactions. Take E = 200 GPa, $A_1 = A_2 = A_3 = 500$ mm^2, P_1 = 300 kN, P_2 = 200 kN

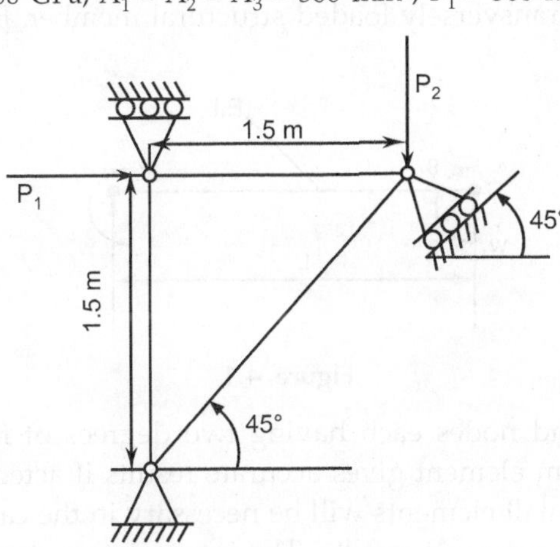

Figure 3.15

Chapter 4

FINITE ELEMENT ANALYSIS OF BEAMS

4.1 INTRODUCTION

Beam by definition is a transversely loaded structural member. Beam element is used in the analysis of beams.

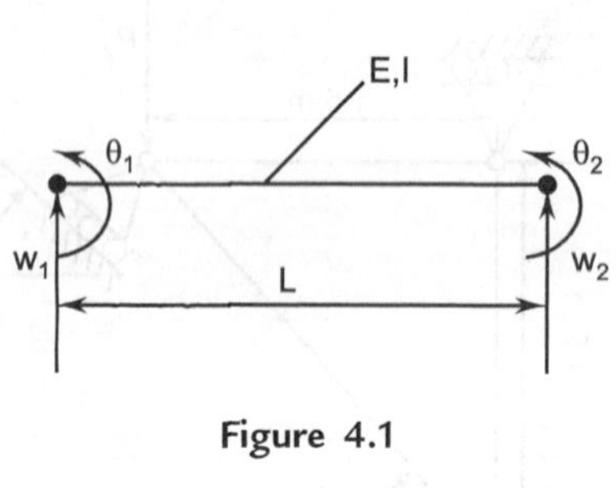

Figure 4.1

This element has two end nodes each having two degrees of freedom, namely transverse displacement and slope. Beam element gives accurate results if acted upon by nodal forces and moments. More number of small elements will be necessary in the case of a beam acted upon by distributed loads in order to get good results. The interpolation equation and element stiffness matrix for beam element are given by

$$w = \begin{bmatrix} N_1 & N_2 & N_3 & N_4 \end{bmatrix} \begin{Bmatrix} w_1 \\ \theta_1 \\ w_2 \\ \theta_2 \end{Bmatrix}$$

$$[K] = \frac{EI}{L^3}\begin{bmatrix} 12 & 6L & -12 & 6L \\ 6L & 4L^2 & -6L & 2L^2 \\ -12 & -6L & 12 & -6L \\ 6L & 2L^2 & -6L & 4L^2 \end{bmatrix}$$

4.2 SIMPLY SUPPORTED BEAMS

PROBLEM 4.1. *For the beam shown in figure 4.2 determine nodal displacements, slopes, reactions. Take E = 210 GPa, I = 4 × 10^{-4} m^4.*

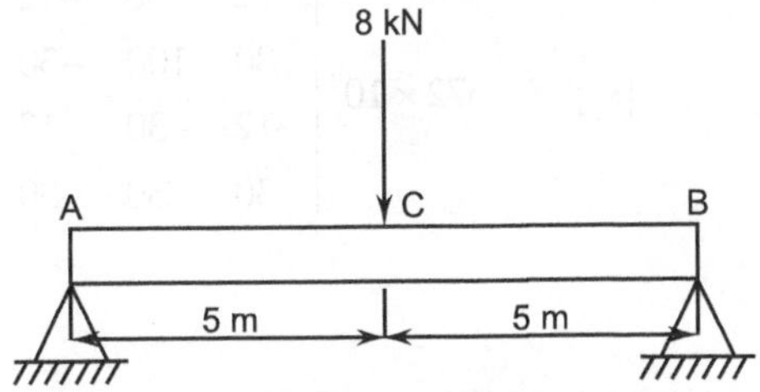

Figure 4.2

SOLUTION. (I) Analytical method. [Refer figure 4.2]

$$L = 10 \text{ m}$$
$$P = 8 \text{ kN}$$

Deflection $\delta_C = -\frac{PL^3}{48EI} = -\frac{8\times10^3\times(10)^3}{48\times210\times10^9\times4\times10^{-4}} = -1.98\times10^{-3} \text{ m} = -1.98 \text{ mm}$

$$|\theta_C| = |\theta_B| = \frac{PL^2}{16EI} = \frac{8\times10^3\times(10)^2}{16\times210\times10^9\times4\times10^{-4}} = 5.95\times10^{-4} \text{ rad}$$

$$\theta_C = 0 \text{ by symmetry.}$$

Reaction, $R_A = R_B = \frac{8}{2} = 4 \text{ kN}$

(II) FEM by hand calculations [Refer figure 4.2(*a*)]

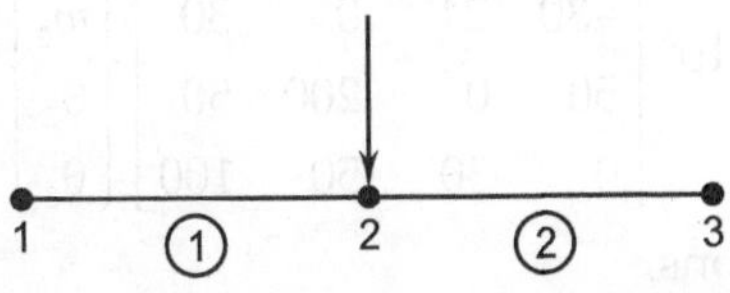

Figure 4.2(a). Finite element model

Element stiffness matrices are,

$$[k_1] = \frac{EI}{L^3}\begin{bmatrix} 12 & 6L & -12 & 6L \\ 6L & 4L^2 & -6L & 2L^2 \\ -12 & -6L & 12 & -6L \\ 6L & 2L^2 & -6L & 4L^2 \end{bmatrix} = \frac{210\times10^9\times4\times10^{-4}}{(5)^3}\begin{bmatrix} 12 & 6(5) & -12 & 6(5) \\ 6(5) & 4(5)^2 & -6(5) & 2(5)^2 \\ -12 & -6(5) & 12 & -6(5) \\ 6(5) & 2(5)^2 & -6(5) & 4(5)^2 \end{bmatrix}$$

$$= 672 \times 10^3 \begin{array}{c} \begin{matrix} w_1 & \theta_1 & w_2 & \theta_2 \end{matrix} \\ \begin{bmatrix} 12 & 30 & -12 & 30 \\ 30 & 100 & -30 & 50 \\ -12 & -30 & 12 & -30 \\ 30 & 50 & -30 & 100 \end{bmatrix} \end{array} \begin{matrix} w_1 \\ \theta_1 \\ w_2 \\ \theta_2 \end{matrix}$$

Since due to symmetry,

$$[k_1] = [k_2]$$

$$[k_2] = 672 \times 10^3 \begin{array}{c} \begin{matrix} w_2 & \theta_2 & w_3 & \theta_3 \end{matrix} \\ \begin{bmatrix} 12 & 30 & -12 & 30 \\ 30 & 100 & -30 & 50 \\ -12 & -30 & 12 & -30 \\ 30 & 50 & -30 & 100 \end{bmatrix} \end{array} \begin{matrix} w_2 \\ \theta_2 \\ w_3 \\ \theta_3 \end{matrix}$$

Global equation is,

$$[K]\{r\} = \{R\}$$

$$672 \times 10^3 \begin{array}{c} \begin{matrix} w_1 & \theta_1 & w_2 & \theta_2 & w_3 & \theta_3 \end{matrix} \\ \begin{bmatrix} 12 & 30 & -12 & 30 & 0 & 0 \\ 30 & 100 & -30 & 50 & 0 & 0 \\ -12 & -30 & 12+12 & -30+30 & -12 & 30 \\ 30 & 50 & -30+30 & 100+100 & -30 & 50 \\ 0 & 0 & -12 & -30 & 12 & -30 \\ 0 & 0 & 30 & 50 & -30 & 100 \end{bmatrix} \end{array} \begin{matrix} w_1 \\ \theta_1 \\ w_2 \\ \theta_2 \\ w_3 \\ \theta_3 \end{matrix} \begin{Bmatrix} w_1 \\ \theta_1 \\ w_2 \\ \theta_2 \\ w_3 \\ \theta_3 \end{Bmatrix} = \begin{Bmatrix} R_1 \\ 0 \\ -8 \times 10^3 \\ 0 \\ R_3 \\ 0 \end{Bmatrix}$$

Using elimination method for applying boundary conditions,

$$w_1 = w_3 = 0$$

The above matrix reduces to

$$672 \times 10^3 \begin{array}{c} \begin{matrix} \theta & w_2 & \theta_2 & \theta_3 \end{matrix} \\ \begin{bmatrix} 100 & -30 & 50 & 0 \\ -30 & 24 & 0 & 30 \\ 50 & 0 & 200 & 50 \\ 0 & 30 & 50 & 100 \end{bmatrix} \end{array} \begin{Bmatrix} \theta_1 \\ w_2 \\ \theta_2 \\ \theta_3 \end{Bmatrix} = \begin{Bmatrix} 0 \\ -8 \times 10^3 \\ 0 \\ 0 \end{Bmatrix}$$

By solving the above equations,

we get,

$$w_2 = -0.002 \text{ m} = -2 \text{ mm},$$

$$\theta_1 = -0.0006 \text{ rad}, \ \theta_2 = 0 \text{ and } \theta_3 = 0.0006 \text{ rad}$$

Reaction calculation

$$672 \times 10^3 (30 \times \theta_1 - 12 \times w_2) = R_1$$

$$672 \times 10^3 [30 \times (-0.0006) - 12(-0.002)] = R_1$$

$$R_1 = 4.032 \text{ kN}$$

$$672 \times 10^3 (-12 \times w_2 - 30 \times \theta_3) = R_3$$

$$672 \times 10^3[(-12 \times (-0.002) - (30 \times 0.0006)] = R_3$$

$$R_3 = 4.032 \text{ kN}$$

(III) Software results

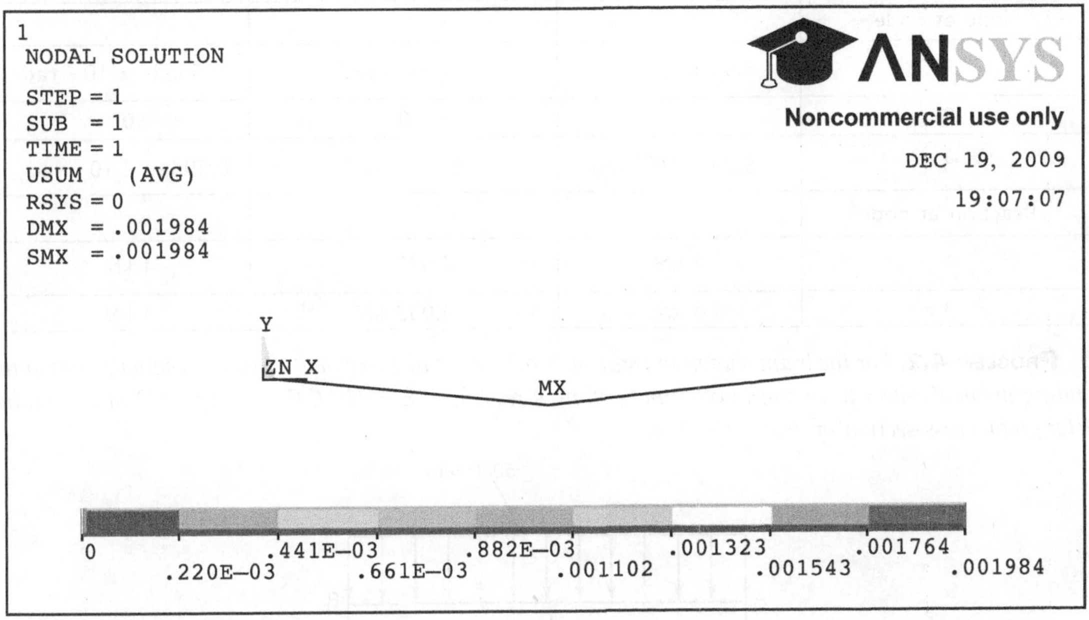

Figure 4.2(b). Deflection pattern for a simply supported beam (Refer to appendix for colour figures)

Deflection values at nodes (in metres)

The following degree of freedom results are in global coordinates

NODE	UX	UY	UZ	USUM
1	0.0000	0.0000	0.0000	0.0000
2	0.0000	-0.19841E-02	0.0000	0.19841E-02
3	0.0000	0.0000	0.0000	0.0000

Rotational deflection values at nodes

The following degree of freedom results are in global coordinates

NODE	ROTZ
1	- 0.59524E-03
2	0.0000
3	0.59524E-03

Reaction values

The following X, Y, Z solutions are in global coordinates

NODE	FX	FY	MZ
1	0.0000	4000.0	
3	0.0000	4000.0	

ANSWERS OF PROBLEM 4.1

Parameter	Analytical method	FEM-hand calculations	Software results
Displacement at node 2	-1.98 mm	- 2 mm	-1.9841 mm
Slope at node			
1	-5.95×10^{-4} rad	-0.0006 rad	-0.59524×10^{-3} rad
2	0	0	0
3	-5.95×10^{-4} rad	0.0006 rad	0.59524×10^{-3} rad
Reaction at node			
1	4 kN	4.032 kN	4 kN
3	4 kN	4.032 kN	4 kN

PROBLEM 4.2. *For the beam shown in figure 4.3, determine displacements, slopes, reactions, maximum bending moment, shear force and maximum bending stress. Take E = 210 GPa, I = 2×10^{-4} m^4. Beam has rectangular cross-section of depth h = 1 m.*

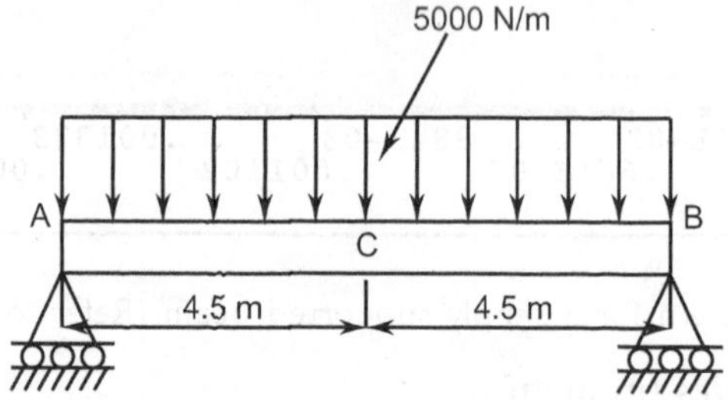

Figure 4.3

SOLUTION. **(I) Analytical method** [Refer figure 4.3]

$$\text{Reaction, } R_A = R_B = \frac{5000 \times 9}{2} = 22500 \text{ N} = 22.5 \text{ kN}$$

$$\delta_C = -\frac{5PL^4}{384EI} = -\frac{5 \times 5000 \times 9^4}{384 \times 210 \times 10^9 \times 2 \times 10^{-4}} = -0.0102 \text{ m} = -10.2 \text{ mm}$$

$$|\theta_A| = |\theta_B| = \frac{PL^3}{24EI} = \frac{5000 \times 9^3}{24 \times 210 \times 10^9 \times 2 \times 10^{-4}} = 3.62 \times 10^{-3} \text{rad}$$

$$\text{Maximum bending moment } M_{max} = \frac{PL^2}{8} = \frac{5000 \times (9)^2}{8} = 50625 \text{ N-m}$$

$$\text{Shear force SF} = \frac{PL}{2} = \frac{5000 \times 9}{2} = 22500 \text{ N}$$

Maximum bending stress,

$$f_{max} = \frac{M_{max}}{I} \times y_{max}$$

$$y_{max} = \frac{h}{2} = \frac{1}{2} = 0.5 \text{ m}$$

$$\therefore \qquad f_{max} = \frac{50625}{2 \times 10^{-4}} \times 0.5 = 126.56 \text{ MPa}$$

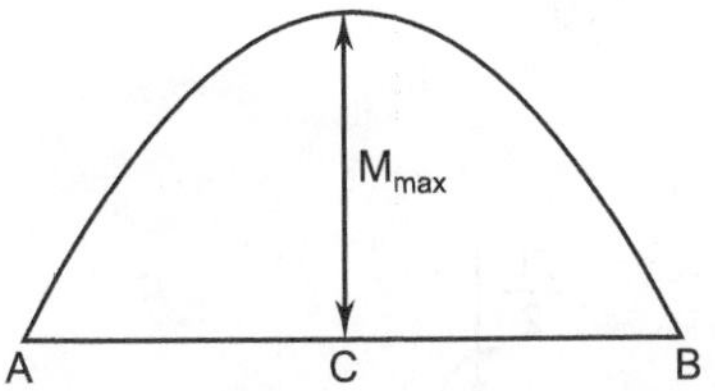

Figure 4.3(a). Bending moment diagram

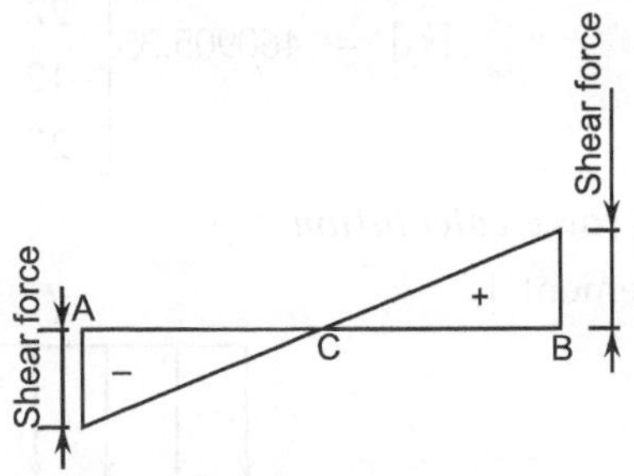

Figure 4.3(b). Shear force diagram

(II) FEM by hand calculations [Refer figure 4.3(*c*)]

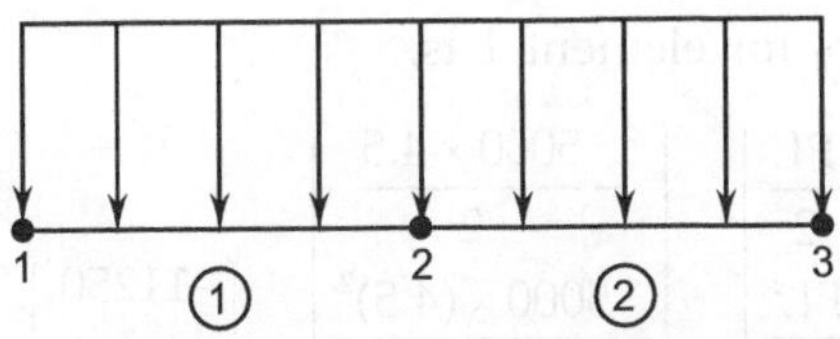

Figure 4.3(c). Finite element model

Stiffness matrices are,

$$[k_1] = \frac{EI}{L^3}\begin{bmatrix} 12 & 6L & -12 & 6L \\ 6L & 4L^2 & -6L & 2L^2 \\ -12 & -6L & 12 & -6L \\ 6L & 2L^2 & -6L & 4L^2 \end{bmatrix}$$

$$= \frac{210 \times 10^9 \times 2 \times 10^{-4}}{(4.5)^3}\begin{bmatrix} 12 & 6(4.5) & -12 & 6(4.5) \\ 6(4.5) & 4(4.5)^2 & -6(4.5) & 2(4.5)^2 \\ -12 & -6(4.5) & 12 & -6(4.5) \\ 6(4.5) & 2(4.5)^2 & -6(4.5) & 4(4.5)^2 \end{bmatrix}$$

$$\begin{matrix} & w_1 & \theta_1 & w_2 & \theta_2 & \end{matrix}$$

$$= 460905.35 \begin{bmatrix} 12 & 27 & -12 & 27 \\ 27 & 81 & 27 & 40.5 \\ -12 & -27 & 12 & -27 \\ 27 & 40.5 & -27 & 81 \end{bmatrix} \begin{matrix} w_1 \\ \theta_1 \\ w_2 \\ \theta_2 \end{matrix}$$

Since due to symmetry, $[k_1] = [k_2]$

$$[k_2] = 460905.35 \begin{array}{c} \begin{array}{cccc} w_1 & \theta_1 & w_2 & \theta_2 \end{array} \\ \left[\begin{array}{cccc} 12 & 27 & -12 & 27 \\ 27 & 81 & -27 & 40.5 \\ -12 & -27 & 12 & -27 \\ 27 & 40.5 & -27 & 81 \end{array}\right] \end{array} \begin{array}{c} \\ w_1 \\ \theta_1 \\ w_2 \\ \theta_2 \end{array}$$

Nodal force calculation

For element 1,

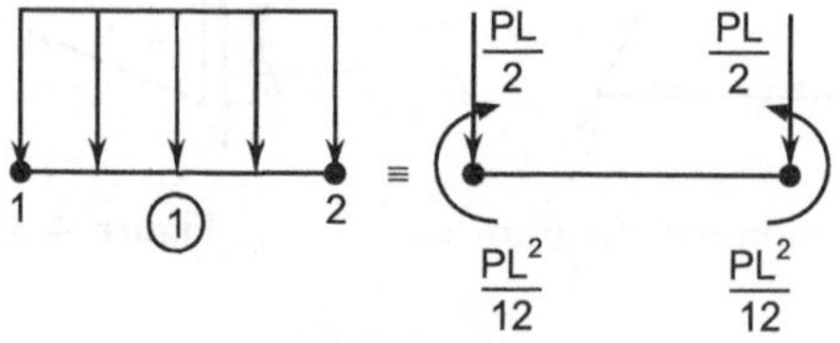

Figure 4.3(d)

Nodal forces and moments for element 1 is,

$$\{F_1\} = \begin{Bmatrix} -\dfrac{PL}{2} \\ -\dfrac{PL^2}{12} \\ -\dfrac{PL}{2} \\ \dfrac{PL^2}{12} \end{Bmatrix} = \begin{Bmatrix} -\dfrac{5000 \times 4.5}{2} \\ -\dfrac{5000 \times (4.5)^2}{12} \\ -\dfrac{5000 \times 4.5}{2} \\ \dfrac{5000 \times (4.5)^2}{12} \end{Bmatrix} = \begin{Bmatrix} -11250 \\ -8437.5 \\ -11250 \\ 8437.5 \end{Bmatrix} \begin{array}{c} f_1 \\ m_1 \\ f_2 \\ m_2 \end{array}$$

For element 2,

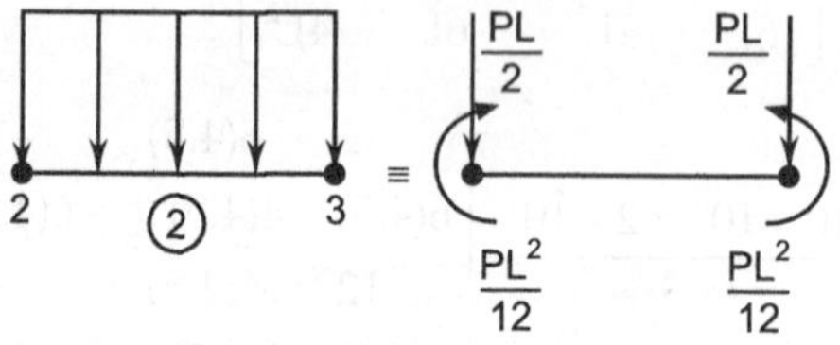

Figure 4.3(e)

Since due to symmetry,

$$\{F_1\} = \{F_2\}$$

$$\therefore \qquad \{F_2\} = \begin{Bmatrix} -11250 \\ -8437.5 \\ -11250 \\ 8437.5 \end{Bmatrix} \begin{array}{c} f_2 \\ m_2 \\ f_3 \\ m_3 \end{array}$$

Global Equation is,

$$460905.35\begin{bmatrix} \overset{w_1}{12} & \overset{\theta_1}{27} & \overset{w_2}{-12} & \overset{\theta_2}{27} & \overset{w_3}{0} & \overset{\theta_3}{0} \\ 27 & 81 & -27 & 40.5 & 0 & 0 \\ -12 & -27 & 12+12 & -27+27 & -12 & 27 \\ 27 & 40.5 & -27+27 & 81+81 & -27 & 40.5 \\ 0 & 0 & -12 & -27 & 12 & -27 \\ 0 & 0 & 27 & 40.5 & -27 & 81 \end{bmatrix}\begin{Bmatrix} w_1 \\ \theta_1 \\ w_2 \\ \theta_2 \\ w_3 \\ \theta_3 \end{Bmatrix} = \begin{Bmatrix} -11250+R_{1y} \\ -8437.5 \\ -11250-11250 \\ 8437.5-8437.5 \\ -11250+R_{3y} \\ 8437.5 \end{Bmatrix}$$

Applying boundary conditions, $w_1 = w_3 = 0$

$$460905.35\begin{bmatrix} 81 & -27 & 40.5 & 0 \\ -27 & 24 & 0 & 27 \\ 40.5 & 0 & 162 & 40.5 \\ 0 & 27 & 40.5 & 81 \end{bmatrix}\begin{Bmatrix} \theta_1 \\ w_2 \\ \theta_2 \\ \theta_3 \end{Bmatrix} = \begin{Bmatrix} -8437.5 \\ -22500 \\ 0 \\ 8437.5 \end{Bmatrix}$$

Solving

$$\theta_1 = -\,0.0036 \text{ rad}$$
$$w_2 = -\,0.0102 \text{ m}$$
$$\theta_2 = 6.0817 \times 10^{-7} = 0$$
$$\theta_3 = 0.0036 \text{ rad}$$

Reactions are calculated from 1st and 5th rows of global matrix.

$$460905.35\begin{bmatrix} 12 & 27 & -12 & 27 & 0 & 0 \end{bmatrix}\begin{Bmatrix} w_1 \\ \theta_1 \\ w_2 \\ \theta_2 \\ w_3 \\ \theta_3 \end{Bmatrix} = -11250 + R_{1y}$$

$$\therefore \quad 11615 = -\,11250 + R_1$$

$$\therefore \quad R_{1y} = 22865 \text{ N} = 22.865 \text{ kN}$$

Similarly from 5th row $R_{3y} = 22.865$ kN

(III) Software results

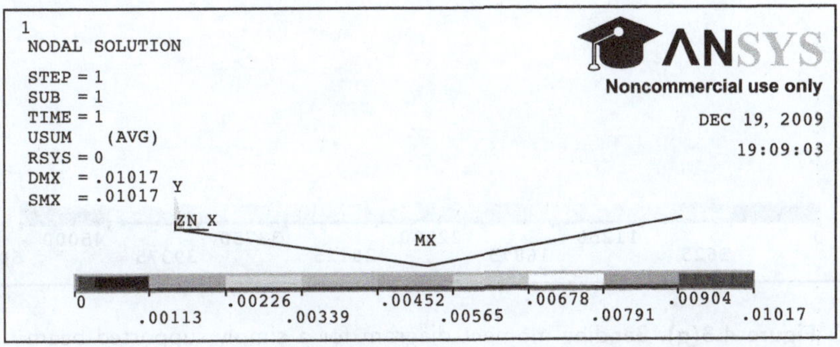

Figure 4.3(f). Deflection pattern for a simply supported beam (Refer to appendix for colour figures)

Deflection values at nodes (in metres)

The following degree of freedom results are in global coordinates

NODE	UX	UY	UZ	USUM
1	0.0000	0.0000	0.0000	0.0000
2	0.0000	-0.10170E-01	0.0000	0.10170E-01
3	0.0000	0.0000	0.0000	0.0000

Slope values at nodes

The following degree of freedom results are in global coordinates

NODE	ROTZ
1	–0.36161E-02
2	0.0000
3	0.36161E-02

Reaction values

The following X, Y, Z solutions are in global coordinates

NODE	FX	FY	MZ
1	0.0000	22500	
3	0.0000	22500	

Total values

VALUE	0.0000	45000	0.0000

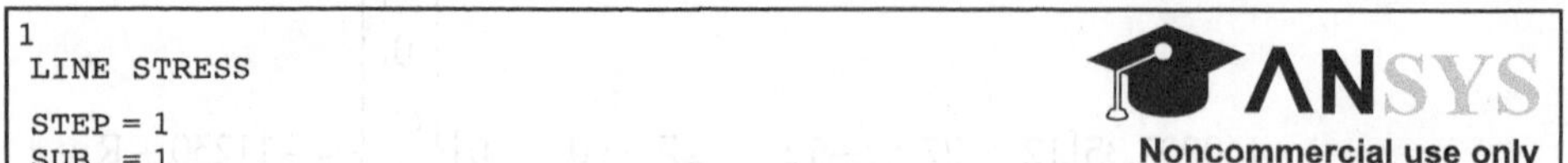

Figure 4.3(g). Bending moment diagram for a simply supported beam (Refer to appendix for colour figures)

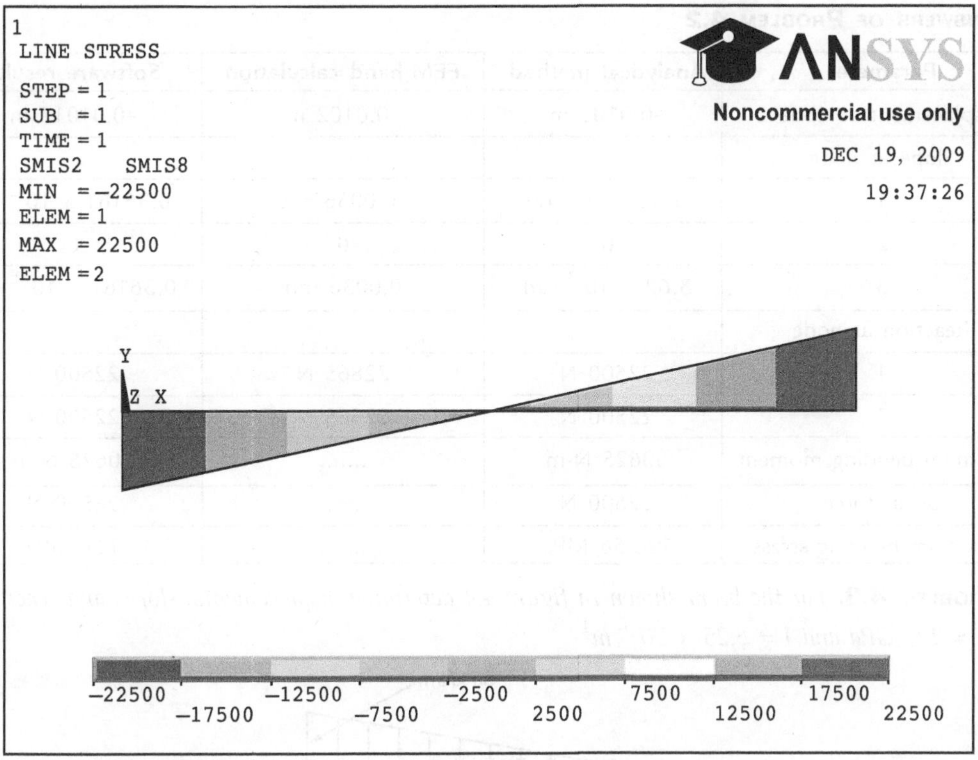

Figure 4.3(h). Shear force diagram for a simply supported beam (Refer to appendix for colour figures)

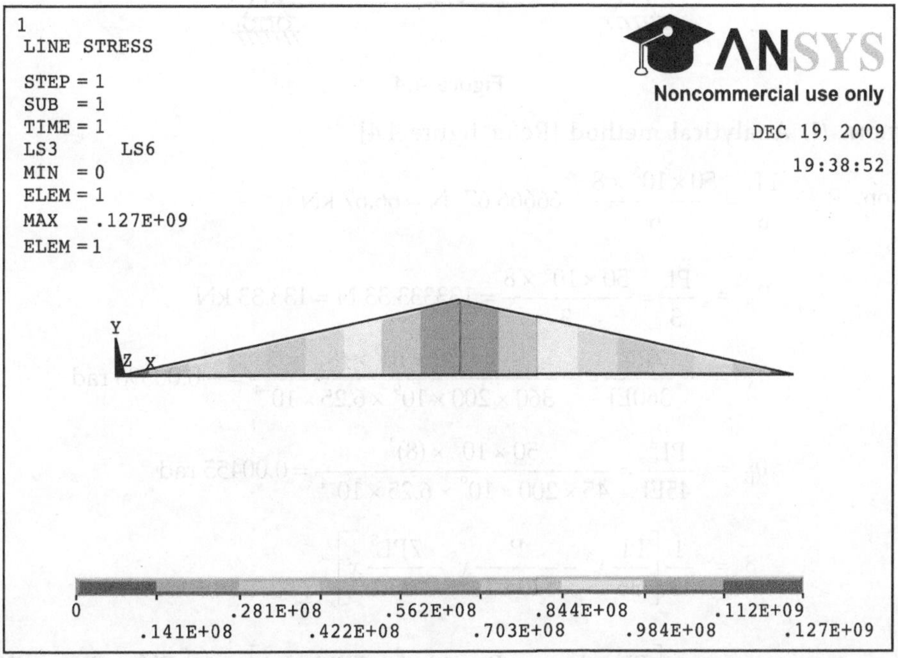

Figure 4.3(i). Bending stress for a simply supported beam (Refer to appendix for colour figures)

Answers of Problem 4.2

Parameter	Analytical method	FEM-hand calculation	Software results
Displacement at node 2	-0.0102 m	-0.0102 m	-0.01017 m
Slope at node			
1	-3.62×10^{-3} rad	-0.0036 rad	-0.36161×10^{-2} rad
2	0	0	0
3	3.62×10^{-3} rad	0.0036 rad	0.36161×10^{-2} rad
Reaction at node			
1	22500 N	22865 N	22500 N
3	22500 N	22865 N	22500 N
Maximum bending moment	50625 N-m	....	50625 N-m
Shear force	22500 N	....	22500 N
Maximum bending stress	126.56 MPa	....	127 MPa

Problem 4.3. *For the beam shown in figure 4.4 determine displacements, slopes and reactions. Take E = 200 GPa and I = 6.25 × 10^{-4} m^4.*

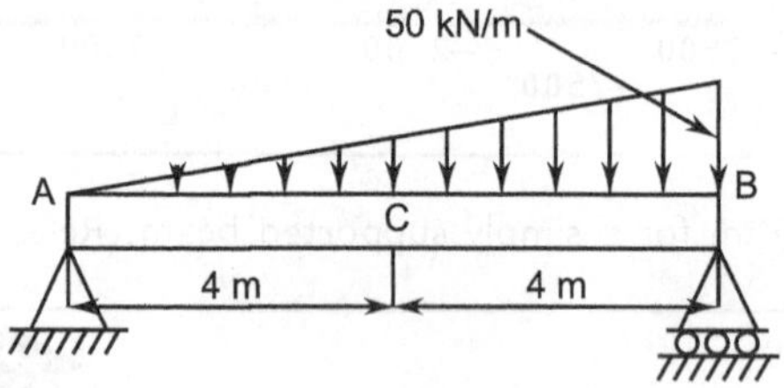

Figure 4.4

Solution. (I) **Analytical method** [Refer figure 4.4]

$$\text{Reaction, } R_A = \frac{PL}{6} = \frac{50 \times 10^3 \times 8}{6} = 66666.67 \text{ N} = 66.67 \text{ kN}$$

$$R_B = \frac{PL}{3} = \frac{50 \times 10^3 \times 8}{3} = 133333.33 \text{ N} = 133.33 \text{ kN}$$

$$\theta_A = -\frac{7PL^3}{360EI} = -\frac{7 \times 50 \times 10^3 \times (8)^3}{360 \times 200 \times 10^9 \times 6.25 \times 10^{-4}} = -0.00398 \text{ rad}$$

$$\theta_B = \frac{PL^3}{45EI} = \frac{50 \times 10^3 \times (8)^3}{45 \times 200 \times 10^9 \times 6.25 \times 10^{-4}} = 0.00455 \text{ rad}$$

$$\delta = \frac{1}{EI}\left[\frac{PL}{36}x^3 - \frac{P}{120 \times L}x^5 - \frac{7PL^3}{360}x\right]_{x=\frac{L}{2}}$$

$$\delta_C = \frac{1}{EI}\left[\frac{PL}{36} \times \frac{L^3}{8} - \frac{P}{120 \times L}\left(\frac{L}{2}\right)^5 - \frac{7PL^3}{360}\left(\frac{L}{2}\right)\right] = \frac{1}{EI}\left[\frac{PL^4}{288} - \frac{PL^4}{3840} - \frac{7PL^4}{720}\right]$$

$$= \frac{PL^4}{EI}\left[\frac{40-3-112}{11520}\right]$$

$$\therefore \quad \delta_C = -\frac{75PL^4}{11520EI} = -\frac{75\times50\times10^3\times(8)^4}{11520\times200\times10^9\times6.25\times10^{-4}} = -0.01067 \text{ m}$$

$$\theta_C = \frac{1}{EI}\left[\frac{PL}{12}x^2 - \frac{P}{24\times L}x^4 - \frac{7PL^3}{360}x\right]_{x=\frac{L}{2}} = \frac{1}{EI}\left[\frac{PL^3}{48} - \frac{PL^3}{384} - \frac{7PL^3}{360}\right]$$

$$\theta_C = \frac{PL^3}{EI}\left[\frac{1}{48} - \frac{1}{384} - \frac{7}{360}\right]$$

$$\therefore \quad \theta_C = -\frac{1.2153\times10^{-3}\times PL^3}{EI} = -\frac{1.2153\times10^{-3}\times50\times10^3\times(8)^3}{200\times10^9\times6.25\times10^{-4}}$$

$$= 2.4889 \times 10^{-4} \text{ rad.}$$

(II) FEM by hand calculations [Refer figure 4.4(*a*)]

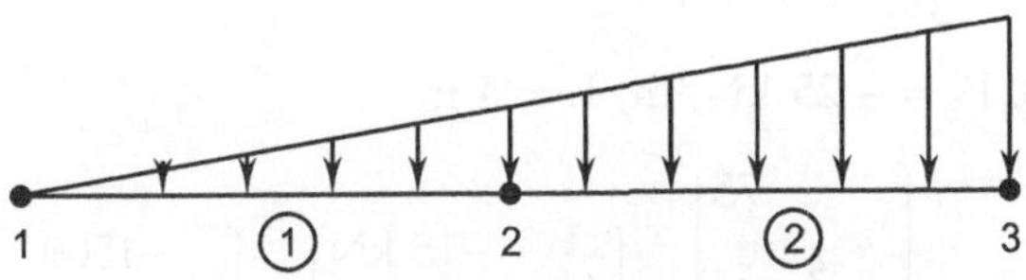

Figure 4.4(a)

Stiffness matrices are,

$$[k_1] = \frac{EI}{L_1^3}\begin{bmatrix} 12 & 6L & -12 & 6L \\ 6L & 4L^2 & -6L & 2L^2 \\ -12 & -6L & 12 & -6L \\ 6L & 2L^2 & -6L & 4L^2 \end{bmatrix} = \frac{200\times10^9\times6.25\times10^{-4}}{(4)^3}\begin{bmatrix} 12 & 6(4) & -12 & 6(4) \\ 6(4) & 4(4)^2 & -6(4) & 2(4)^2 \\ -12 & -6(4) & 12 & -6(4) \\ 6(4) & 2(4)^2 & -6(4) & 4(4)^2 \end{bmatrix}$$

$$= 195.3125\times10^4 \begin{bmatrix} 12 & 24 & -12 & 24 \\ 24 & 64 & -24 & 32 \\ -12 & -24 & 12 & -24 \\ 24 & 32 & -24 & 64 \end{bmatrix} \begin{matrix} w_1 \\ \theta_1 \\ w_2 \\ \theta_2 \end{matrix} \quad (\text{columns } w_1,\ \theta_1,\ w_2,\ \theta_2)$$

$$[k_2] = \frac{EI}{L_2^3}\begin{bmatrix} 12 & 6L & -12 & 6L \\ 6L & 4L^2 & -6L & 2L^2 \\ -12 & -6L & 12 & -6L \\ 6L & 2L^2 & -6L & 4L^2 \end{bmatrix} = \frac{200\times10^9\times6.25\times10^{-4}}{(4)^3}\begin{bmatrix} 12 & 6(4) & -12 & 6(4) \\ 6(4) & 4(4)^2 & -6(4) & 2(4)^2 \\ -12 & -6(4) & 12 & -6(4) \\ 6(4) & 2(4)^2 & -6(4) & 4(4)^2 \end{bmatrix}$$

$$= 195.3125\times10^4 \begin{bmatrix} 12 & 24 & -12 & 24 \\ 24 & 64 & -24 & 32 \\ -12 & -24 & 12 & -24 \\ 24 & 32 & -24 & 64 \end{bmatrix} \begin{matrix} w_2 \\ \theta_2 \\ w_3 \\ \theta_3 \end{matrix} \quad (\text{columns } w_2,\ \theta_2,\ w_3,\ \theta_3)$$

Global Stiffness Matrix is,

$$[K] = 195.3125\times10^4 \begin{array}{c} \begin{array}{cccccc} w_1 & \theta_1 & w_2 & \theta_2 & w_3 & \theta_3 \end{array} \\ \left[\begin{array}{cccccc} 12 & 24 & -12 & 24 & 0 & 0 \\ 24 & 64 & -24 & 32 & 0 & 0 \\ -12 & -24 & 12+12 & -24+24 & -12 & 24 \\ 24 & 32 & 0 & 64+64 & -24 & 32 \\ 0 & 0 & -12 & -24 & 12 & -24 \\ 0 & 0 & 24 & 32 & -24 & 64 \end{array}\right] \end{array} \begin{array}{c} \\ w_1 \\ \theta_1 \\ w_2 \\ \theta_2 \\ w_3 \\ \theta_3 \end{array}$$

Load vector $\{F\} = \dfrac{L}{20}\begin{Bmatrix} 7P_1+3P_2 \\ \dfrac{L}{3}(3P_1+2P_2) \\ 3P_1+7P_2 \\ -\dfrac{L}{3}(2P_1+3P_2) \end{Bmatrix}$

For element 1, $P_1 = 0$, $P_2 = -25$ kN/m, L = 4 m

$$\{F_1\} = \frac{4}{20}\begin{Bmatrix} -75 \\ \dfrac{4}{3}(50) \\ -175 \\ -\dfrac{4}{3}(-75) \end{Bmatrix} = \begin{Bmatrix} -15\ \text{kN} \\ -13.33\ \text{kN-m} \\ -35\ \text{kN} \\ 20\ \text{kN-m} \end{Bmatrix} = \begin{Bmatrix} -15000\ \text{N} \\ -13333\ \text{N-m} \\ -35000\ \text{N} \\ 20000\ \text{N-m} \end{Bmatrix}$$

For element 2, $P_1 = -25$ kN/m, $P_2 = -50$ kN/m, L = 4 m

$$\{F_2\} = \frac{4}{20}\begin{Bmatrix} -175-150 \\ \dfrac{4}{3}(-75-100) \\ -75-350 \\ -\dfrac{4}{3}(-50-150) \end{Bmatrix} = \begin{Bmatrix} -65000\ \text{N} \\ -46667\ \text{N-m} \\ -85000\ \text{N} \\ 53333\ \text{N-m} \end{Bmatrix}$$

Global load vector is,

$$\{F\} = \begin{Bmatrix} -15000 \\ -13333 \\ -100000 \\ -26667 \\ -85000 \\ 53333 \end{Bmatrix}$$

Global equation is,

$$195.3125 \times 10^4 \begin{array}{c} \begin{array}{cccccc} w_1 & \theta_1 & w_2 & \theta_2 & w_3 & \theta_3 \end{array} \\ \begin{bmatrix} 12 & 24 & -12 & 24 & 0 & 0 \\ 24 & 64 & -24 & 32 & 0 & 0 \\ -12 & -24 & 12+12 & -24+24 & -12 & 24 \\ 24 & 32 & 0 & 64+64 & -24 & 32 \\ 0 & 0 & -12 & -24 & 12 & -24 \\ 0 & 0 & 24 & 32 & -24 & 64 \end{bmatrix} \end{array} \begin{Bmatrix} w_1 \\ \theta_1 \\ w_2 \\ \theta_2 \\ w_3 \\ \theta_3 \end{Bmatrix} = \begin{Bmatrix} -15000 + R_1 \\ -13333 \\ -100000 \\ -26667 \\ -85000 + R_3 \\ 53333 \end{Bmatrix}$$

Applying boundary conditions, $w_1 = w_3 = 0$

The above matrix reduces to,

$$195.3125 \times 10^4 \begin{bmatrix} 64 & -24 & 32 & 0 \\ -24 & 24 & 0 & 24 \\ 32 & 0 & 128 & 32 \\ 0 & 24 & 32 & 64 \end{bmatrix} \begin{Bmatrix} \theta_1 \\ w_2 \\ \theta_2 \\ \theta_3 \end{Bmatrix} = \begin{Bmatrix} -13333 \\ -100000 \\ -26667 \\ 53333 \end{Bmatrix}$$

Solving,

$$\begin{Bmatrix} \theta_1 \\ w_2 \\ \theta_2 \\ \theta_3 \end{Bmatrix} = \begin{Bmatrix} -0.00398 \text{ rad} \\ -0.01067 \text{ m} \\ -0.00025 \text{ rad} \\ 0.00455 \text{ rad} \end{Bmatrix}$$

Reaction calculation

$$1953125\ (12 \times w_1 + 24 \times \theta_1 - 12 \times w_2 + 24 \times \theta_2) = -15000 + R_1$$

$$R_1 = 66796.875 \text{ N} = 66.79 \text{ kN}$$

$$1953125\ [-12\ (-0.01067) - 24(-0.00025) - 24\ (0.00455)] = -85000 + R_3$$

$$R_3 = 133515.63 \text{ N} = 133.52 \text{ kN}$$

(III) Software results

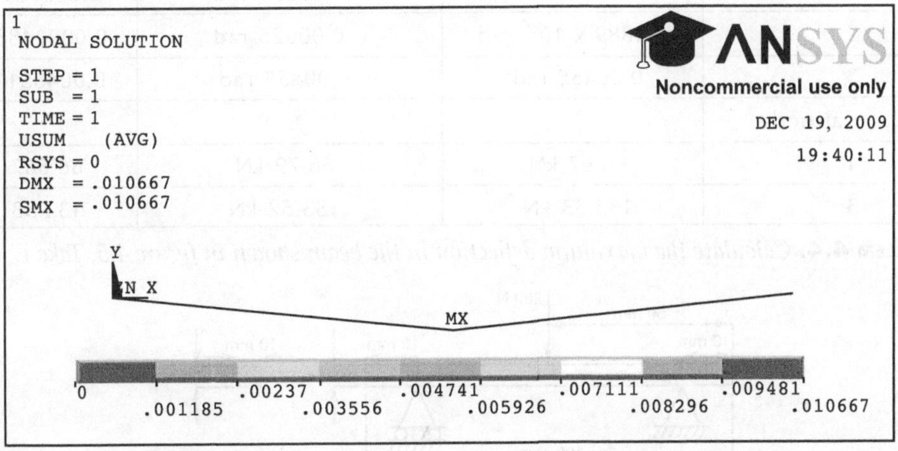

Figure 4.4(b). Deflection pattern for a simply supported beam (Refer to appendix for colour figures)

Deflection values at nodes (in metres)

The following degree of freedom results are in global coordinates

NODE	UX	UY	UZ	USUM
1	0.0000	0.0000	0.0000	0.0000
2	0.0000	-0.10667E-01	0.0000	0.10667E-01
3	0.0000	0.0000	0.0000	0.0000

Maximum absolute values

NODE	0	2	0	2
VALUE	0.0000	-0.10667E-01	0.0000	0.10667E-01

Slope values at nodes

The following degree of freedom results are in Global coordinates

NODE	ROTZ
1	- 0.39822E-02
2	- 0.24889E-03
3	0.45511E-02

Reaction values

The following X, Y, Z solutions are in global coordinates

NODE	FX	FY	MZ
1	0.0000	66667	
3	0.0000	0.13333E + 06	

Total values

VALUE	0.0000	0.20000E + 06	0.0000

Answers of Problem 4.3

Parameter	Analytical method	FEM-hand calculation	Software results
Displacement at node 2	-0.01067 m	-0.01067 m	-0.010667 m
Slope at node			
1	-0.00398 rad	-0.00398 rad	-0.0039822 rad
2	-2.4889×10^{-4} rad	-0.00025 rad	-0.00024889 rad
3	0.00455 rad	0.00455 rad	0.0045511 rad
Reaction at node			
1	66.67 kN	66.79 kN	66.667 kN
3	133.33 kN	133.52 kN	133.33 kN

Problem 4.4. *Calculate the maximum deflection in the beam shown in figure 4.5. Take E = 200 GPa.*

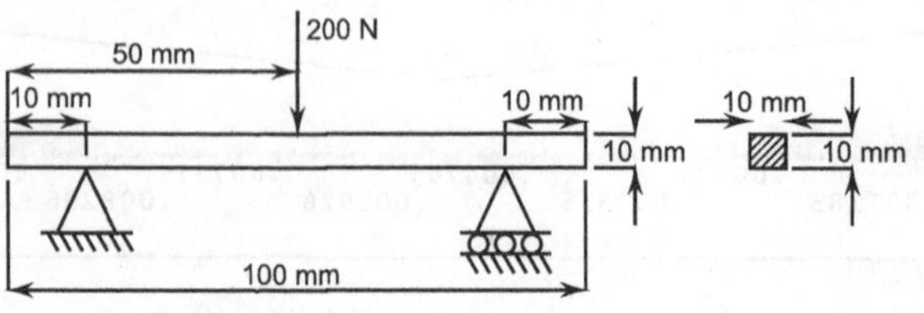

Figure 4.5

SOLUTION. (I) **Analytical method** [Refer figure 4.5(*a*)]

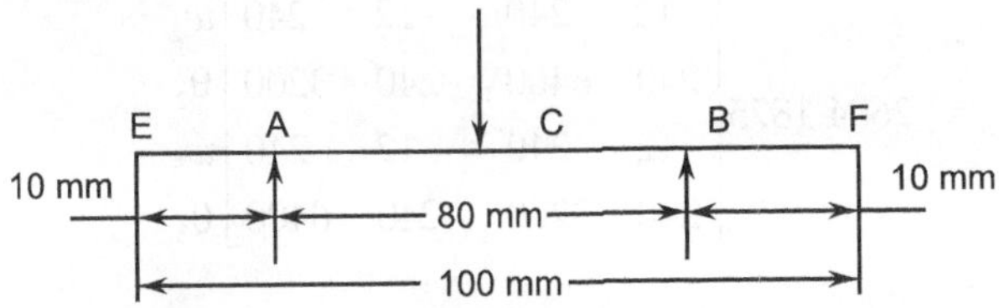

Figure 4.5(a)

$$I = \frac{bh^3}{12} = \frac{0.01 \times (0.01)^3}{12} = 8.333 \times 10^{-10} \text{ m}^4$$

$$\delta_C = \frac{PL^3}{48EI} = -\frac{200 \times (0.08)^3}{48EI} = -\frac{2.133 \times 10^{-3}}{EI} = -\frac{2.133 \times 10^{-3}}{200 \times 10^9 \times 8.33 \times 10^{-10}}$$

$$= -1.2803 \times 10^{-5} \text{ m} = -0.0128 \text{ mm}$$

$$\theta_C = 0$$

$$|\theta_A| = |\theta_B| = \frac{PL^2}{16EI} = \frac{200 \times (0.08)^2}{16EI} = \frac{0.08}{EI} = \frac{0.08}{200 \times 10^9 \times 8.33 \times 10^{-10}} = 4.802 \times 10^{-4} \text{rad}$$

$$\delta_E = \delta_F = \theta_B \times BF = 4.802 \times 10^{-4} \times 10 = 4.802 \times 10^{-3} \text{ mm}$$

(II) FEM by hand calculations [Refer figure 4.5(*b*)]

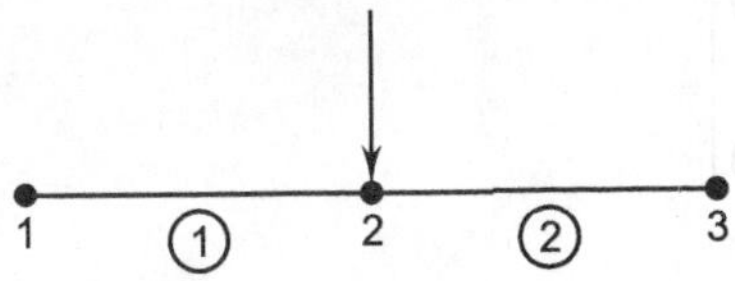

Figure 4.5(b)

For beam,

$$I = \frac{bh^3}{12} = \frac{10 \times (10)^3}{12} = 833.34 \text{ mm}^4$$

For element 1 and 2, L = 40 mm

$$[k_1] = [k_2] = \frac{200 \times 10^3 \times 833.34}{(40)^3} \begin{bmatrix} 12 & 6 \times 40 & -12 & 6 \times 40 \\ 6 \times 40 & 4 \times (40)^2 & -6 \times 40 & 2 \times (40)^2 \\ -12 & -6 \times 40 & 12 & -6 \times 40 \\ 6 \times 40 & 2 \times (40)^2 & -6 \times 40 & 4 \times (40)^2 \end{bmatrix}$$

$$\begin{matrix} & & w_1 & \theta_1 & w_2 & \theta_2 & \end{matrix}$$

$$[k_1] = 2604.1875 \begin{bmatrix} 12 & 240 & -12 & 240 \\ 240 & 6400 & -240 & 3200 \\ -12 & -240 & 12 & -240 \\ 240 & 3200 & -240 & 6400 \end{bmatrix} \begin{matrix} w_1 \\ \theta_1 \\ w_2 \\ \theta_2 \end{matrix}$$

$$[k_2] = 2604.1875\begin{array}{c} \begin{matrix} w_2 & \theta_2 & w_3 & \theta_3 \end{matrix} \\ \begin{bmatrix} 12 & 240 & -12 & 240 \\ 240 & 6400 & -240 & 3200 \\ -12 & -240 & 12 & -240 \\ 240 & 3200 & -240 & 6400 \end{bmatrix} \end{array}\begin{matrix} w_2 \\ \theta_2 \\ w_3 \\ \theta_3 \end{matrix}$$

Global stiffness matrix is,

$$[K] = 2604.1875\begin{array}{c} \begin{matrix} w_1 & \theta_1 & w_2 & \theta_2 & w_3 & \theta_3 \end{matrix} \\ \begin{bmatrix} 12 & 240 & -12 & 240 & 0 & 0 \\ 240 & 6400 & -240 & 3200 & 0 & 0 \\ -12 & -240 & 12+12 & -240+240 & -12 & 240 \\ 240 & 3200 & -240+240 & 6400+6400 & -240 & 3200 \\ 0 & 0 & -12 & -240 & 12 & -240 \\ 0 & 0 & 240 & 3200 & -240 & 6400 \end{bmatrix} \end{array}\begin{matrix} w_1 \\ \theta_1 \\ w_2 \\ \theta_2 \\ w_3 \\ \theta_3 \end{matrix}$$

$$\{F\} = \begin{Bmatrix} R_1 \\ 0 \\ 0 \\ -200 \\ 0 \\ R_3 \\ 0 \end{Bmatrix}$$

Global equation is,

$$2604.1875\begin{array}{c} \begin{matrix} w_1 & \theta_1 & w_2 & \theta_2 & w_3 & \theta_3 \end{matrix} \\ \begin{bmatrix} 12 & 240 & -12 & 240 & 0 & 0 \\ 240 & 6400 & -240 & 3200 & 0 & 0 \\ -12 & -240 & 24 & 0 & -12 & 240 \\ 240 & 3200 & 0 & 12800 & -24 & 3200 \\ 0 & 0 & -12 & -240 & 12 & -240 \\ 0 & 0 & 240 & 3200 & -240 & 6400 \end{bmatrix} \end{array}\begin{Bmatrix} w_1 \\ \theta_1 \\ w_2 \\ \theta_2 \\ w_3 \\ \theta_3 \end{Bmatrix} = \begin{Bmatrix} R_1 \\ 0 \\ -200 \\ 0 \\ R_3 \\ 0 \end{Bmatrix}$$

Boundary conditions are $w_1 = w_3 = 0$

$$2604.1875\begin{bmatrix} 6400 & -240 & 3200 & 0 \\ -240 & 24 & 0 & 240 \\ 3200 & 0 & 12800 & 3200 \\ 0 & 240 & 3200 & 6400 \end{bmatrix}\begin{Bmatrix} \theta_1 \\ w_2 \\ \theta_2 \\ \theta_3 \end{Bmatrix} = \begin{Bmatrix} 0 \\ -200 \\ 0 \\ 0 \end{Bmatrix}$$

Solving the matrix and equations,

$$\theta_1 = -4.8 \times 10^{-4} \text{ rad and } \theta_2 = 0$$

$$w_1 = -0.0128 \text{ mm}$$

$$\theta_3 = 4.8 \times 10^{-4} \text{ rad}$$

At $\theta_2 = 0$, max deflection between supports is 0.0128 mm

Deflection at ends (overhang) = $4.8 \times 10^{-4} \times 10 = 4.8 \times 10^{-3}$ mm

(III) Software results

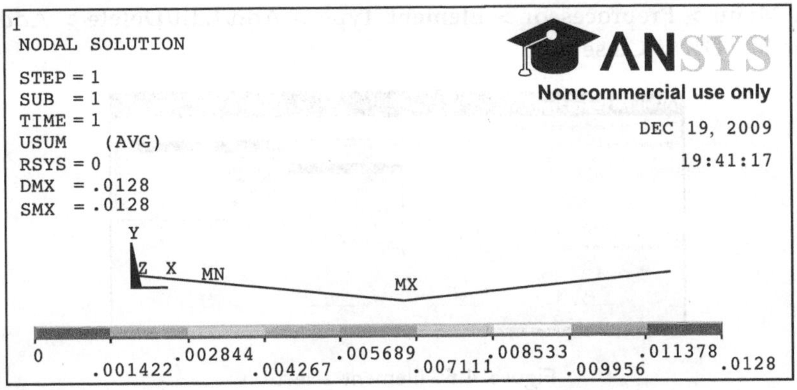

Figure 4.5(c). Deflection pattern for a simply supported beam (Refer to appendix for colour figures)

Deflection values at nodes (in mm)

The following degree of freedom results are in global coordinates

NODE	UX	UY	UZ	USUM
1	0.0000	0.48000E-02	0.0000	0.48000E-02
2	0.0000	0.0000	0.0000	0.0000
3	0.0000	-0.12800E-01	0.0000	0.12800E-01
4	0.0000	0.0000	0.0000	0.0000
5	0.0000	0.48000E-02	0.0000	0.48000E-02

Maximum absolute values

NODE	0	3	0	3
VALUE	0.0000	-0.12800E-01	0.0000	0.12800E-01

Slope values at nodes

The following degree of freedom results are in global coordinates

NODE	ROT Z
1	-0.48000E-03
2	-0.48000E-03
3	0.0000
4	0.48000E-03
5	0.48000E-03

ANSWERS OF PROBLEM 4.4

Parameter	Analytical method	FEM-hand calculation	Software results
Deflection at applied load	-0.0128 mm	-0.0128 mm	-0.0128 mm
Deflection at ends (overhang)	4.802×10^{-3} mm	4.8×10^{-3} mm	4.8×10^{-3} mm
Slope at hinged support	-4.802×10^{-4} rad	-4.8×10^{-4} rad	-4.8×10^{-4} rad
Slope at roller support	4.802×10^{-4} rad	4.8×10^{-4} rad	4.8×10^{-4} rad

Procedure for solving the problems using ANSYS® 12.0 academic teaching software.

For Problem 4.2

PREPROCESSING

1. **Main Menu > Preprocessor > Element Type > Add/Edit/Delete > Add >Beam > 2D elastic 3 > OK > Close**

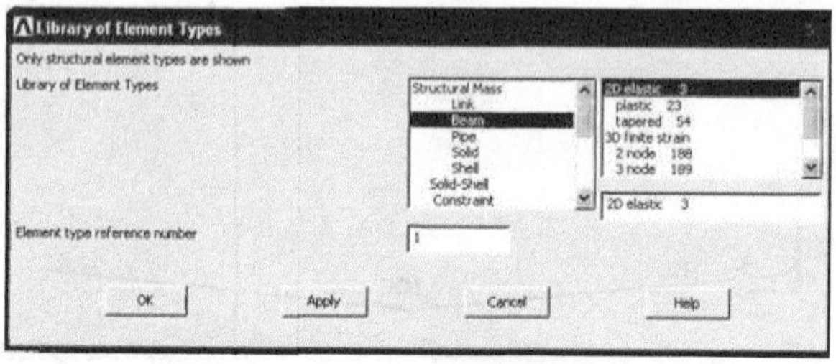

Figure 4.6. Element selection

2. **Main Menu > Preprocessor > Real Constants > Add/Edit/Delete > Add > OK**

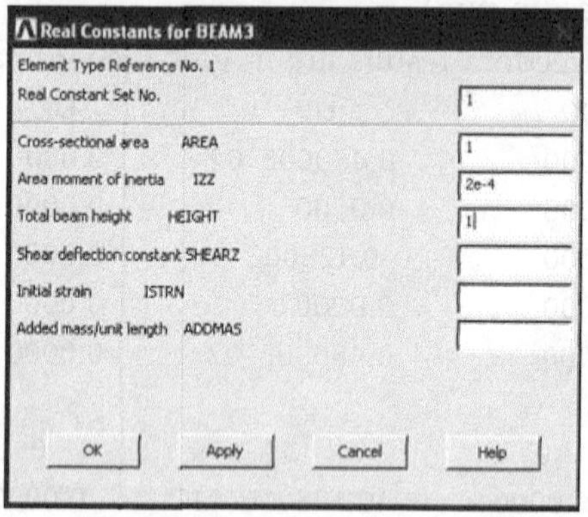

Figure 4.7. Enter the area, moment of inertia and height of beam

Cross-sectional area AREA > **Enter 1**

Area moment of inertia IZZ > **Enter 2e-4**

Total beam height HEIGHT > **Enter 1 > OK > Close**

Enter the material properties.

3. **Main Menu > Preprocessor > Material Props > Material Models**

   Material Model Number 1, click **Structural > Linear > Elastic > Isotropic**

   Enter **EX = 210E9 and PRXY = 0.3** > **OK**

   (**Close** the define material model behaviour window.)

   Create the nodes and elements as shown in the below table and figure 4.8.

4. **Main Menu > Preprocessor > Modeling > Create > Nodes > In Active CS** Enter the coordinates of node 1 > **Apply** Enter the coordinates of node 2 > **Apply** Enter the coordinates of node 3 > **OK.**

Node locations		
Node number	X coordinate	Y coordinate
1	0	0
2	4.5	0
3	9	0

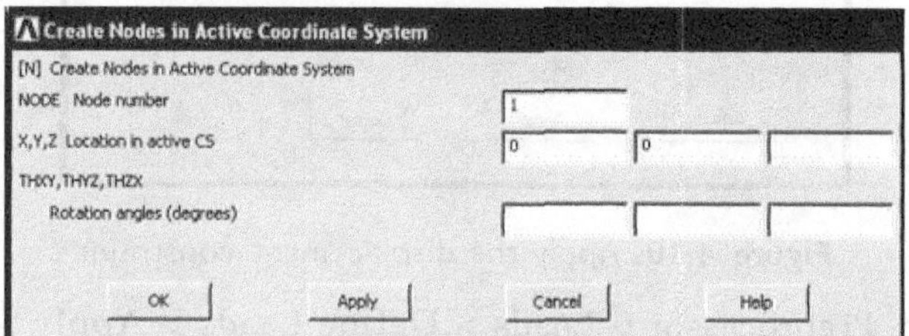

Figure 4.8. Enter the node coordinates

5. **Main Menu > Preprocessor > Modeling > Create > Elements > Auto Numbered > Thru nodes** Pick the 1st and 2nd node > **Apply** Pick the 2nd and 3rd node > **OK**

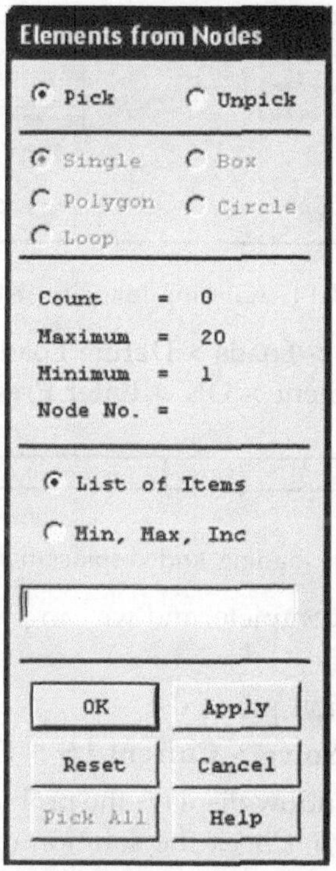

Figure 4.9. Pick the nodes to create elements

Apply the displacement boundary conditions and loads.

6. **Main Menu > Preprocessor > Loads > Define Loads > Apply > Structural >Displacement > On Nodes** Pick the 1st node and 3rd node > **Apply > Select UX and UY and Enter displacement value = 0 > OK**

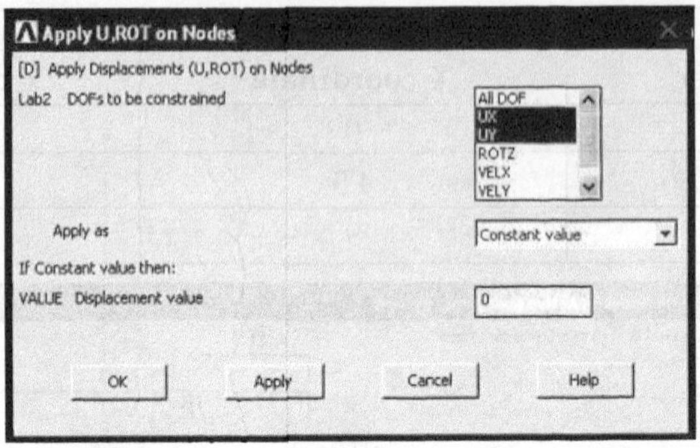

Figure 4.10. Apply the displacement constraint

7. **Main Menu > Preprocessor > Loads > Define Loads > Apply > Structural >Pressure > On Beams** Pick the 1st element **> OK > Enter Pressure value at node I = 5000 > OK**

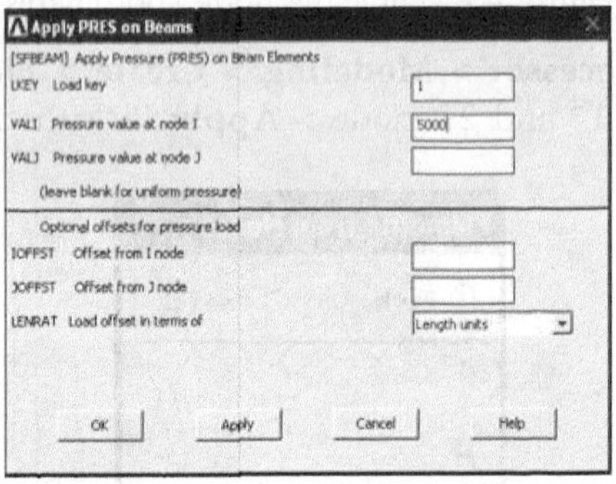

Figure 4.11. Applying loads on element 1

8. **Main Menu > Preprocessor > Loads > Define Loads > Apply > Structural > Pressure > On Beams** Pick the 2nd element **> OK > Enter Pressure value at node I = 5000 > OK**

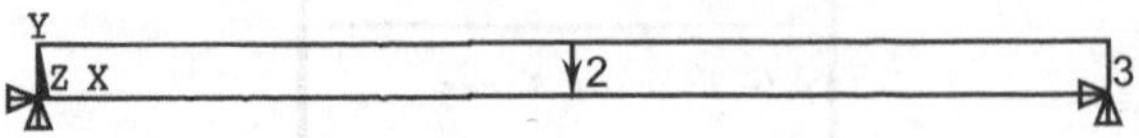

Figure 4.12. Model with loading and displacement boundary conditions.

The model-building step is now complete, and we can proceed to the solution. First to be safe, save the model.

Solution. The interactive solution proceeds.

9. **Main Menu > Solution > Solve > Current LS > OK**

   The **/STATUS Command** window displays the problem parameters and **the SolveCurrent Load Step** window is shown. Check the solution options in the **/STATUS** window and if all is OK, select **File > Close**

   In the Solve **Current Load Step** window, Select **OK**, and when the solution is complete, **close** the **'Solution is Done!'** window.

POSTPROCESSING

We can now plot the results of this analysis and also list the computed values.

10. **Main Menu > General Postproc > Plot Results > Contour Plot > Nodal Solu > DOF Solution > Displacement vector sum > OK**

    This result is shown in figure 4.3(*f*).

11. **Main Menu > General Postproc > List Results > Nodal Solution > Select Rotation vector sum > OK**

12. **Main Menu > General Postproc > List Results > Reaction Solu > OK**

    To find the **bending moment diagram** following procedure is followed.

13. **Main Menu > General Postproc > Element Table > Define Table > Add**

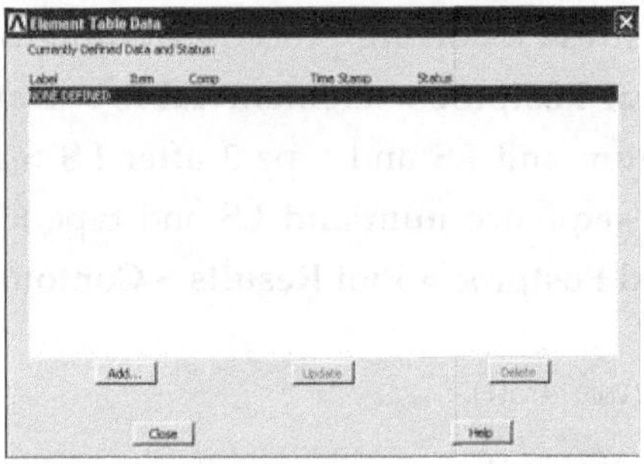

Figure 4.13. Define the element table

Select **By sequence num and SMISC** and type **6 after SMISC** as shown in figure 4.14.

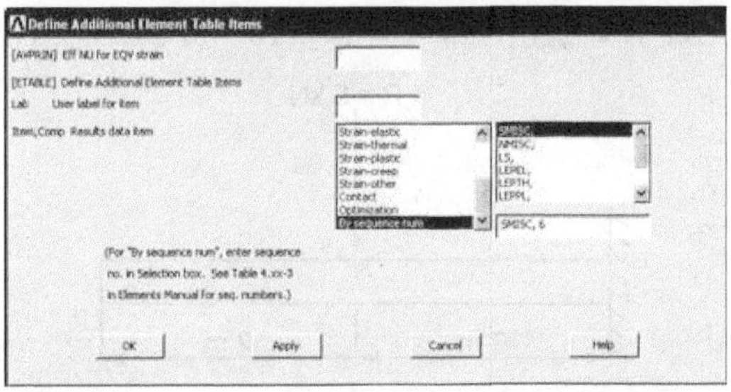

Figure 4.14. Selecting options in element table

> **> APPLY**

Then again select **By sequence num and SMISC** and type **12 after SMISC > OK**

14. **Main Menu > General Postproc > Plot Results > Contour Plot > Line Elem Res >** Select **SMIS 6 and SMIS 12 in the rows of LabI and LabJ** respectively as shown in figure 4.15 **> OK**

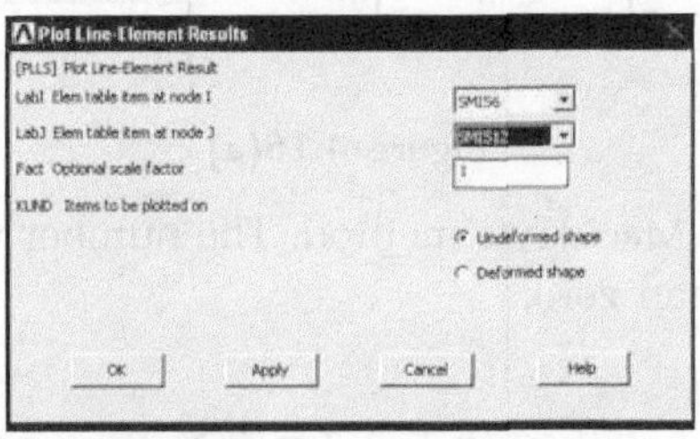

Figure 4.15. Selecting options for finding out bending moment

This result is shown in figure 4.3(*g*).

To find the **shear force diagram** following procedure is followed.

15. **Main Menu > General Postproc > Element Table > Define Table > Add**
    Select **By sequence num and SMISC** and type **2 after SMISC >APPLY**
    Then again select **By sequence num and SMISC** and type **8 after SMISC > OK**
16. **Main Menu > General Postproc > Plot Results > Contour Plot > Line Elem Res >** Select **SMIS 2 and SMIS 8 > OK**
    This result is shown in figure 4.3(*h*)
    To find the **bending stress** following procedure is followed.
17. **Main Menu > General Postproc > Element Table > Define Table > Add**
    Select **By sequence num and LS** and type **3 after LS > APPLY**
    Then again select **By sequence num and LS** and type **6 after LS > OK**
18. **Main Menu > General Postproc > Plot Results > Contour Plot > Line Elem Res >** Select **LS 3 and LS 6 > OK**

This result is shown in figure 4.3(*i*).

4.3 CANTILEVER BEAMS

PROBLEM 4.5. *Beam subjected to concentrated load. For the beam shown in figure 4.16, determine the deflections and reactions. Let E = 210 GPa, I = 2 × 10^{-4} m^4. Take 2 elements.*

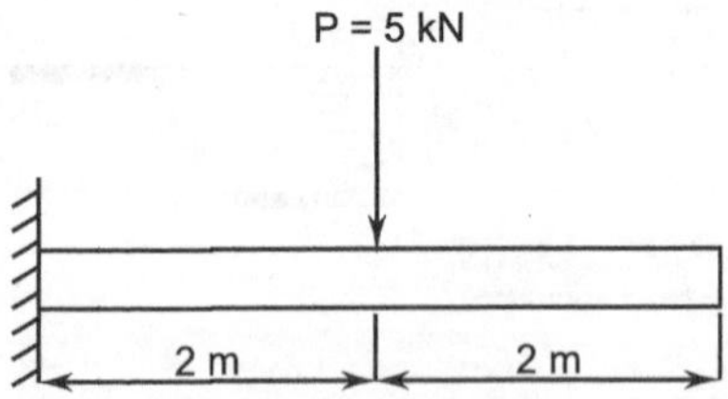

Figure 4.16

SOLUTION. (I) **Analytical Method** [Refer figure 4.16(*a*)]

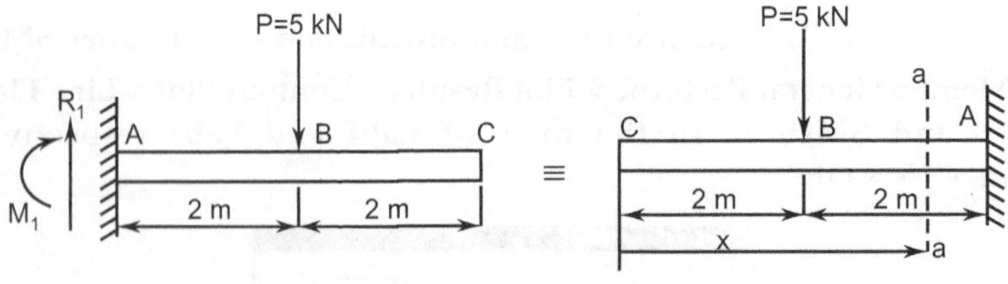

Figure 4.16(a)

The solution is obtained by Maculay's method. The number within the brackets < > is to be neglected whenever it is less than zero.

At section *a–a*

$$EIy'' = -P < x - 2 >$$

$$EIy' = \frac{-P<x-2>^2}{2} + C_1$$

$$EIy = \frac{-P<x-2>^3}{6} + C_1x + C_2$$

$$\text{At } x = 4 \quad y' = 0 \Rightarrow C_1 = 2P$$

$$\text{At } x = 4 \quad y = 0 \Rightarrow C_2 = \frac{-20P}{3}$$

$$\therefore \quad y' = \frac{1}{EI}\left[\frac{-P<x-2>^2}{2} + 2P\right]$$

$$y = \frac{1}{EI}\left[\frac{-P<x-2>^3}{2} + 2Px - \frac{20P}{3}\right]$$

$$y'_B = y'_{x=2} = \frac{1}{EI}(2P) = \frac{2\times 5000}{210\times 10^9 \times 2\times 10^{-4}} = 2.381\times 10^{-4} \text{ rad}$$

$$y'_C = y'_{x=0} = \frac{1}{EI}(2P) = \frac{2\times 5000}{210\times 10^9 \times 2\times 10^{-4}} = 2.381\times 10^{-4} \text{ rad}$$

Similarly,

$$y_B = y_{x=2} = \frac{1}{EI}\left(4P - \frac{20P}{3}\right) = \frac{1}{210\times 10^9\times 2\times 10^{-4}}\left(4\times 5000 - \frac{20\times 5000}{3}\right) = -3.1746\times 10^{-4} \text{ m}$$

$$y_C = y_{x=0} = \frac{1}{EI}\left(-\frac{20P}{3}\right) = \frac{1}{210\times 10^9\times 2\times 10^{-4}}\left(-\frac{20\times 5000}{3}\right) = -7.9365\times 10^{-4} \text{ m}$$

$$\Sigma F_y = 0 \Rightarrow R_1 = 5 \text{ kN}$$

$$\Sigma M = 0 \Rightarrow M_1 = 10 \text{ kN-m}$$

(II) FEM by hand calculations

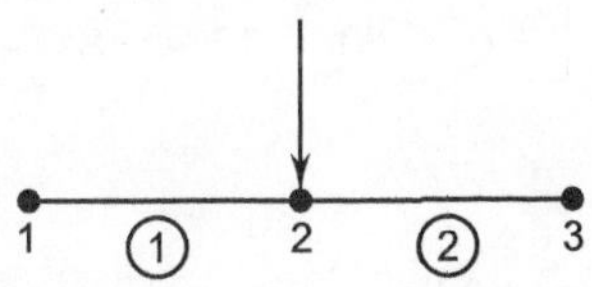

Figure 4.16(b). Finite element model

Element stiffness matrix for element 1 is,

$$[k_1] = \frac{EI}{L^3}\begin{bmatrix} 12 & 6L & -12 & 6L \\ 6L & 4L^2 & -6L & 2L^2 \\ -12 & -6L & 12 & -6L \\ 6L & 2L^2 & -6L & 4L^2 \end{bmatrix} = \frac{210\times 10^9\times 2\times 10^{-4}}{(2)^3}\begin{bmatrix} 12 & 6\times 2 & -12 & 6\times 2 \\ 6\times 2 & 4(2)^2 & -6(2) & 2(2)^2 \\ -12 & -6(2) & 12 & -6(2) \\ 6\times 2 & 2(2)^2 & -6(2) & 4(2)^2 \end{bmatrix}$$

$$= 5.25 \times 10^6 \begin{bmatrix} 12 & 12 & -12 & 12 \\ 12 & 16 & -12 & 8 \\ -12 & -12 & 12 & -12 \\ 12 & 8 & -12 & 16 \end{bmatrix} \begin{matrix} w_1 \\ \theta_1 \\ w_2 \\ \theta_2 \end{matrix}$$

(columns: w_1, θ_1, w_2, θ_2)

Element stiffness matrix for element 2 is,

$$[k_2] = \frac{210 \times 10^9 \times 2 \times 10^{-4}}{(2)^3} \begin{bmatrix} 12 & 12 & -12 & 12 \\ 12 & 16 & -12 & 8 \\ -12 & -12 & 12 & -12 \\ 12 & 8 & -12 & 16 \end{bmatrix} = 5.25 \times 10^6 \begin{bmatrix} 12 & 12 & -12 & 12 \\ 12 & 16 & -12 & 8 \\ -12 & -12 & 12 & -12 \\ 12 & 8 & -12 & 16 \end{bmatrix} \begin{matrix} w_2 \\ \theta_2 \\ w_3 \\ \theta_3 \end{matrix}$$

(columns: w_2, θ_2, w_3, θ_3)

Global stiffness matrix is,

$$[K] = 5.25 \times 10^6 \begin{bmatrix} 12 & 12 & -12 & 12 & 0 & 0 \\ 12 & 16 & -12 & 8 & 0 & 0 \\ -12 & -12 & 12+12 & -12+12 & -12 & 12 \\ 12 & 8 & -12+12 & 16+16 & -12 & 8 \\ 0 & 0 & -12 & -12 & 12 & -12 \\ 0 & 0 & 12 & 8 & -12 & 16 \end{bmatrix}$$

$$[K] = 5.25 \times 10^6 \begin{bmatrix} 12 & 12 & -12 & 12 & 0 & 0 \\ 12 & 16 & -12 & 8 & 0 & 0 \\ -12 & -12 & 24 & 0 & -12 & 12 \\ 12 & 8 & 0 & 32 & -12 & 8 \\ 0 & 0 & -12 & -12 & 12 & -12 \\ 0 & 0 & 12 & 8 & -12 & 16 \end{bmatrix} \begin{matrix} w_1 \\ \theta_1 \\ w_2 \\ \theta_2 \\ w_3 \\ \theta_3 \end{matrix}$$

(columns: w_1, θ_1, w_2, θ_2, w_3, θ_3)

The Global equations are,

$$[K]\{r\} = \{R\}$$

$$5.25 \times 10^6 \begin{bmatrix} 12 & 12 & -12 & 12 & 0 & 0 \\ 12 & 16 & -12 & 8 & 0 & 0 \\ -12 & -12 & 24 & 0 & -12 & 12 \\ 12 & 8 & 0 & 32 & -12 & 8 \\ 0 & 0 & -12 & -12 & 12 & -12 \\ 0 & 0 & 12 & 8 & -12 & 16 \end{bmatrix} \begin{matrix} w_1 \\ \theta_1 \\ w_2 \\ \theta_2 \\ w_3 \\ \theta_3 \end{matrix} \begin{Bmatrix} w_1 \\ \theta_1 \\ w_2 \\ \theta_2 \\ w_3 \\ \theta_3 \end{Bmatrix} = \begin{Bmatrix} 0 + R_1 \\ 0 + M_1 \\ -5 \times 10^3 \\ 0 \\ 0 \\ 0 \end{Bmatrix}$$

(columns: w_1, θ_1, w_2, θ_2, w_3, θ_3; rows 1–2 and columns 1–2 struck out)

By using elimination method for applying boundary conditions, $w_1 = \theta_1 = 0$

The above matrix reduces to,

$$5.25 \times 10^6 \begin{bmatrix} 24 & 0 & -12 & 12 \\ 0 & 32 & -12 & 8 \\ -12 & -12 & 12 & -12 \\ 12 & 8 & -12 & 16 \end{bmatrix} \begin{bmatrix} w_2 \\ \theta_2 \\ w_3 \\ \theta_3 \end{bmatrix} = \begin{bmatrix} -5 \times 10^3 \\ 0 \\ 0 \\ 0 \end{bmatrix}$$

Solving the above matrix and equations, we get,

Deflections and slopes as

$$w_2 = -0.3175 \times 10^{-3} \text{ m}$$
$$\theta_2 = -0.2381 \times 10^{-3} \text{ rad}$$
$$w_3 = -0.7937 \times 10^{-3} \text{ m}$$
$$\theta_3 = -0.2381 \times 10^{-3} \text{ rad}$$

Reaction Calculation

$$5.25 \times 10^6 (12w_1 + 12\,\theta_1 - 12\,w_2 + 12\theta_2) = R_1$$
$$5.25 \times 10^6 [12 \times 0 + 12 \times 0 - 12 (-0.3175 \times 10^{-3}) + 12 (-0.2381 \times 10^{-3}) = R_1$$
$$\therefore \quad R_1 = 5002.2 \text{ N} \approx 5 \text{ kN}$$
$$5.25 \times 10^6(-12w_2 + 8\theta_2) = M_1$$
$$5.25 \times 10^6 [-12 (-0.3175 \times 10^{-3}) + 8 (-0.238 \times 10^{-3})] = M_1$$
$$\therefore \quad M_1 = 10002.3 \text{ N-m} \approx 10 \text{ kN-m}$$

(III) Software results

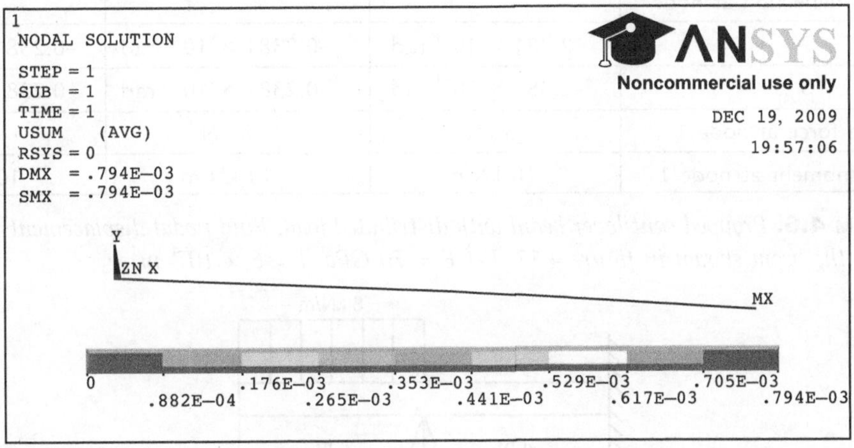

Figure 4.16(c). Deflection pattern for a cantilever beam (Refer to appendix for colour figures)

Deflection values at nodes (in metres)

The following degree of freedom results are in global coordinates

NODE	UX	UY	UZ	USUM
1	0.0000	0.0000	0.0000	0.0000
2	0.0000	-0.31746E-03	0.0000	0.31746E-03
3	0.0000	-0.79365E-03	0.0000	0.79365E-03

Maximum absolute values

NODE	0	3	0	3
VALUE	0.0000	- 0.79365E-03	0.0000	0.79365E-03

Rotational deflection values at nodes

The following degree of freedom results are in Global coordinates

NODE	ROTZ
1	0.0000
2	- 0.23810E-03
3	- 0.23810E-03

Reaction values

The following X, Y, Z solutions are in Global coordinates

NODE	FX	FY	MZ
1	0.0000	5000.0	10000.

Answers of Problem 4.5

Parameter	Analytical method	FEM-hand calculation	Software results
Deflection at node			
2	-3.1746×10^{-4} m	-0.3175×10^{-3} m	-0.31746×10^{-3} m
3	-7.9365×10^{-4} m	-0.7937×10^{-3} m	-0.79365×10^{-3} m
Rotational deflection at node			
2	-2.381×10^{-4} rad	-0.2381×10^{-3} rad	-0.2381×10^{-3} rad
3	-2.381×10^{-4} rad	-0.2381×10^{-3} rad	-0.2381×10^{-3} rad
Reaction force at node 1	5 kN	5 kN	5 kN
Reaction moment at node 1	10 kN-m	10 kN-m	10 kN-m

Problem 4.6. *Propped cantilever beam with distributed load. Find nodal displacements and support reactions for the beam shown in figure 4.17. Let $E = 70$ GPa, $I = 6 \times 10^{-4}$ m^4.*

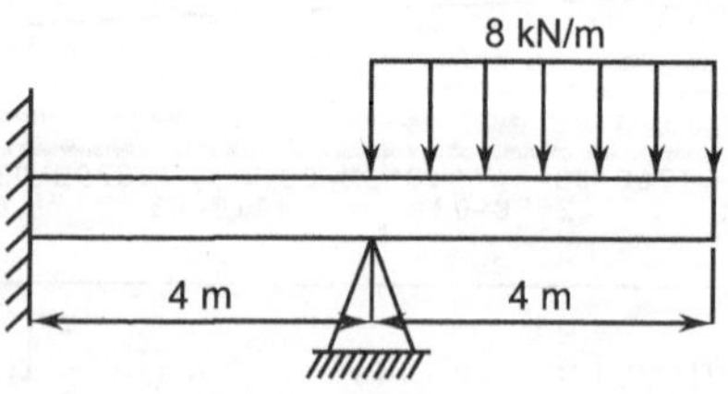

Figure 4.17

SOLUTION. (I) **Analytical method** [Refer figure 4.17(*a*)]

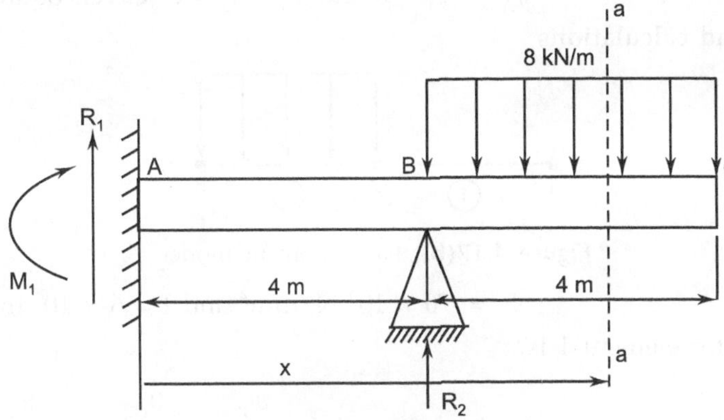

Figure 4.17(a)

The solution is obtained by Maculay's method. The number within the brackets < > is to be neglected whenever it is less than zero.

$$\Sigma F_y = 0 \Rightarrow R_1 = -R_2 + 8 \times 10^3 \times 4 = 32000 - R_2$$

$$\Sigma M = 0 \Rightarrow M_1 = 4R_2 - (8 \times 10^3 \times 4) \times 6 = (4R_2 - 192000) \text{ N-m}$$

At section *a-a*

$$M_x = M_1 + R_1 x + R_2 <x-4> - \frac{8 \times 10^3}{2} <x-4>^2$$

$$EIy'' = (4R_2 - 192000) + (32000 - R_2)\, x + R_2 <x-4> - 4 \times 10^3 <x-4>^2$$

$$EIy' = 4R_2 x - 192000x + 32000\frac{x^2}{2} - R_2 \times \frac{x^2}{2} + \frac{R_2}{2} <x-4>^2 - \frac{4 \times 10^3 <x-4>^3}{3} + C_1 \quad ...(i)$$

$$EIy = 4R_2 \times \frac{x^2}{2} - 192000\frac{x^2}{2} + 32000\frac{x^3}{6} - R_2 \times \frac{x^3}{6} + \frac{R_2 <x-4>^3}{6} - \frac{4 \times 10^3 <x-4>^4}{4} + C_1 x + C_2 \quad ...(ii)$$

Boundary conditions are,

$$\text{At } x = 0 \quad y = 0 \Rightarrow C_2 = 0$$

$$\text{At } x = 0 \quad y' = 0 \Rightarrow C_1 = 0$$

$$\text{At } x = 4 \quad y = 0 \Rightarrow R_2 = 56008 \text{ N}$$

$$\therefore \quad R_1 = 32000 - R_2 = -24008 \text{ N}$$

$$M_1 = 4R_2 - 192000 = 32032 \text{ N-m (Clockwise) (negative)}$$

Substituting in (*i*) and (*ii*)

$$EIy' = 4 \times 56008x - 192000x + 32000\frac{x^2}{2} - 56008\frac{x^2}{2} + \frac{56008}{2} <x-4>^2 - \frac{4 \times 10^3}{3} <x-4>^3$$

$$y'\big|_{x=4} = -0.00152 \text{ rad}$$

and

$$y'\big|_{x=8} = -0.00355 \text{ rad}$$

$$EIy = 4 \times 56008\frac{x^2}{2} - 192000\frac{x^2}{2} + 32000\frac{x^3}{6} - \frac{56008x^3}{6} + \frac{56008 <x-4>^3}{6} - \frac{4 \times 10^3}{4} <x-4>^4$$

$$y\big|_{x=8} = y_C = y_3 = -\,0.0122 \text{ m}$$

$$y_A = y_1 = 0 \text{ and } y_B = y_2 = 0 \text{ (Given boundary conditions)}$$

(II) FEM by hand calculations

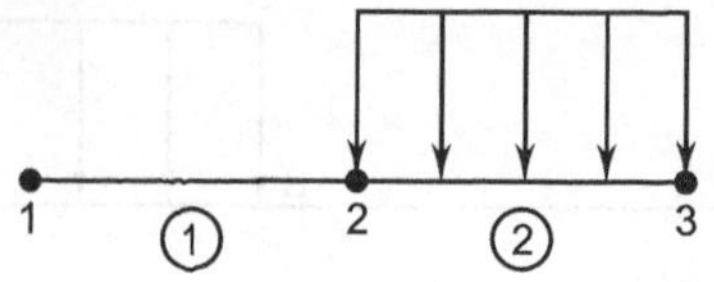

Figure 4.17(b). Finite element model

$$E = 70 \times 10^3 \text{ N/mm}^2 \text{ and } I = 6 \times 10^8 \text{ mm}^4$$

Stiffness matrix for element 1 is,

$$[k_1] = \frac{EI}{L^3}\begin{bmatrix} 12 & 6L & -12 & 6L \\ 6L & 4L^2 & -6L & 2L^2 \\ -12 & -6L & 12 & -6L \\ 6L & 2L^2 & -6L & 4L^2 \end{bmatrix} = \frac{70\times10^3\times6\times10^8}{(4000)^3}\begin{array}{c} \begin{array}{cccc} w_1 & \theta_1 & w_2 & \theta_2 \end{array} \\ \left[\begin{array}{cccc} 12 & 24000 & -12 & 24000 \\ 24000 & 64\times10^6 & -24000 & 32\times10^6 \\ -12 & -24000 & 12 & -24000 \\ 24000 & 32\times10^6 & -24000 & 64\times10^6 \end{array}\right] \end{array}\begin{array}{c} \\ w_1 \\ \theta_1 \\ w_2 \\ \theta_2 \end{array}$$

$$[k_1] = 656.25\begin{array}{c} \begin{array}{cccc} w_1 & \theta_1 & w_2 & \theta_2 \end{array} \\ \left[\begin{array}{cccc} 12 & 24000 & -12 & 24000 \\ 24000 & 64\times10^6 & -24000 & 32\times10^6 \\ -12 & -24000 & 12 & -24000 \\ 24000 & 32\times10^6 & -24000 & 64\times10^6 \end{array}\right] \end{array}\begin{array}{c} \\ w_1 \\ \theta_1 \\ w_2 \\ \theta_2 \end{array}$$

$$[k_2] = 656.25\begin{array}{c} \begin{array}{cccc} w_2 & \theta_2 & w_3 & \theta_3 \end{array} \\ \left[\begin{array}{cccc} 12 & 24000 & -12 & 24000 \\ 24000 & 64\times10^6 & -24000 & 32\times10^6 \\ -12 & -24000 & 12 & -24000 \\ 24000 & 32\times10^6 & -24000 & 64\times10^6 \end{array}\right] \end{array}\begin{array}{c} \\ w_2 \\ \theta_2 \\ w_3 \\ \theta_3 \end{array}$$

Nodal force calculation

For element 2,

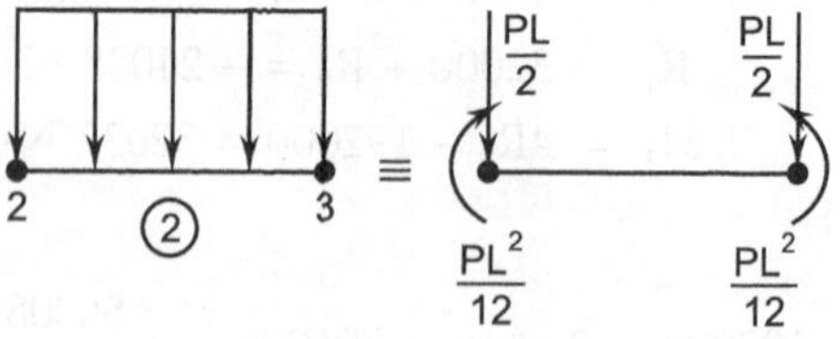

Figure 4.17(c)

$$P = 8 \text{ N/mm}$$

$$L = 4000 \text{ mm}$$

$$\frac{PL}{2} = \frac{8\times4000}{2} = 16000 \text{ N}$$

$$\frac{PL^2}{12} = \frac{8 \times (4000)^2}{12} = 10.667 \times 10^6 \text{ N - mm}$$

The Nodal Forces and moments for element 2 is,

$$= \begin{Bmatrix} -\dfrac{PL}{2} \\ -\dfrac{PL^2}{12} \\ -\dfrac{PL}{2} \\ \dfrac{PL^2}{12} \end{Bmatrix} = \begin{Bmatrix} -16000 \\ -10.667 \times 10^6 \\ -16000 \\ 10.667 \times 10^3 \end{Bmatrix}$$

The Global equations are,

$$[K]\{r\} = \{R\}$$

$$656.25 \begin{bmatrix} \overset{w_1}{12} & \overset{\theta_1}{24000} & \overset{w_2}{-12} & \overset{\theta_2}{24000} & \overset{w_3}{0} & \overset{\theta_3}{0} \\ 24000 & 64 \times 10^6 & -24000 & 32 \times 10^6 & 0 & 0 \\ -12 & -24000 & 12 & -24000 + 24000 & -12 & 24000 \\ 24000 & 32 \times 10^6 & -24000 + 24000 & 64 \times 10^6 + 64 \times 10^6 & -24000 & 32 \times 10^6 \\ 0 & 0 & -12 & -24000 & 12 & -24000 \\ 0 & 0 & 24000 & 32 \times 10^6 & -24000 & 64 \times 10^6 \end{bmatrix} \begin{matrix} w_1 \\ \theta_1 \\ w_2 \\ \theta_2 \\ w_3 \\ \theta_3 \end{matrix} \begin{Bmatrix} w_1 \\ \theta_1 \\ w_2 \\ \theta_2 \\ w_3 \\ \theta_3 \end{Bmatrix} = \begin{Bmatrix} 0 + R_1 \\ 0 + M_1 \\ 16000 + R_2 \\ -10.667 \times 10^6 \\ -16000 \\ 10.667 \times 10^6 \end{Bmatrix}$$

By using elimination method for applying boundary conditions,

$$w_1 = \theta_1 = w_2 = 0,$$

the above matrix reduces to

$$656.25 \begin{bmatrix} 128 \times 10^6 & -24000 & 32 \times 10^6 \\ -24000 & 12 & -24000 \\ 32 \times 10^6 & -24000 & 64 \times 10^6 \end{bmatrix} \begin{Bmatrix} \theta_2 \\ w_3 \\ \theta_3 \end{Bmatrix} = \begin{Bmatrix} -10.667 \times 10^6 \\ -16000 \\ 10.667 \times 10^6 \end{Bmatrix}$$

Solving above matrix and equations,

we get

$$w_2 = -12.19 \text{ mm} = -0.01219 \text{ m}$$

$$\theta_3 = -0.00355 \text{ rad}$$

$$\theta_2 = -0.00152 \text{ rad}$$

Reaction Calculation

$$656.25\,(12w_1 + 24000\theta_1 - 12w_2 + 24000\theta_2) = R_1$$

$$656.25\,[12 \times 0 + 24000 \times 0 - 12 \times 0 + 24000\,(-0.00152)] = R_1$$

$$\therefore \quad R_1 = -23.94 \text{ kN} \approx -24 \text{ kN}$$

$$656.25\,(24000w_1 + 64 \times 10^6\,\theta_1 - 24000w_2 + 32 \times 10^6\theta_2) = M_1$$

$$656.25\,[0 + 0 - 0 + 32 \times 10^6(-0.00152)] = M_1$$

$$\therefore \quad M_1 = -31.92 \text{ kN-m} \approx -32 \text{ kN-m}$$

$$656.25\ (-12w_1 - 24000\theta_1 + 24w_2 + 0 \times \theta_2 - 12w_3 + 24000\theta_3) = R_2 - 16000$$

$$656.25\ [(0 - 0 + 0 + 0 - 12\ (-12.19) + 24000\ (-0.00355)] = R_2 - 16000$$

$$\therefore \quad R_2 = 56.08 \text{ kN} \approx 56 \text{ kN}$$

(III) Software results

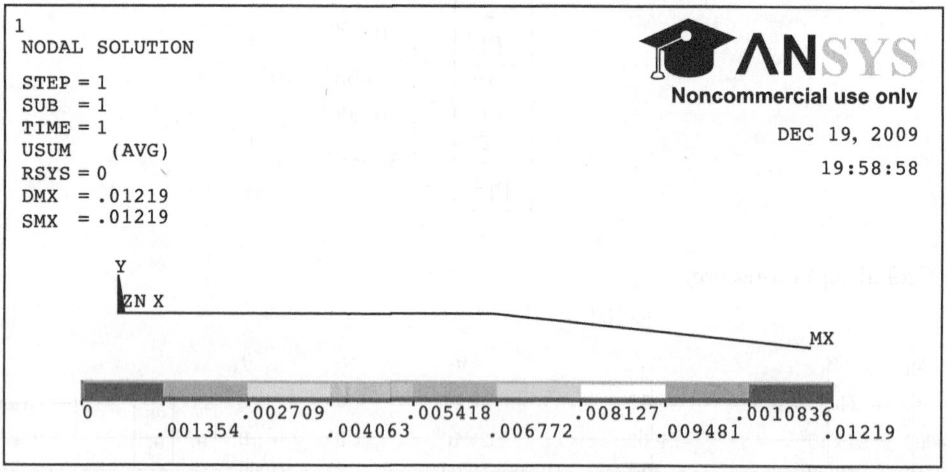

Figure 4.17(d). Deflection pattern for a cantilever beam (Refer to appendix for colour figures)

Deflection values at nodes (in metres)

The following degree of freedom results are in Global coordinates

NODE	UX	UY	UZ	USUM
1	0.0000	0.0000	0.0000	0.0000
2	0.0000	0.0000	0.0000	0.0000
3	0.0000	-0.12190E-01	0.0000	0.12190E-01

Rotational deflection values at nodes

The following degree of freedom results are in Global coordinates

NODE	ROT Z
1	0.0000
2	– 0.15238E-02
3	– 0.35556E-02

Reaction values

The following X, Y, Z solutions are in Global coordinates

NODE	FX	FY	MZ
1	0.0000	-24000	-32000
2	0.0000	56000	

Answers of Problem 4.6

Parameter	Analytical method	FEM-hand calculation	Software results
Deflection at node 3	-0.0122 m	-0.01219 m	-0.01219 m
Rotational deflection at node			
2	-0.00152 rad	-0.00152 rad	-0.001524 rad
3	-0.00355 rad	-0.00355 rad	-0.00355 rad
Reaction force at			
1	-24 kN	-24 kN	-24 kN
2	56 kN	56 kN	56 kN
Reaction moment at node 1	-32 kN-m	-32 kN-m	-32 kN-m

Problem 4.7. *Propped cantilever beam with varying load. For the beam shown in figure 4.18, determine the nodal displacements, slopes, reactions, maximum bending moment, shear force and maximum bending stress. Take E = 200 GPa.*

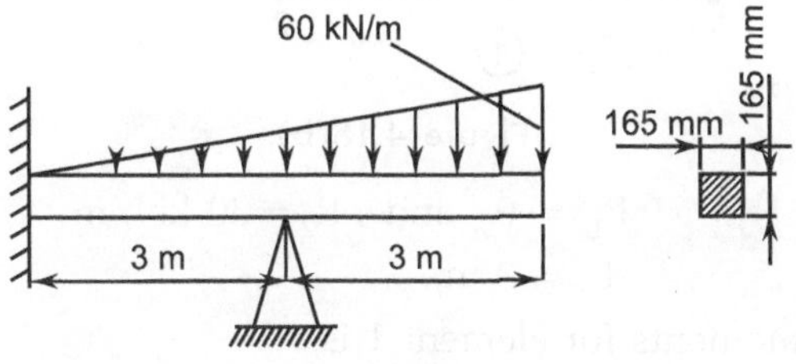

Figure 4.18

Solution. (I) FEM hand calculations

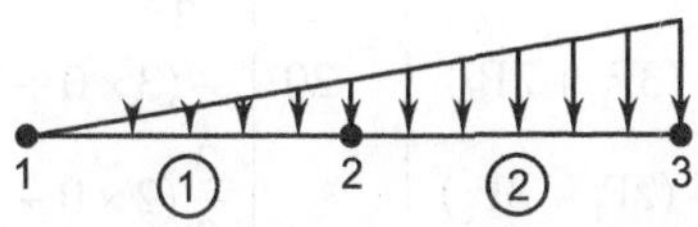

Figure 4.18(a). Finite element model

$$I = \frac{bh^3}{12} = \frac{165 \times (165)^3}{12} = 61766718.75 \text{ mm}^4 = 6.18 \times 10^{-5} \text{ m}^4$$

Stiffness matrices for element 1 and 2 are,

$$[k] = \frac{EI}{L^3}\begin{bmatrix} 12 & 6L & -12 & 6L \\ 6L & 4L^2 & -6L & 2L^2 \\ -12 & -6L & 12 & -6L \\ 6L & 2L^2 & -6L & 4L^2 \end{bmatrix}$$

$$[k_1] = \frac{200 \times 10^9 \times 6.18 \times 10^{-5}}{(3^3)}\begin{bmatrix} 12 & 18 & -12 & 18 \\ 18 & 36 & -18 & 18 \\ -12 & -18 & 12 & -18 \\ 18 & 18 & -18 & 36 \end{bmatrix}$$

$$= 457.78\times 10^{3}\begin{array}{c} \begin{matrix} \quad w_1 & \quad\theta_1 & \quad w_2 & \quad\theta_2 \end{matrix} \\ \begin{bmatrix} 12 & 18 & -12 & 18 \\ 18 & 36 & -18 & 18 \\ -12 & -18 & 12 & -18 \\ 18 & 18 & -18 & 36 \end{bmatrix} \end{array}\begin{matrix} \\ w_1 \\ \theta_1 \\ w_2 \\ \theta_2 \end{matrix}$$

$$[k_2] = 457.78\times 10^{3}\begin{array}{c} \begin{matrix} \quad w_2 & \quad\theta_2 & \quad w_3 & \quad\theta_3 \end{matrix} \\ \begin{bmatrix} 12 & 18 & -12 & 18 \\ 18 & 36 & -18 & 18 \\ -12 & -18 & 12 & -18 \\ 18 & 18 & -18 & 36 \end{bmatrix} \end{array}\begin{matrix} \\ w_2 \\ \theta_2 \\ w_3 \\ \theta_3 \end{matrix}$$

Nodal force calculation

For element 1,

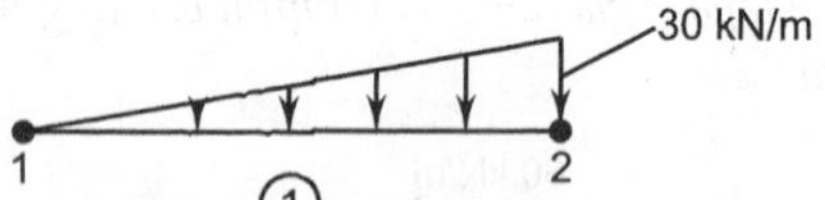

Figure 4.18(b)

$$P_1 = 0 \quad \text{and} \quad P_2 = 30 \text{ kN/m}$$

$$L = 3 \text{ m}$$

The Nodal forces and moments for element 1 is,

$$F_1 = \frac{L}{20}\begin{Bmatrix} -(7P_1+3P_2) \\ -\dfrac{L}{3}(3P_1+2P_2) \\ -(3P_1+7P_2) \\ \dfrac{L}{3}(2P_1+3P_2) \end{Bmatrix} = \frac{3}{20}\begin{Bmatrix} -(7\times 0+3\times 30\times 10^3) \\ -\dfrac{3}{3}(3\times 0+2\times 30\times 10^3) \\ -(3\times 0+7\times 30\times 10^3) \\ \dfrac{3}{3}(2\times 0+3\times 30\times 10^3) \end{Bmatrix} = \begin{Bmatrix} -13.5\times 10^3 \\ -9\times 10^3 \\ -31.5\times 10^3 \\ 13.5\times 10^3 \end{Bmatrix}\begin{matrix} f_1 \\ m_1 \\ f_2 \\ m_2 \end{matrix}$$

For element 2,

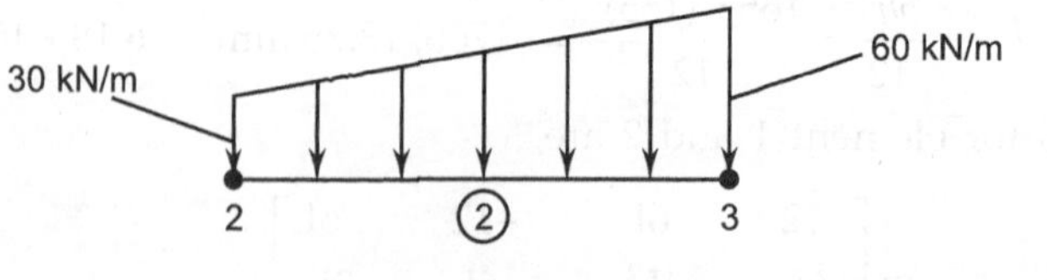

Figure 4.18(c)

$$P_1 = 30 \times 10^3 \text{ N/m and } P_2 = 60 \times 10^3 \text{ N/m}$$

$$L = 3 \text{ m}$$

The nodal forces and moments for element 2 is,

$$F_2 = \frac{L}{20}\begin{Bmatrix} -(7P_1+3P_2) \\ -\frac{L}{3}(3P_1+2P_2) \\ -(3P_1+7P_2) \\ -\frac{L}{3}(2P_1+3P_2) \end{Bmatrix} = \frac{3}{20}\begin{Bmatrix} -(7\times 30\times 10^3 + 3\times 60\times 10^3) \\ -\frac{3}{3}(3\times 30\times 10^3 + 2\times 60\times 10^3) \\ -(3\times 30\times 10^3 + 7\times 60\times 10^3) \\ -\frac{3}{3}(2\times 30\times 10^3 + 3\times 60\times 10^3) \end{Bmatrix} = \begin{Bmatrix} -58.5\times 10^3 \\ -31.5\times 10^3 \\ -76.5\times 10^3 \\ 36\times 10^3 \end{Bmatrix} \begin{matrix} f_2 \\ m_2 \\ f_3 \\ m_3 \end{matrix}$$

The combined nodal forces and moments matrix is,

$$[F] = \begin{Bmatrix} -13.5\times 10^3 \\ -9\times 10^3 \\ (-31.5-58.5)\times 10^3 \\ (13.5-31.5)\times 10^3 \\ -76.5\times 10^3 \\ 36\times 10^3 \end{Bmatrix} = \begin{Bmatrix} -13.5\times 10^3 \\ -9\times 10^3 \\ -90\times 10^3 \\ -18\times 10^3 \\ -76.5\times 10^3 \\ 36\times 10^3 \end{Bmatrix} \begin{matrix} f_1 \\ m_1 \\ f_2 \\ m_2 \\ f_3 \\ m_3 \end{matrix}$$

The global equations are,

$$457.78\times \not{10^3}\; \begin{matrix} w_1 & \theta_1 & w_2 & \theta_2 & w_3 & \theta_3 \end{matrix} \begin{bmatrix} 12 & 18 & -12 & 18 & 0 & 0 \\ 18 & 36 & -18 & 18 & 0 & 0 \\ -12 & -18 & 12+12 & -18+18 & -12 & 18 \\ 18 & 18 & -18+18 & 36+36 & -18 & 18 \\ 0 & 0 & -12 & -18 & 12 & -18 \\ 0 & 0 & 18 & 18 & -18 & 36 \end{bmatrix} \begin{matrix} w_1 \\ \theta_1 \\ w_2 \\ \theta_2 \\ w_3 \\ \theta_3 \end{matrix} \begin{Bmatrix} w_1 \\ \theta_1 \\ w_2 \\ \theta_2 \\ w_3 \\ \theta_3 \end{Bmatrix} = \not{10^3} \begin{Bmatrix} -13.5+R_1 \\ -9+M_1 \\ -90+R_2 \\ -18 \\ -76.5 \\ 36 \end{Bmatrix}$$

By using elimination method for applying boundary conditions, $w_1 = \theta_1 = w_2 = 0$ the above matrix reduces to,

$$457.78\begin{bmatrix} 72 & -18 & 18 \\ -18 & 12 & -18 \\ 18 & -18 & 36 \end{bmatrix}\begin{Bmatrix} \theta_2 \\ w_3 \\ \theta_3 \end{Bmatrix} = \begin{Bmatrix} -18 \\ -76.5 \\ 36 \end{Bmatrix}$$

By solving above matrix and equations,

we get $\theta_2 = -0.0128$ rad, $w_3 = -0.0811$ m and $\theta_3 = -0.0319$ rad

Reaction calculation

$$457.78\times 10^3\,(12w_1 + 18\theta_1 - 12w_2 + 18\theta_2) = -13.5\times 10^3 + R_1$$

$$457.78\times 10^3[12\times 0 + 18\times 0 - 12\times 0 + 18\times(-0.0128)] = -13.5\times 10^3 + R_1$$

$$\therefore \quad R_1 = -91972.5 \text{ N}$$

$$457.78\times 10^3\,(0 + 0 - 0 + 18\theta_2) = -9\times 10^3 + M_1$$

$$457.78\times 10^3\,[18(-0.0128)] = -9\times 10^3 + M_1$$

$$\therefore \quad M_1 = -96472.5 \text{ N-m}$$

$$457.78 \times 10^3[0 - 0 + 0 + 0 - 12w_3 + 18\theta_3) = -90 \times 10^3 + R_2$$

$$457.78 \times 10^3[-12\,(-0.0811) + 18\,(-0.0319)] = -90 \times 10^3 + R_2$$

$$\therefore \qquad R_2 = 272654.22 \text{ N}$$

(II) Software results

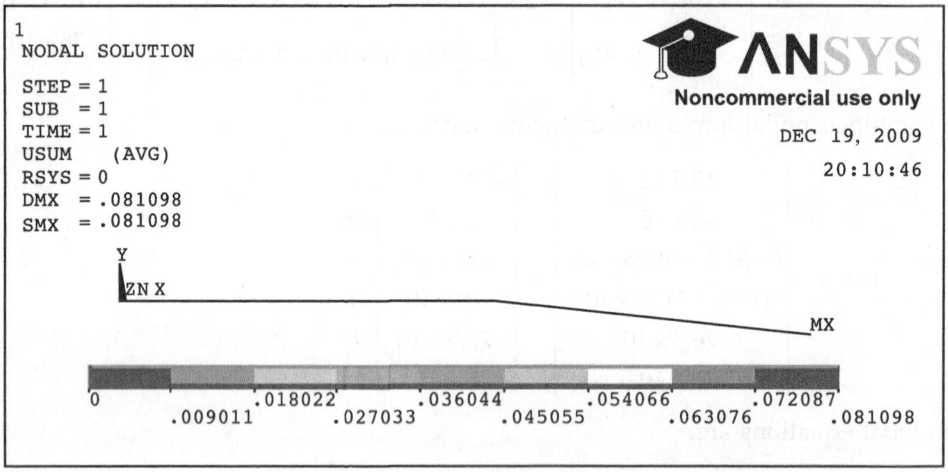

Figure 4.18(d). Deflection pattern for a cantilever beam (Refer to appendix for colour figures)

Deflection values at nodes (in metres)

The following degree of freedom results are in Global coordinates

NODE	UX	UY	UZ	USUM
1	0.0000	0.0000	0.0000	0.0000
2	0.0000	0.0000	0.0000	0.0000
3	0.0000	-0.81098E-01	0.0000	0.81098E-01

Maximum absolute values

NODE	0	3	0	3
VALUE	0.0000	-0.81098E-01	0.0000	0.81098E-01

Rotational deflection values at nodes

The following degree of freedom results are in Global coordinates

NODE	ROTZ
1	0.0000
2	-0.12834E-01
3	-0.31948E-01

Reaction values

The following X, Y, Z solutions are in Global coordinates

NODE	FX	FY	MZ
1	0.0000	-92250	-96750
2	0.0000	0.27225E + 06	

Total values

VALUE	0.0000	0.18000E + 06	-96750

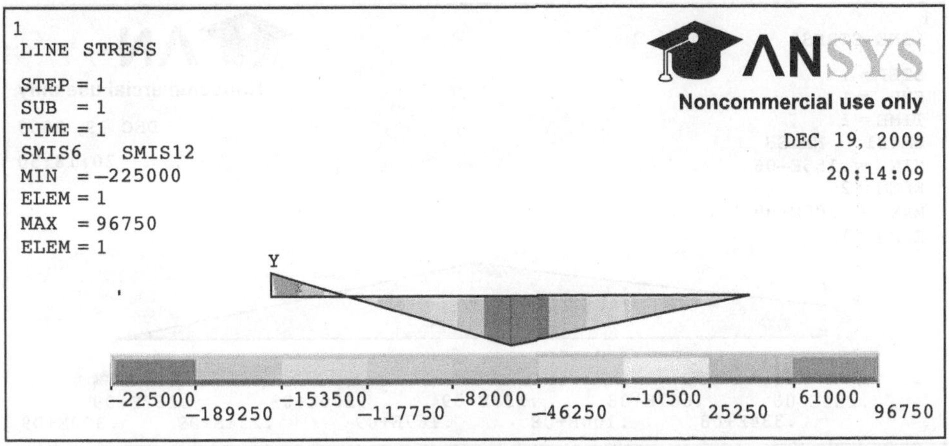

Figure 4.18(e) Bending moment diagram for a propped cantilever beam (Refer to appendix for colour figures)

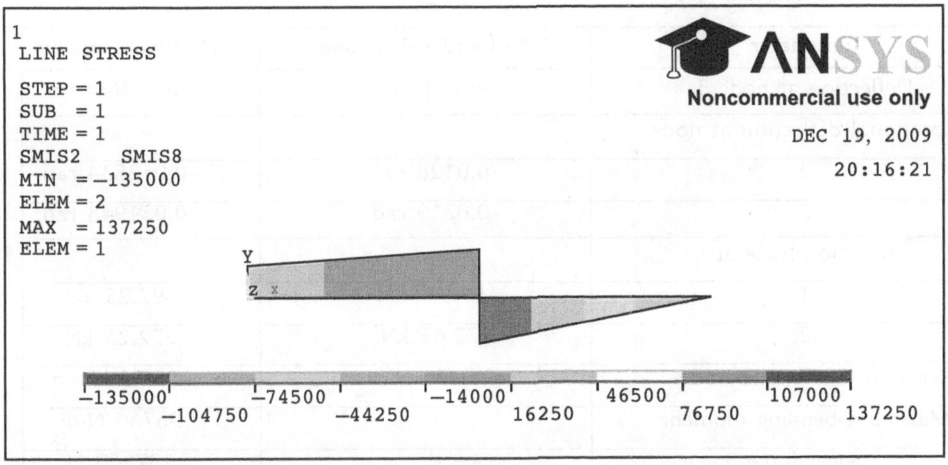

Figure 4.18(f). Shear force diagram for a propped cantilever beam (Refer to appendix for colour figures)

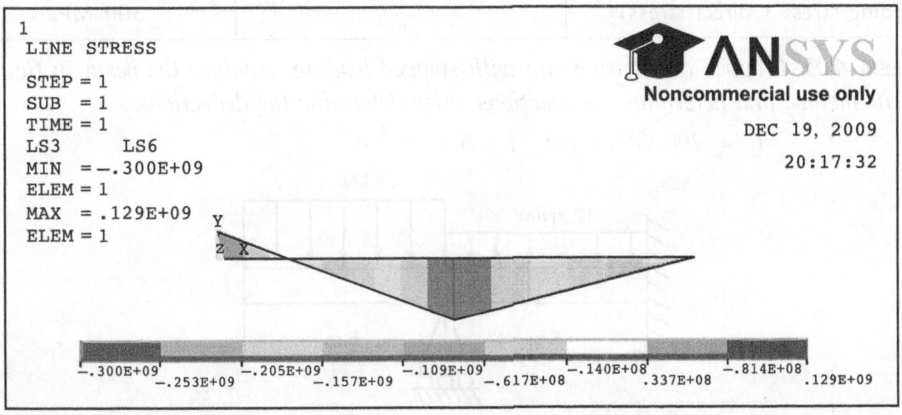

Figure 4.18(g). Bending stress for a propped cantilever beam (Refer to appendix for colour figures)

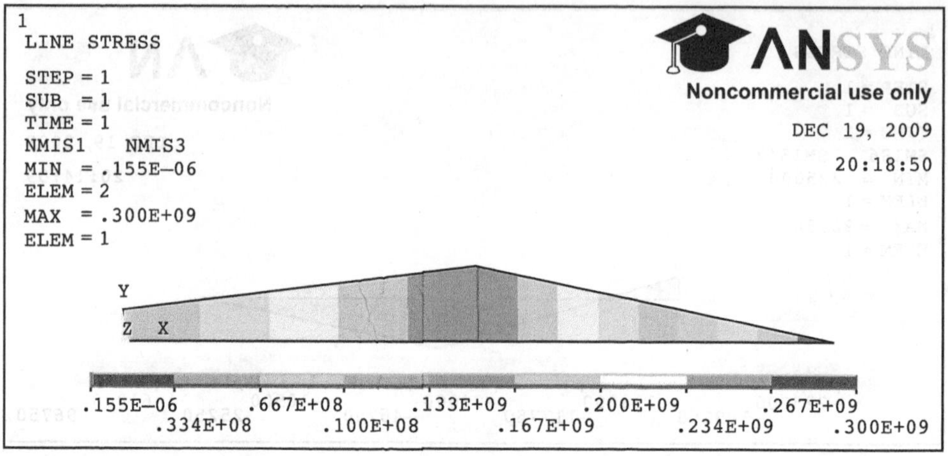

Figure 4.18(h). Maximum stress for a propped cantilever beam (Refer to appendix for colour figures)

ANSWERS OF PROBLEM 4.7

Parameter	FEM-hand calculation	Software results
Deflection at node 3	-0.0811 m	-0.081098 m
Rotational deflection at node		
2	-0.0128 rad	-0.012834 rad
3	-0.0319 rad	-0.031948 rad
Reaction force at		
1	-91.97 kN	-92.25 kN
2	272.65 kN	272.25 kN
Reaction moment at node 1	-96.47 kN-m	-96.75 kN-m
Maximum bending moment	......	96750 N-m
Shear force	......	137250 N
Maximum bending stress	......	129 MPa
Maximum stress (bending stress + direct stress)	......	300 MPa

PROBLEM 4.8. *Propped cantilever beam with stepped loading. Analyze the beam in figure 4.19 by finite element method and determine the reactions. Also determine the deflections.*

Given $E = 200 \text{ GPa}$ and $I = 5 \times 10^{-4} \text{ m}^4$.

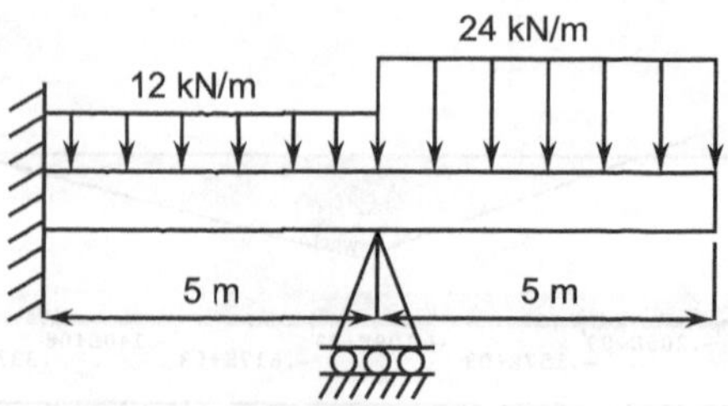

Figure 4.19

SOLUTION. (I) FEM by hand calculations

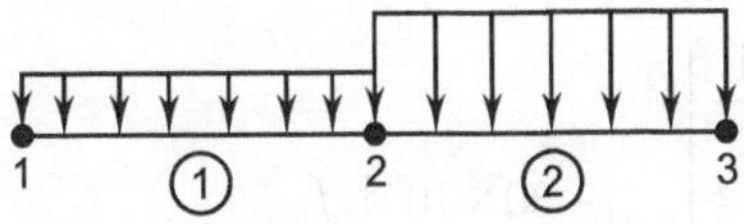

Figure 4.19(a). Finite element model

Stiffness matrix for element 1 is,

$$[k] = \frac{EI}{L^3}\begin{bmatrix} 12 & 6L & -12 & 6L \\ 6L & 4L^2 & -6L & 2L^2 \\ -12 & -6L & 12 & -6L \\ 6L & 2L^2 & -6L & 4L^2 \end{bmatrix}$$

$$[k_1] = \frac{200\times10^9\times5\times10^{-4}}{(5)^3}\begin{bmatrix} 12 & 30 & -12 & 30 \\ 30 & 100 & -30 & 50 \\ -12 & -30 & 12 & -30 \\ 30 & 50 & -30 & 100 \end{bmatrix}$$

$$= 800\times10^3 \begin{bmatrix} \overset{w_2}{12} & \overset{\theta_2}{30} & \overset{w_3}{-12} & \overset{\theta_3}{30} \\ 30 & 100 & -30 & 50 \\ -12 & -30 & 12 & -30 \\ 30 & 50 & -30 & 100 \end{bmatrix}\begin{matrix} w_2 \\ \theta_2 \\ w_3 \\ \theta_3 \end{matrix}$$

Nodal force calculation

For element 1,

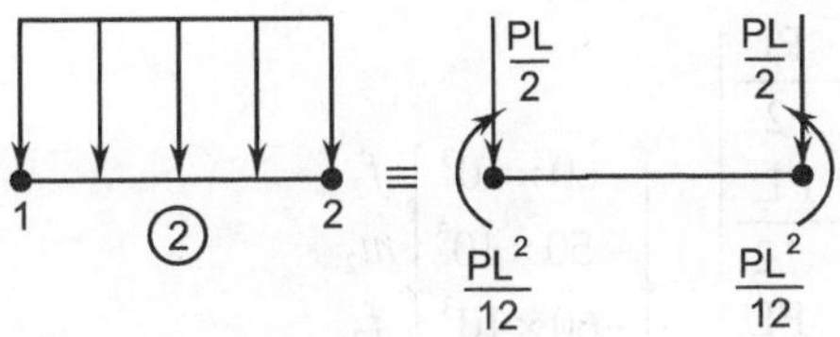

Figure 4.19(b)

$$P = 12\text{ kN/m} = 12\times10^3\text{ N/m}$$

$$L = 5\text{ m}$$

$$\frac{PL}{2} = \frac{12\times10^3\times5}{2} = 30\times10^3\text{N}$$

$$\frac{PL^2}{12} = \frac{12\times10^3\times(5)^2}{12} = 25\times10^3\text{N-m}$$

The nodal forces and moments for element 1 is,

$$[F_1] = \begin{Bmatrix} -\dfrac{PL}{2} \\ -\dfrac{PL^2}{12} \\ -\dfrac{PL}{2} \\ \dfrac{PL^2}{12} \end{Bmatrix} = \begin{Bmatrix} -30 \times 10^3 \\ -25 \times 10^3 \\ -30 \times 10^3 \\ 25 \times 10^3 \end{Bmatrix} \begin{matrix} f_1 \\ m_1 \\ f_2 \\ m_2 \end{matrix}$$

For element 2,

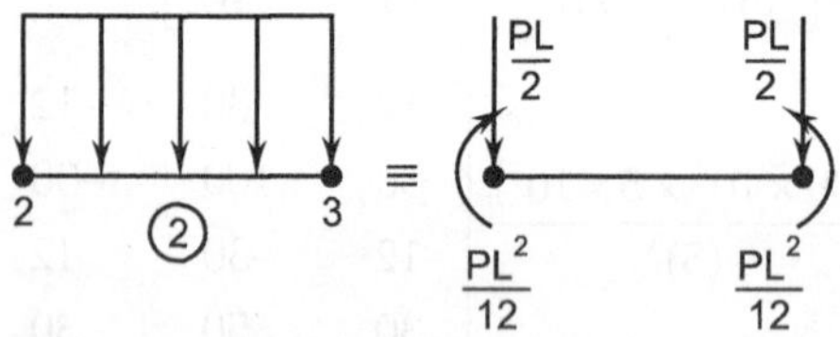

Figure 4.19(c)

$$P = 24 \text{ kN/m} = 24 \times 10^3 \text{ N/m and } L = 5 \text{ m}$$

$$\frac{PL}{2} = \frac{24 \times 10^3 \times 5}{2} = 60 \times 10^3 \text{N}$$

$$\frac{PL^2}{12} = \frac{24 \times 10^3 \times (5)^2}{12} = 50 \times 10^3 \text{ N-m}$$

The nodal forces and moments for element 2 is,

$$[F_2] = \begin{Bmatrix} -\dfrac{PL}{2} \\ -\dfrac{PL^2}{12} \\ -\dfrac{PL}{2} \\ \dfrac{PL^2}{12} \end{Bmatrix} = \begin{Bmatrix} -60 \times 10^3 \\ -50 \times 10^3 \\ -60 \times 10^3 \\ 50 \times 10^3 \end{Bmatrix} \begin{matrix} f_2 \\ m_2 \\ f_3 \\ m_3 \end{matrix}$$

The combined nodal forces and moments is,

$$[F] = \begin{Bmatrix} -30 \times 10^3 \\ -25 \times 10^3 \\ -30 \times 10^3 - 60 \times 10^3 \\ 25 \times 10^3 - 50 \times 10^3 \\ -60 \times 10^3 \\ 50 \times 10^3 \end{Bmatrix} = \begin{Bmatrix} -30 \times 10^3 \\ -25 \times 10^3 \\ -90 \times 10^3 \\ -25 \times 10^3 \\ -60 \times 10^3 \\ 50 \times 10^3 \end{Bmatrix} \begin{matrix} f_1 \\ m_1 \\ f_2 \\ m_2 \\ f_3 \\ m_3 \end{matrix}$$

The Global equations are,

$$[K]\,\{r\} = \{R\}$$

$$
\begin{matrix} & w_1 & \theta_1 & w_2 & \theta_2 & w_3 & \theta_3 \end{matrix}
$$

$$
800 \times 10^3 \begin{bmatrix} 12 & 30 & -12 & 30 & 0 & 0 \\ 30 & 100 & -30 & 50 & 0 & 0 \\ -12 & -30 & 12+12 & -30+30 & -12 & 30 \\ 30 & 50 & -30+30 & 100+100 & -30 & 50 \\ 0 & 0 & -12 & -30 & 12 & -30 \\ 0 & 0 & 30 & 50 & -30 & 100 \end{bmatrix} \begin{Bmatrix} w_1 \\ \theta_1 \\ w_2 \\ \theta_2 \\ w_3 \\ \theta_3 \end{Bmatrix} = \begin{Bmatrix} -30 \times 10^3 + R_1 \\ 25 \times 10^3 + M_1 \\ -90 \times 10^3 + R_2 \\ -25 \times 10^3 \\ -60 \times 10^3 \\ 50 \times 10^3 \end{Bmatrix}
$$

By using elimination method for applying boundary conditions,

$$w_1 = \theta_1 = w_2 = 0,$$

the above matrix reduces to,

$$
800 \times 10^3 \begin{bmatrix} 200 & -30 & 50 \\ -30 & 12 & -30 \\ 50 & -30 & 100 \end{bmatrix} \begin{Bmatrix} \theta_2 \\ w_3 \\ \theta_3 \end{Bmatrix} = \begin{Bmatrix} -25 \times 10^3 \\ -60 \times 10^3 \\ 50 \times 10^3 \end{Bmatrix}
$$

By solving above matrix and equations,

we get, $\theta_2 = -0.003438$ rad, $w_3 = -0.035938$ m and $\theta_3 = -0.008438$ rad

Reaction Calculation

$$800 \times 10^3 (12w_1 + 30\theta_1 - 12w_2 + 30\theta_2) = -30 \times 10^3 + R_1$$

$$800 \times 10^3 [0 + 0 - 0 + 30 (-0.003438)] = -30 \times 10^3 + R_1$$

$$\therefore \quad R_1 = -52512 \text{ N}$$

$$800 \times 10^3 (0 + 0 - 0 + 50\theta_2) = -25 \times 10^3 + M_1$$

$$800 \times 10^3 [50 (-0.003438)] = -25 \times 10^3 + M_1$$

$$\therefore \quad M_1 = -112520 \text{ N-m}$$

$$800 \times 10^3 [0 + 0 + 0 + 0 - 12w_3 + 30\theta_3) = -90 \times 10^3 + R_2$$

$$800 \times 10^3 [-12 (-0.035938) + 30 (-0.008438)] = -90 \times 10^3 + R_2$$

$$\therefore \quad R_2 = 232492.8 \text{ N}$$

(II) Software results

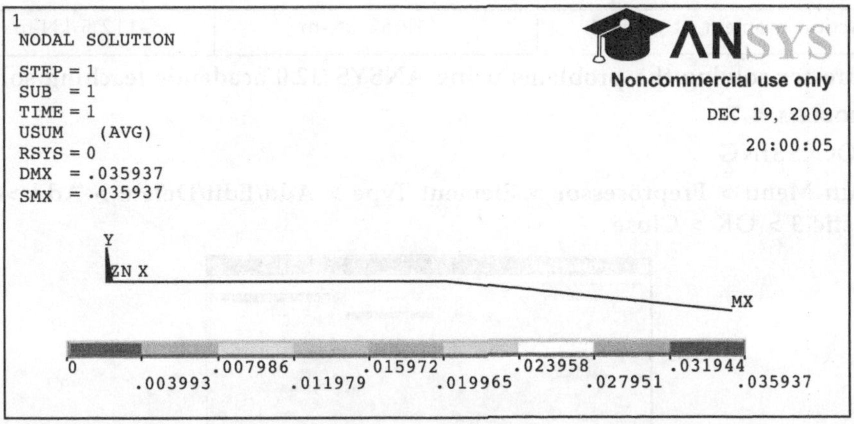

Figure 4.19(d). Deflection pattern for a cantilever beam (Refer to appendix for colour figures)

Deflection values at nodes (in metres)

The following degree of freedom results are in Global coordinates

NODE	UX	UY	UZ	USUM
1	0.0000	0.0000	0.0000	0.0000
2	0.0000	0.0000	0.0000	0.0000
3	0.0000	-0.35938E-01	0.0000	0.35938E-01

Maximum absolute values

NODE	0	3	0	3
VALUE	0.0000	-0.35938E-01	0.0000	0.35938E-01

Rotational deflection values at nodes

The following degree of freedom results are in coordinates

NODE	ROTZ
1	0.0000
2	- 0.34375E-02
3	- 0.84375E-02

Reaction values

The following X, Y, Z solutions are in Global coordinates

NODE	FX	FY	MZ
1	0.0000	- 52500	- 0.11250E + 06
2	0.0000	0.23250E + 06	

ANSWERS OF PROBLEM 4.8

Parameter	FEM-hand calculation	Software results
Deflection at node 3	-0.035938 m	-0.035938 m
Rotational deflection at node		
2	-0.003438 rad	-0.0034375 rad
3	-0.008438 rad	-0.0084375 rad
Reaction force at		
1	-52.512 kN	-52.5 kN
2	232.493 kN	232.25 kN
Reaction moment at node 1	-112.52 kN-m	-112.5 kN-m

Procedure for solving the problems using ANSYS®12.0 academic teaching software.

FOR PROBLEM 4.7

PREPROCESSING

1. **Main Menu > Preprocessor > Element Type > Add/Edit/Delete > Add >Beam > 2D elastic 3 > OK > Close**

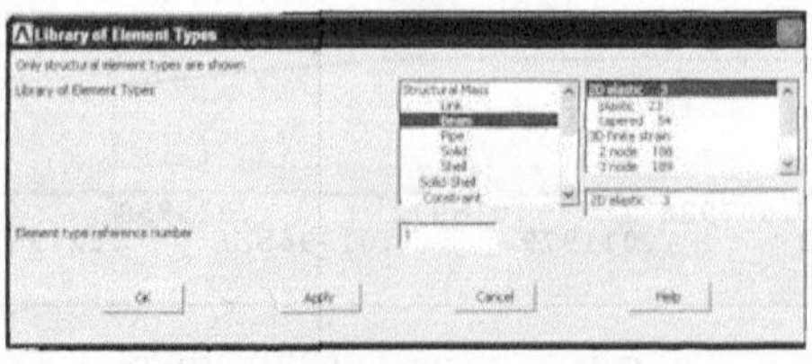

Figure 4.20. Element selection

2. **Main Menu > Preprocessor > Sections >Beam> Common sections,** following dialog box appears

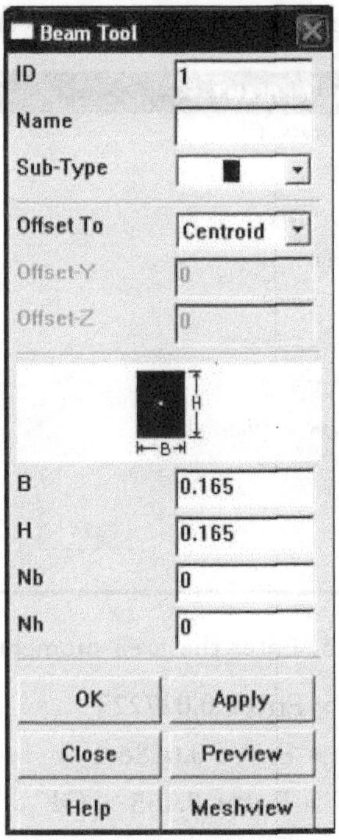

Figure 4.21. Choose cross-section of the beam

In that dialog box select **Sub-Type** Choose **Square Cross-Section,** then Enter values of B = 0.165 and H = 0.165 as shown in figure 4.21.

Click on **Preview > OK**

The following figure appears on the screen.

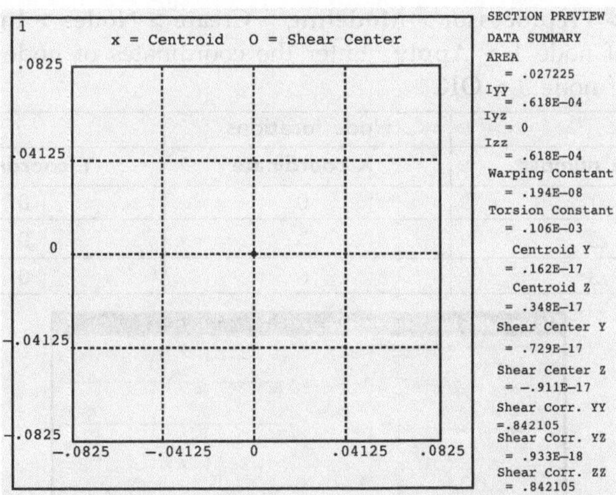

Figure 4.22. Details of geometrical properties of the beam

From figure 4.22 note down the values of Area A = 0.027225 m^2 and Moment of inertia $I_{zz} = 0.618 \times 10^{-4}$ m^4.

3. **Main Menu > Preprocessor > Real Constants > Add/Edit/Delete > Add > OK**

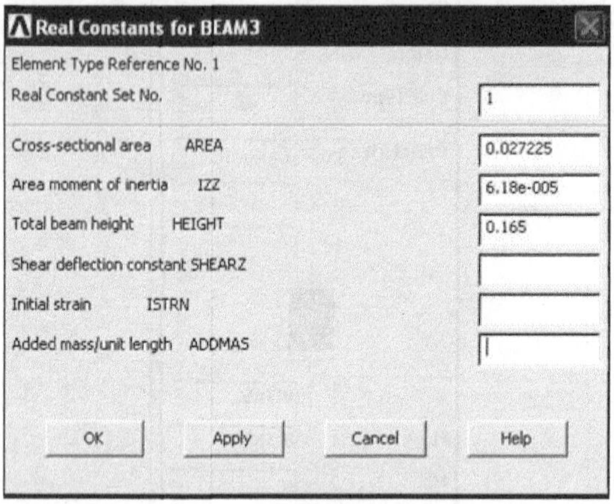

Figure 4.23. Enter the area moment of inertia

Cross-sectional area AREA > **Enter 0.027225**

Area moment of inertia I_{ZZ} > **Enter 0.618e-4**

Total beam height HEIGHT > **Enter 0.165 > OK > Close**

Enter the material properties.

4. **Main Menu > Preprocessor > Material Props > Material Models**

   Material Model Number 1, click **Structural > Linear > Elastic > Isotropic**

   Enter **EX = 200E9 and PRXY = 0.3 > OK**

   (**Close** the Define Material Model Behaviour window.)

   Create the nodes and elements as shown in the figure.

5. **Main Menu > Preprocessor > Modeling > Create > Nodes > In Active CS** Enter the coordinates of node 1 > **Apply** Enter the coordinates of node 2 > **Apply** Enter the coordinates of node 3 > **OK.**

	Node locations	
Node number	X coordinate	Y coordinate
1	0	0
2	3	0
3	6	0

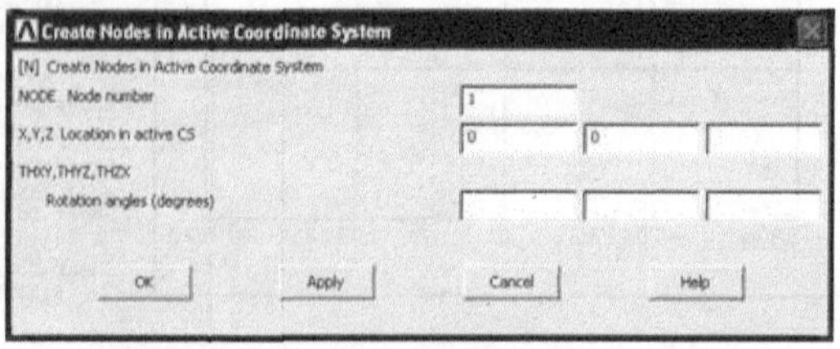

Figure 4.24. Enter the node coordinates

6. **Main Menu > Preprocessor > Modeling > Create > Elements > Auto Numbered > Thru nodes** Pick the 1st and 2nd node > **Apply** Pick the 2nd and 3rd node > **OK**

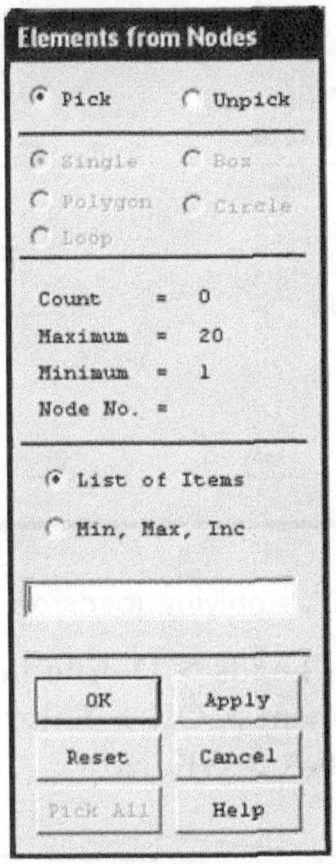

Figure 4.25. Pick the nodes to create elements

Apply the displacement boundary conditions and loads.

7. **Main Menu > Preprocessor > Loads > Define Loads > Apply > Structural > Displacement > On Nodes** Pick the 1st node > **Apply > All DOF = 0 > OK**
8. **Main Menu > Preprocessor > Loads > Define Loads > Apply > Structural > Displacement > On Nodes** Pick the 2nd node > **Apply >Select UX and UY = 0 > OK**

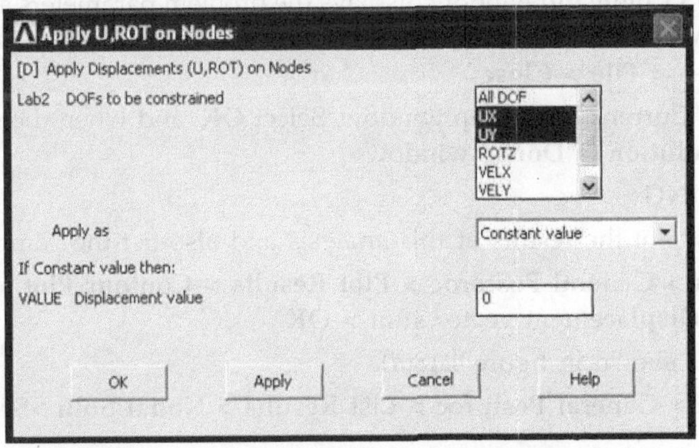

Figure 4.26. Applying boundary conditions on node 2

9. **Main Menu > Preprocessor > Loads > Define Loads > Apply > Structural > Pressure > On Beams** Pick the 1st element > **OK > Enter Pressure value at node I = 0 and Pressure value at node J = 30e3 > OK**

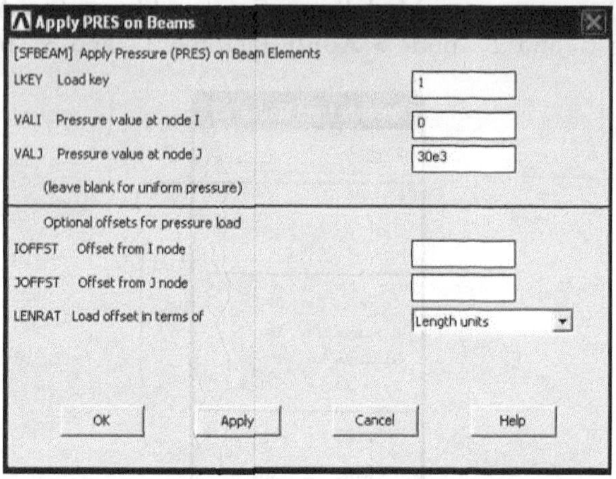

Figure 4.27. Applying loads on element 1

10. **Main Menu > Preprocessor > Loads > Define Loads > Apply > Structural > Pressure > On Beams** Pick the 2nd element **> OK > Enter Pressure value at node I = 30e3 and Pressure value at node J = 60e3 > OK**

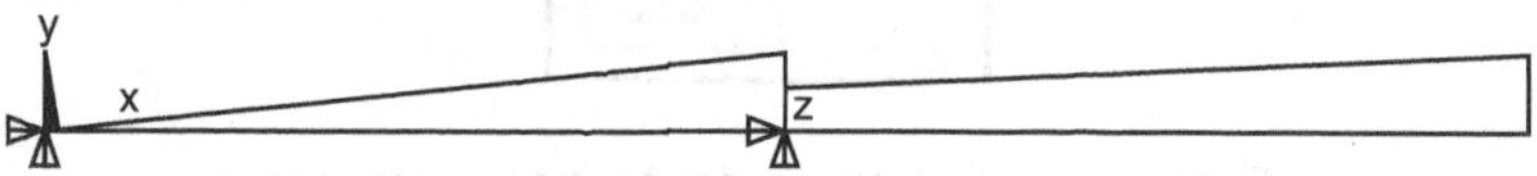

Figure 4.28. Model with loading and displacement boundary conditions

The model-building step is now complete, and we can proceed to the solution. First to be safe, save the model.

SOLUTION. The interactive solution proceeds.

11. **Main Menu > Solution > Solve > Current LS > OK**

    The **/STATUS Command** window displays the problem parameters and **the SolveCurrent Load Step** window is shown. Check the solution options in the **/STATUS**window and if all is OK, select **File > Close**

    In the Solve **Current Load Step** window, Select **OK**, and when the solution is complete, **close** the **'Solution is Done!'** window.

POSTPROCESSING

We can now plot the results of this analysis and also list the computed values.

12. **Main Menu > General Postproc > Plot Results > Contour Plot > Nodal Solu >DOF Solution > Displacement vector sum > OK**

    This result is shown in figure 4.18(*d*).

13. **Main Menu > General Postproc > List Results > Nodal Solu >Select Rotation vector sum > OK**

14. **Main Menu > General Postproc > List Results > Reaction Solu > OK**

    To find the **bending moment diagram** following procedure is followed.

15. **Main Menu > General Postproc > Element Table > Define Table > Add**

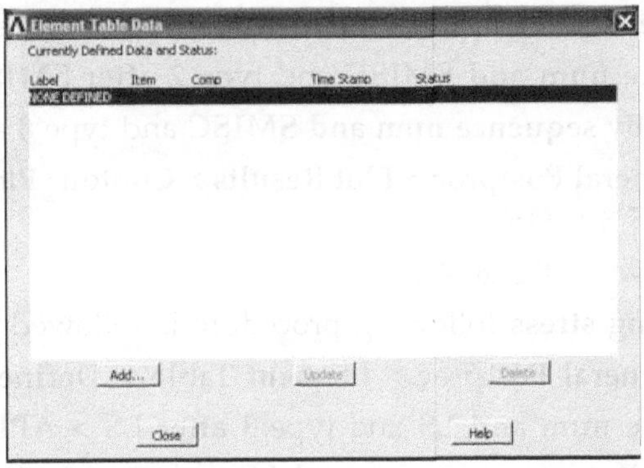

Figure 4.29. Define the element table

Select **By sequence num and SMISC** and type **6 after SMISC** as shown in figure 4.30.

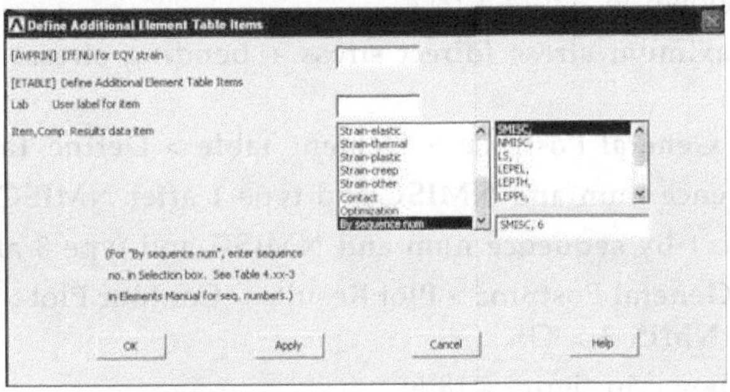

Figure 4.30. Selecting options in element table

> **APPLY**

Then again select **By sequence num and SMISC** and type **12 after SMISC >OK**

16. **Main Menu > General Postproc > Plot Results > Contour Plot>Line Elem Res >** Select **SMIS 6 and SMIS 12 in the rows of LabI and LabJ** respectively as shown in figure 4.31 **> OK**

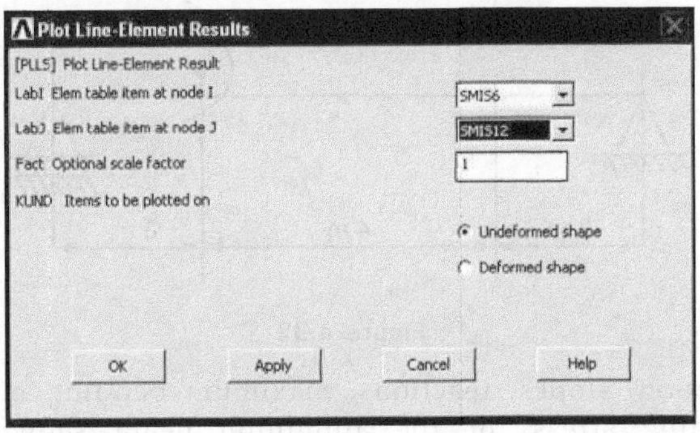

Figure 4.31. Selecting options for finding out bending moment

This result is shown in figure 4.18(*e*).

To find the **shear force diagram** following procedure is followed.

17. **Main Menu > General Postproc > Element Table > Define Table > Add**
    Select **By sequence num and SMISC** and type **2 after SMISC >APPLY**
    Then again select **By sequence num and SMISC** and type **8 after SMISC > OK > Close**
18. **Main Menu > General Postproc > Plot Results > Contour Plot > Line Elem Res >** Select **SMIS 2 and SMIS 8 > OK**
    This result is shown in figure 4.18(*f*).
    To find the **bending stress** following procedure is followed.
19. **Main Menu > General Postproc > Element Table > Define Table > Add**
    Select **By sequence num and LS** and type **3 after LS > APPLY**
    Then again select **By sequence num and LS** and type **6 after LS > OK**
20. **Main Menu > General Postproc > Plot Results > Contour Plot > Line Elem Res >** Select **LS 3 and LS 6 > OK**
    This result is shown in figure 4.18(*g*).
    To find the **maximum stress (direct stress + bending stress)** following procedure is followed.
21. **Main Menu > General Postproc > Element Table > Define Table > Add**
    Select **By sequence num and NMISC** and type **1 after NMISC >APPLY**
    Then again select **By sequence num and NMISC** and type **3 after NMIS > OK**
22. **Main Menu > General Postproc > Plot Results > Contour Plot > Line Elem Res >** Select **NMISC 1 and NMIS 3 > OK**
    This result is shown in figure 4.18(*h*).

EXERCISES

1. For the beam shown in figure 4.32, determine the deflections, slopes, reactions, maximum bending moment, shear force and maximum bending stress. Take E = 210 GPa, I = 7 × 10^{-4} m^4.

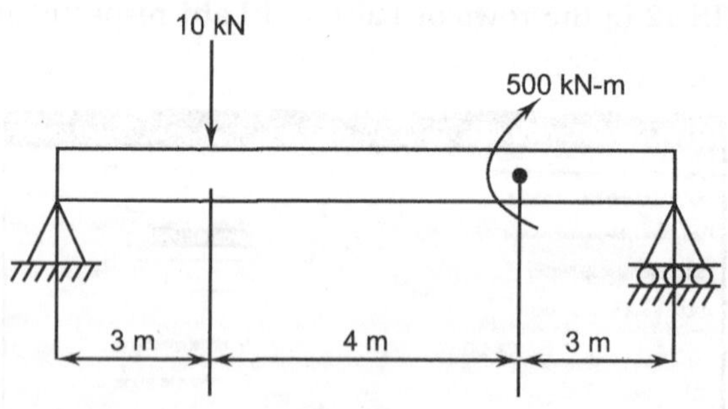

Figure 4.32

2. Find the deflection, slopes, reactions, maximum bending moment, shear force and maximum bending stress for the aluminum beam shown in figure 4.33. Take E = 200 GPa, I = 3 × 10^{-4} m^4.

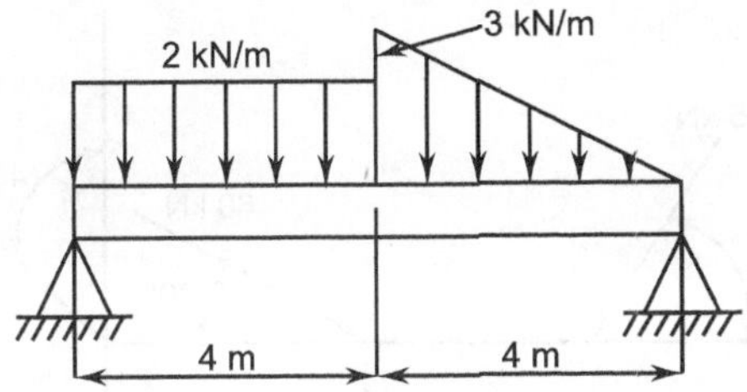

Figure 4.33

3. Find the deflection at the load and the slopes at the ends for the shaft shown in figure 4.34. Also find the maximum bending moment, maximum bending stress and reactions developed in the bearings. Consider the shaft to be simply supported at bearings A and B. Take E = 200 GPa.

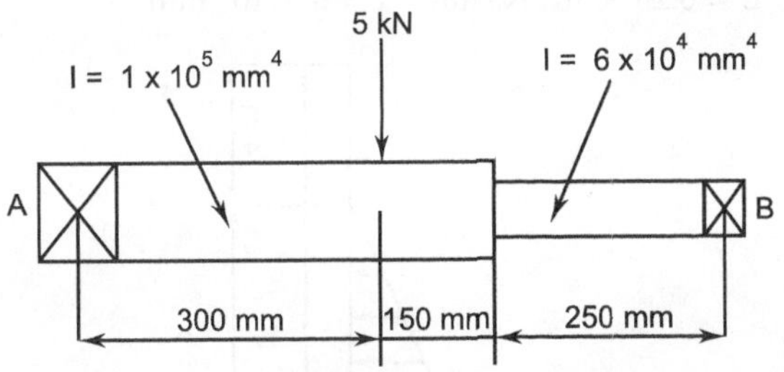

Figure 4.34

4. Find the deflection of the beam shown in figure 4.35 under self weight. Take E = 200 GPa, mass density ρ = 7800 kg/m^3.

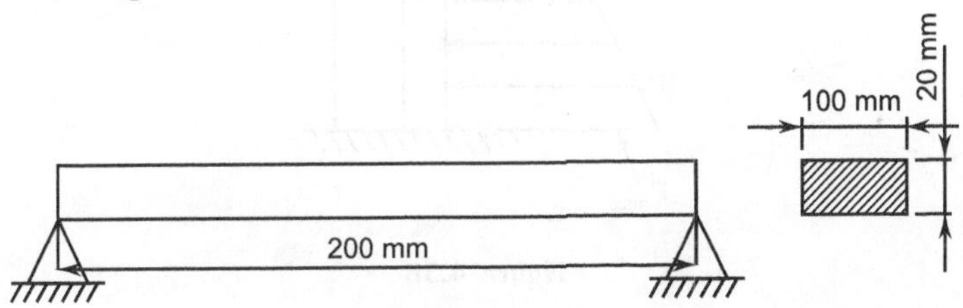

Figure 4.35

5. Find the deflection and bending stress distribution for the cantilever beam shown in figure 4.36 under combined loading. Take E = 200 GPa.

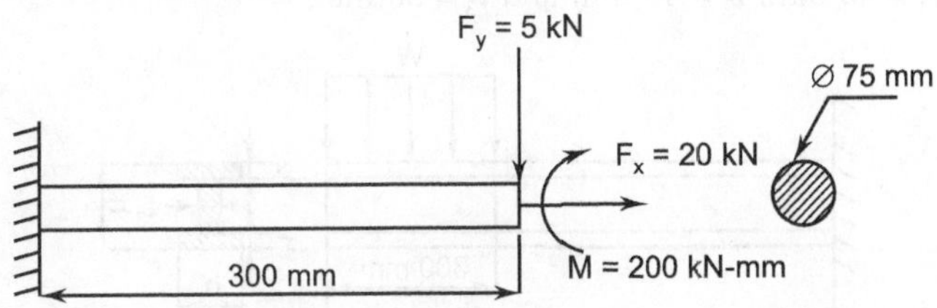

Figure 4.36

[Hint: I = $\pi d^4/64$]

6. For the beam shown in figure 4.37, determine the deflections at nodes and reaction. Also plot the bending moment diagram, shear force diagram and find the bending stress. Take E = 200 GPa and I = 8 × 10^{-4}m^4.

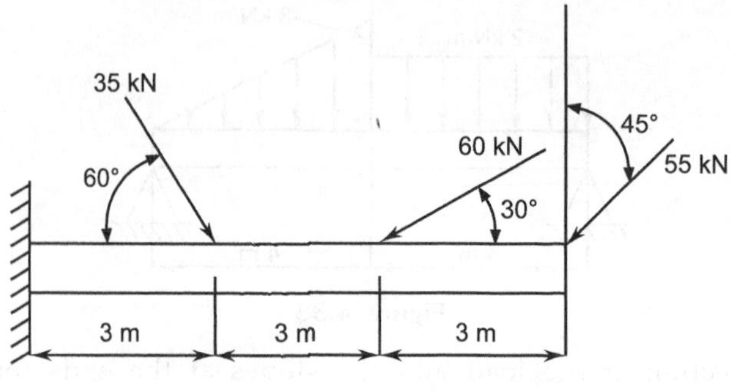

Figure 4.37

7. A cantilever beam is shown in figure 4.38. Using two beam elements determine the nodal deflection and reaction. Take $E = 0.25 \times 10^5$ N/mm^2, $I = 8 \times 10^9$ mm^4.

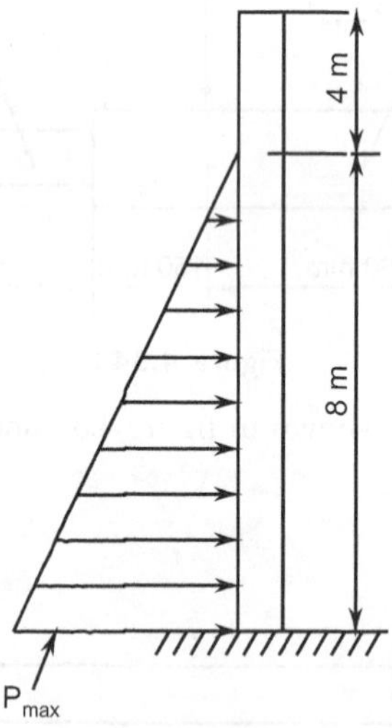

Figure 4.38

[**Hint:** $P_{max} = \rho \times g \times h$]

8. Determine the deflection, reaction and bending stress for the beam shown in figure 4.39. Also plot the bending moment and shear force diagram. Take E = 207 GPa, W = 150 N/mm, h = 800 mm, b = 400 mm, t_1 = 40 mm, t_2 = 40 mm and t_3 = 50 mm.

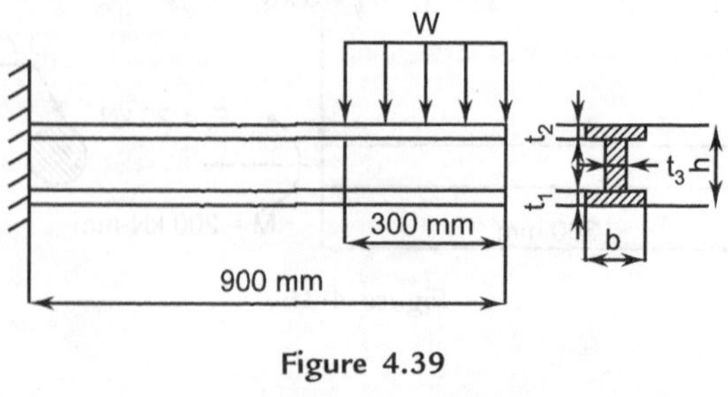

Figure 4.39

Chapter 5

Stress Analysis of a Rectangular Plate With a Circular Hole

5.1 INTRODUCTION

Two dimensional problems in structural analysis are dealt in this chapter. Hand calculations, even with two elements, become too long and hence are not given for these problems; only analytical method solutions and software solutions have been provided.

Two dimensional problems can either be plane stress or plain strain problems. Method of analysis is same for both, except that stress strain matrix is different in two cases.

Plane bodies that are flat and of constant thickness subjected to in-plane loading fall under the category of plane stress problems. Stress components σ_z, τ_{xz} and τ_{yz} assume zero values in these problems.

Some of the elements used in the analysis of two dimensional problems are constant strain triangle(CST), linear strain triangle(LST), linear quadrilateral, isoparametric quadrilateral, etc. each of these elements has two degrees of freedom per node namely translation in x and y directions.

Stress within the element may be calculated using the equation,

$$\{\sigma\} = [D][B]\{q\}$$

The stress analysis of a rectangular plate with a circular hole problem is assumed as a two dimensional plane stress problem. Plane stress is defined to be a state of stress in which the normal stress and the shear stress directed perpendicular to the plane are assumed to be negligible.

The above problem can be categorized into 3 sub cases.

Sub Case 1

A rectangular plate with a very small circular hole at the centre with one vertical edge fixed and the other vertical edge is acted upon by a horizontal tensile load in the form of pressure.

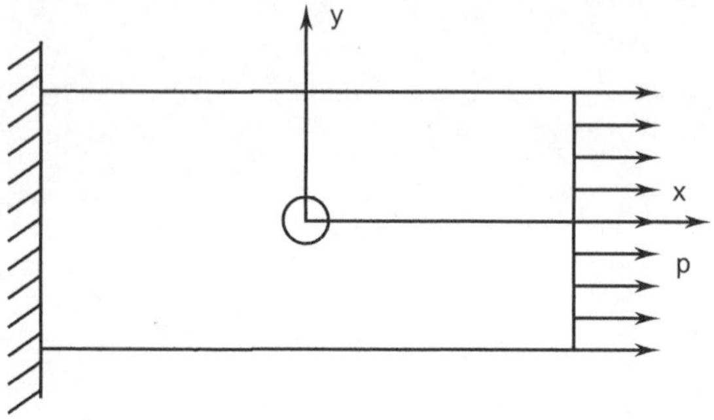

Figure 5.1. Rectangular plate with a very small circular hole subjected to tensile load at one edge

Sub Case 2

A rectangular plate with a small circular hole at the centre and a horizontal tensile force in the form of pressure is acting on both the vertical edges of the plate.

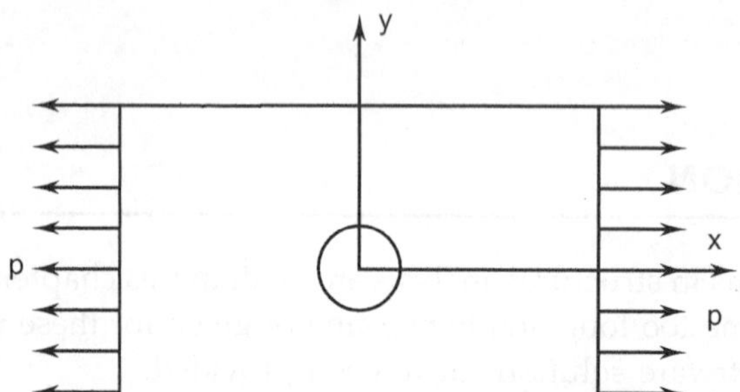

Figure 5.2. Rectangular plate with a hole subjected to tensile load at both the edges

The above problem is solved by exploiting the symmetric geometry and symmetric loading boundary conditions. Now we can draw the above figure 5.2 as below for the analysis purpose (refer figure 5.3).

Place the origin of *x-y* coordinates at the centre of the hole and pull on both ends of the plate. Then points on the centerlines will not move perpendicular to them but move along the centerlines. This indicates the appropriate displacement conditions to use as shown in figure 5.3.

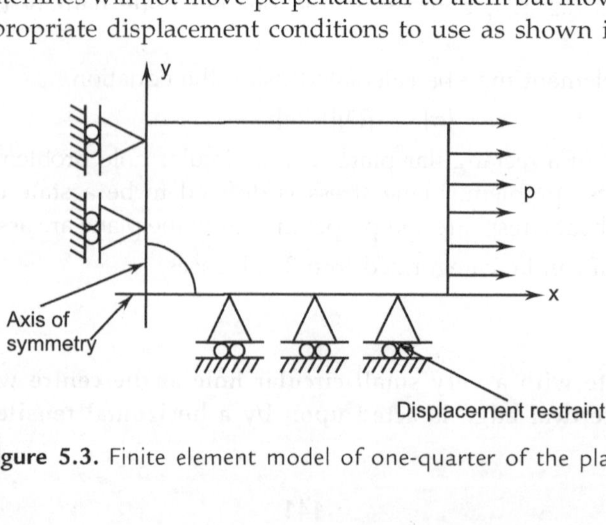

Figure 5.3. Finite element model of one-quarter of the plate

Sub Case 3

A rectangular plate with a large circular hole at the centre and a uniform pressure acts on the boundary of the hole.

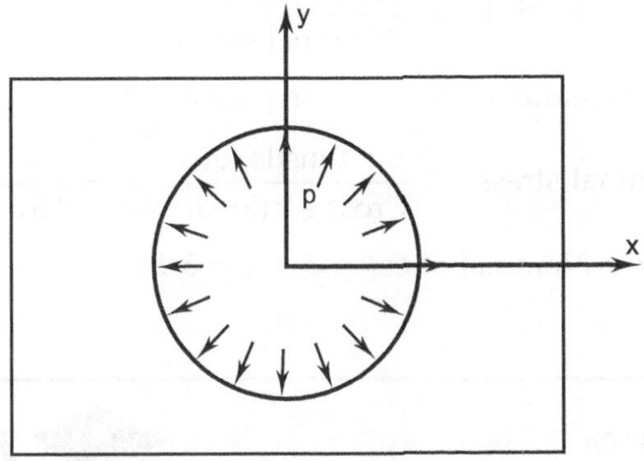

Figure 5.4. Rectangular plate with a hole subjected to uniform pressure at the boundary of the hole

The above problem can be solved considering one quarter of the plate and by exploiting the symmetric geometry and loading conditions. The finite element model is shown below.

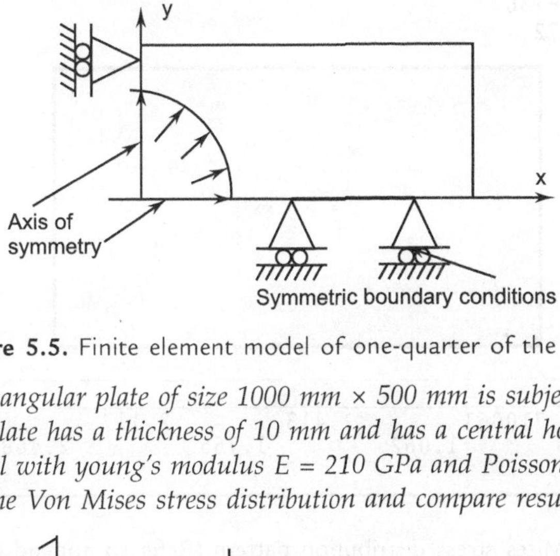

Figure 5.5. Finite element model of one-quarter of the plate

PROBLEM 5.1. *A rectangular plate of size 1000 mm × 500 mm is subjected to uniform pressure as shown in figure 5.6. The plate has a thickness of 10 mm and has a central hole 50 mm in diameter. The material of the plate is steel with young's modulus E = 210 GPa and Poisson's ratio, ν = 0.3. Assume a case of plane stress. Plot the Von Mises stress distribution and compare result with analytical method.*

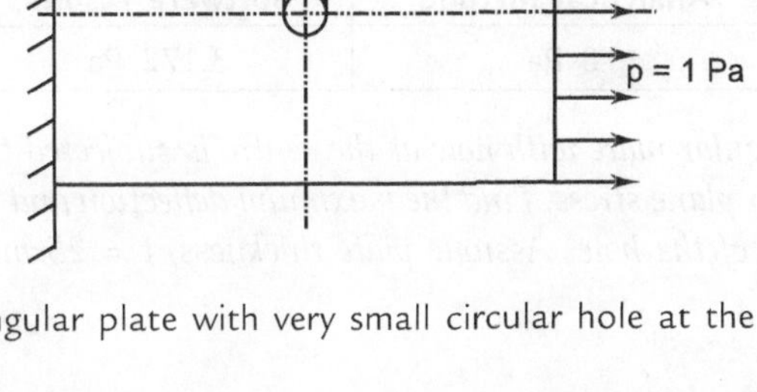

Figure 5.6. Rectangular plate with very small circular hole at the centre of the plate

SOLUTION. (I) Analytical method

Comparing the above case with the infinite plate with a very small circular hole, for this, the stress concentration factor is (SCF) = 3

$$\text{SCF} = \frac{\text{Maximum stress}}{\text{Nominal stress}}$$

Hence, Tensile force = Pressure × cross-sectional area = 1 × 0.5 × 0.01 = 0.005 N

$$\text{Nominal stress} = \frac{\text{Tensile force}}{\text{Cross-sectional area}} = \frac{0.005}{0.5 \times 0.01} = 1\,\text{N/m}^2$$

Maximum stress = SCF × Nominal stress = 3 × 1 = **3 Pa.**

(II) Software results

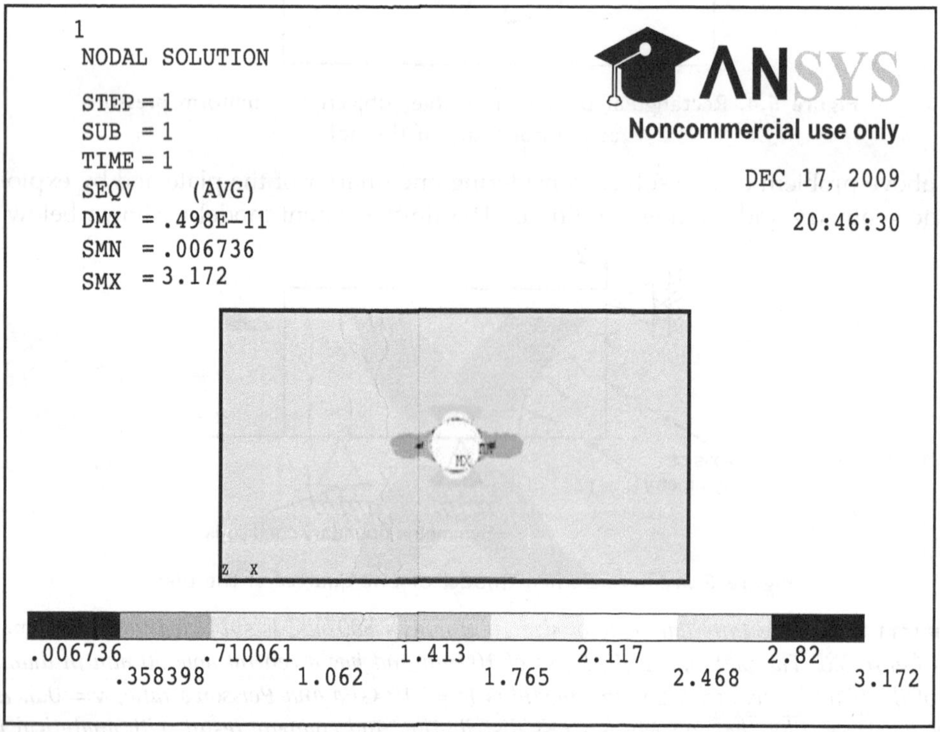

Figure 5.6(a). Von Mises stress distribution pattern (Refer to appendix for colour figures)

From the software, we got, Maximum stress (Von Mises stress) = **3.172 Pa**

ANSWERS OF PROBLEM 5.1

Parameter	Analytical method	Software results	Percentage of error
Maximum stress	3 Pa	3.172 Pa	5.42

PROBLEM 5.2. *A rectangular plate with hole at the centre is subjected to uniform pressure as shown in figure 5.7. The plate is under plane stress. Find the maximum deflection and maximum stress distribution. Also find the deformed shape of the hole. Assume plate thickness, t = 25 mm, E = 207 GPa, v = 0.3.*

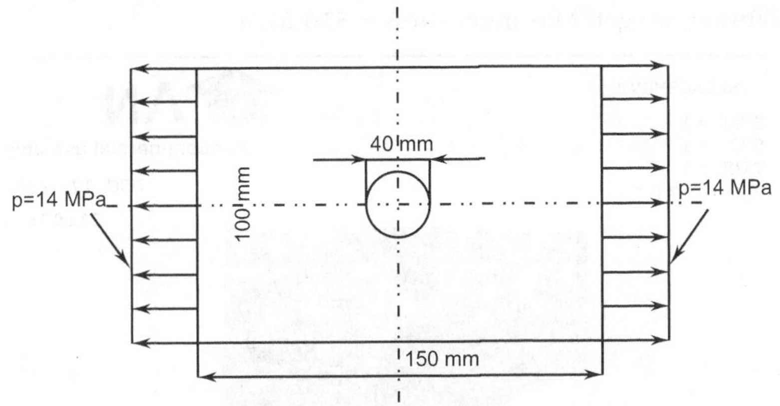

Figure 5.7. Rectangular plate with a hole with symmetrical loading

SOLUTION. (I) Analytical method

$$\text{Geometric factor} = \frac{\text{Diameter of hole}}{\text{Width of plate}} = \frac{d}{w} = \frac{40}{100} = 0.4$$

From the design data handbook,

For $\frac{d}{w}$ of 0.4 the stress concentration factor (SCF) = 2.25

$$\text{SCF} = \frac{\text{Maximum stress}}{\text{Nominal stress}}$$

Hence,

Tensile force = Pressure × Cross-sectional area = 14 × 100 × 25 = 35000 N

$$\text{Nominal stress} = \frac{\text{Tensile force}}{(w-d)t} = \frac{35000}{(100-40)25} = 23.33 \text{ MPa}$$

Maximum stress = SCF × Nominal stress = 2.25 × 23.333 = **52.5 MPa**

(II) Software results

For the analysis using software, one quarter of the plate is modeled and analyzed.

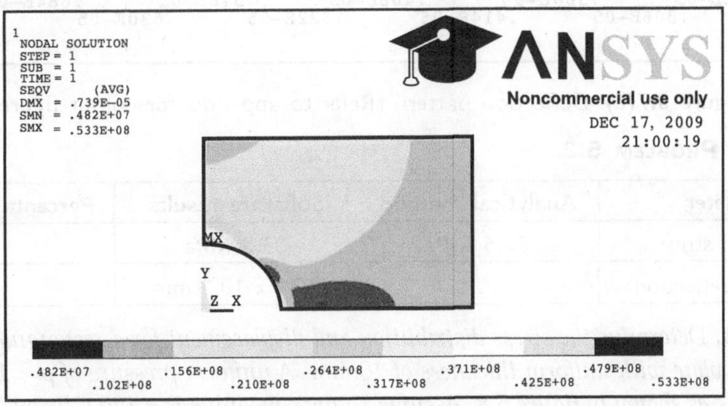

Figure 5.7(a). Von Mises stress distribution pattern (Refer to appendix for colour figures)

From the software, we got, Maximum stress = **53.3 MPa**

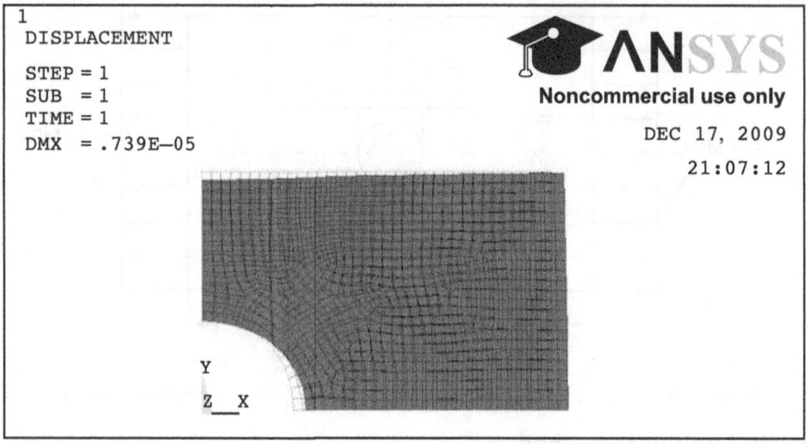

Figure 5.7(b). Deformed shape of the hole (Refer to appendix for colour figures)

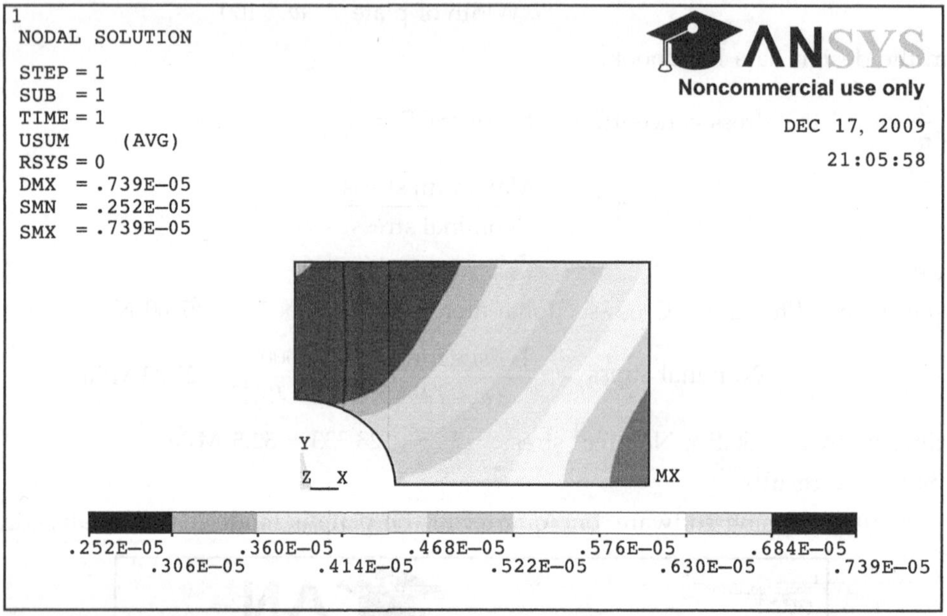

Figure 5.7(c). Deflection pattern (Refer to appendix for colour figures)

Answers of Problem 5.2

Parameter	Analytical method	Software results	Percentage of error
Maximum stress	52.5 MPa	53.3 MPa	1.5
Maximum deflection	......	7.39×10^{-3} mm	......

Problem 5.3. *Determine the stress distribution and displacement for a rectangular plate with a hole at the centre of the plate with uniform thickness of 10 mm. A uniform pressure of p = 10 MPa acts on the boundary of the hole as shown in figure 5.8. Assume Young's modulus E = 120 GPa and the Poisson's ratio is 0.28. Assume plane stress condition.*

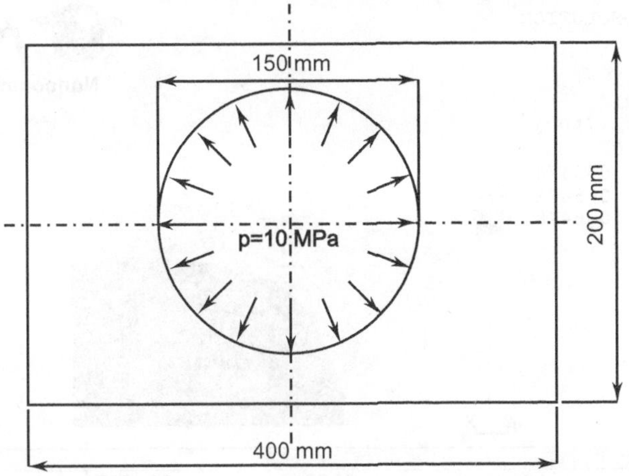

Figure 5.8. Rectangular plate with a hole subjected to uniform pressure at the boundary of the hole

Solution. **(I) Software results**

For the analysis using software, one quarter of the plate is modelled and analyzed.

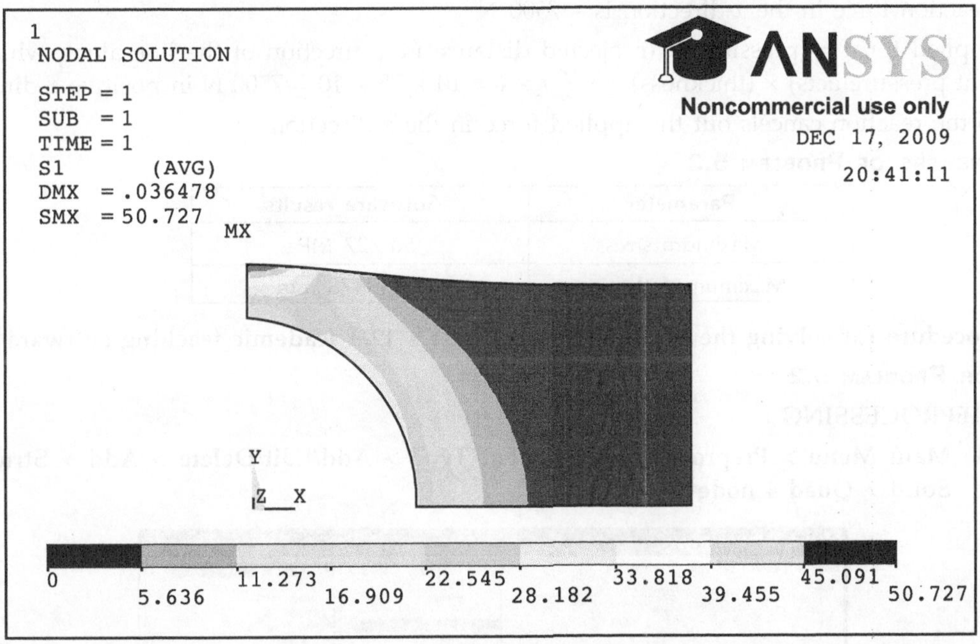

Figure 5.8(a). 1st principal stress distribution pattern (Refer to appendix for colour figures)

From the software, we got, Maximum stress = **50.727 MPa.**

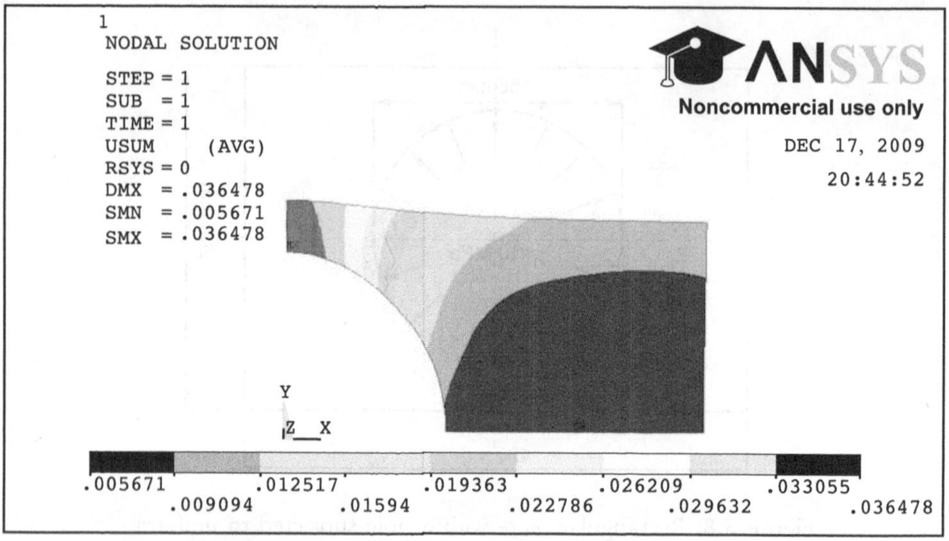

Figure 5.8(b). Deflection pattern (Refer to appendix for colour figures)

Validation of the results

The reactions at the supports must balance the applied forces. Therefore, From software, the total reaction force in the *x*-direction is – 7500 N

Applied force = (pressure) × (projected distance in *x*-direction of the line along which the constant pressure acts) × (thickness) = $p \times r \times t = 10 \times 75 \times 10 = 7500$ N in positive *x*-direction.

So the reaction cancels out the applied force in the *x*-direction.

ANSWERS OF PROBLEM 5.3

Parameter	Software results
Maximum stress	50.727 MPa
Maximum deflection	0.036478 mm

Procedure for solving the problem using ANSYS®12.0 academic teaching software

FOR PROBLEM 5.2

PREPROCESSING :

1. **Main Menu > Preprocessor > Element Type > Add/Edit/Delete > Add > Structural Solid > Quad 4 node 42 > OK**

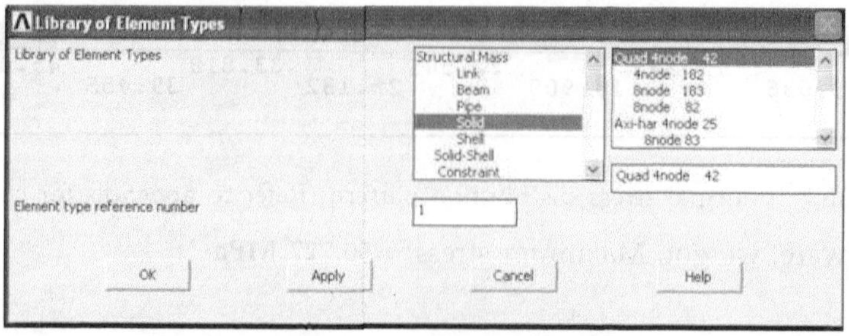

Figure 5.9. Element selection

Select the option where you define the plate thickness.

2. **Options (Element behaviour K3) > Plane strs w/thk > OK > Close**

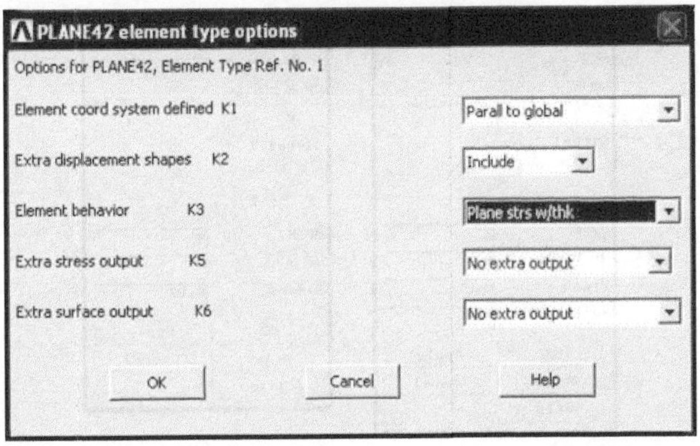

Figure 5.10. Element options

3. **Main Menu > Preprocessor > Real Constants > Add/Edit/Delete > Add > OK**

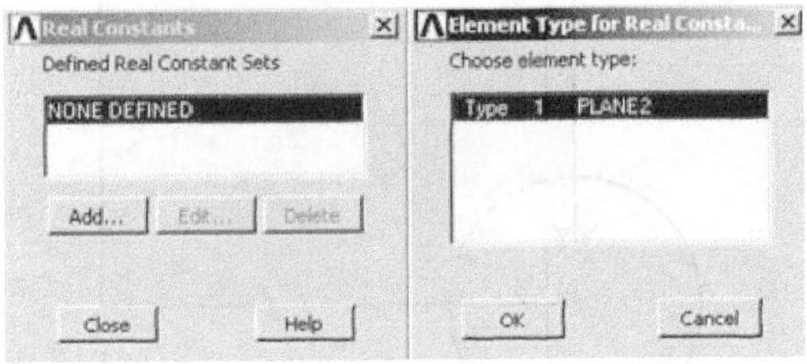

Figure 5.11. Real constants

(Enter the plate thickness of 0.025 m) > **Enter 0.025 > OK > Close**

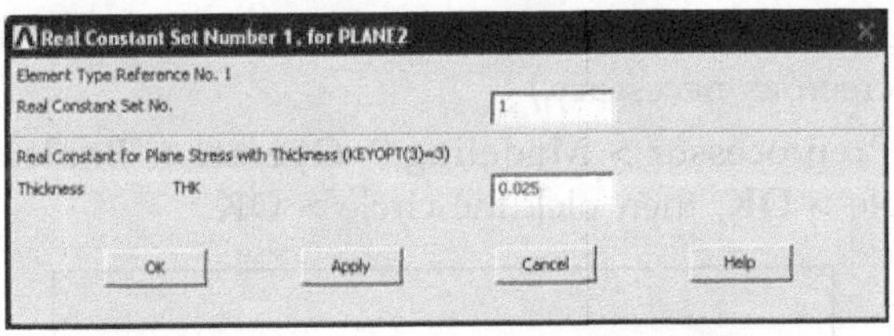

Figure 5.12. Enter the plate thickness

Enter the material properties.

4. **Main Menu > Preprocessor > Material Props > Material Models**

   Material Model Number 1, click **Structural > Linear > Elastic > Isotropic**

   Enter **EX = 2.07E11** and **PRXY = 0.3 > OK** (**Close** the Define Material Model Behaviour window.)

   Create the geometry for the upper right quadrant of the plate by subtracting a 0.04 m diameter circle from a 0.075 × 0.05 m rectangle. Generate the rectangle first.

5. **Main Menu > Preprocessor > Modeling > Create > Areas > Rectangle > By 2 Corners**

   Enter (lower left corner) **WP X = 0.0, WP Y = 0.0 and Width = 0.075, Height = 0.05 > OK**

6. **Main Menu > Preprocessor > Modeling > Create > Areas > Circle > Solid Circle** Enter **WP X = 0.0, WP Y = 0.0 and Radius = 0.02 > OK**

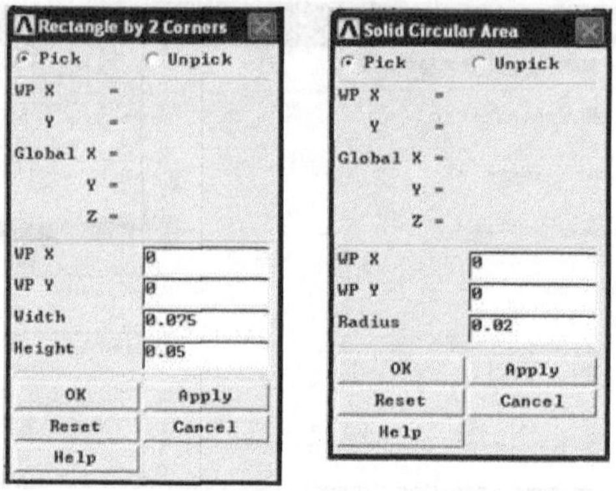

Figure 5.13. Create areas

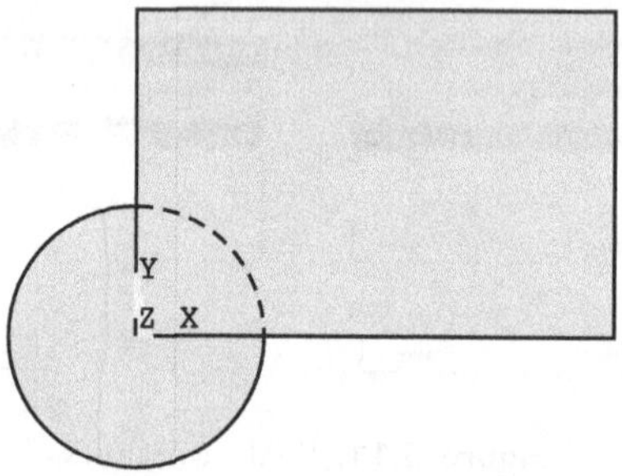

Figure 5.14. Rectangle and circle

Now **subtract** the circle from the rectangle. (Read the messages in the window at the bottom of the screen as necessary.)

7. **Main Menu > Preprocessor > Modeling > Operate > Booleans > Subtract > Areas>** Pick the rectangle **> OK,** then pick the circle **> OK**

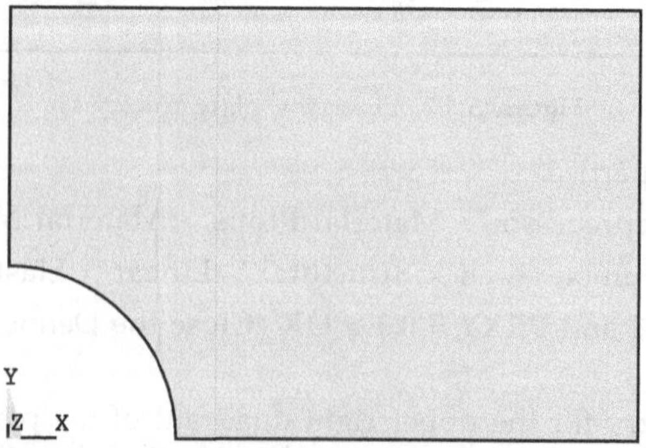

Figure 5.15. Geometry for quadrant of plate

Create a mesh of triangular elements over the quadrant area.

8. **Main Menu > Preprocessor > Meshing > Mesh Tool**

The **Mesh Tool** dialog box appears. In that dialog box click on the **Smart Size** and move the slider available below the **Smart Size to 2** (*i.e.,* towards **Fine** side). Then close the Mesh Tool box.

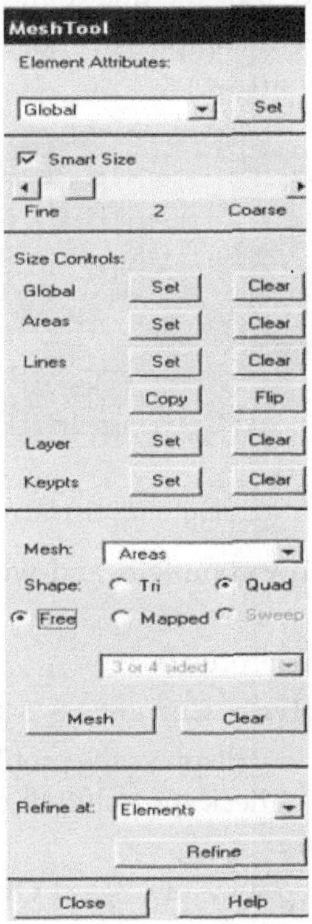

Figure 5.16. Mesh Tool box

9. **Main Menu > Preprocessor > Meshing > Mesh > Areas > Free** Pick the quadrant > **OK**

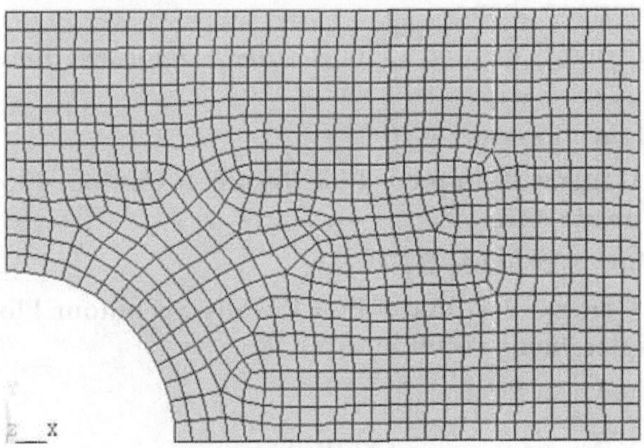

Figure 5.17. Quad element mesh

Apply the displacement boundary conditions and loads.

10. **Main Menu > Preprocessor > Loads > Define Loads > Apply > Structural > Displacement > On Lines** Pick the left edge of the quadrant > **OK > UX = 0 > OK**

11. **Main Menu > Preprocessor > Loads > Define Loads > Apply > Structural > Displacement > On Lines** Pick the bottom edge of the quadrant **> OK > UY = 0 > OK**
12. **Main Menu > Preprocessor > Loads > Define Loads > Apply > Structural > Pressure > On Lines.** Pick the right edge of the quadrant **> OK > Pressure = –14E6 > OK**

    (A positive pressure would be a compressive load, so we use a negative pressure. The pressure is shown as a single arrow.)

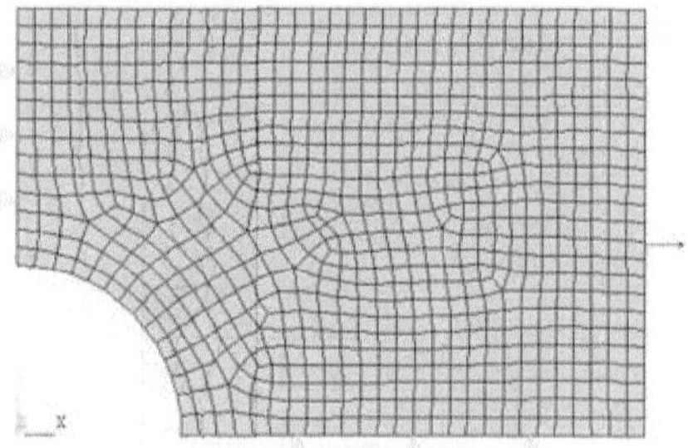

Figure 5.18. Model with loading and displacement boundary conditions

The model-building step is now complete, and we can proceed to the solution. First to be safe, save the model.

SOLUTION. The interactive solution proceeds.

13. **Main Menu > Solution > Solve > Current LS > OK**

    The **/STATUS Command** window displays the problem parameters and **the Solve Current Load Step** window is shown. Check the solution options in the **/STATUS** window and if all is OK, select **File > Close.**

    In the Solve **Current Load Step** window, Select **OK**, and when the solution is complete, **close** the **'Solution is Done!'** window.

POSTPROCESSING

We can now plot the results of this analysis and also list the computed values. First examine the deformed shape.

14. **Main Menu > General Postproc > Plot Results > Deformed Shape > Def. + Undeformed > OK**

    This result is shown in figure 5.7(*b*).
15. **Main Menu > General Postproc > Plot Results > Contour Plot > Nodal Solu >Stress > Von Mises stress > OK**

    This result is shown in figure 5.7(*a*).
16. **Main Menu > General Postproc > Plot Results > Contour Plot > Nodal Solu > DOF Solution > Displacement vector sum > OK**

    This result is shown in figure 5.7(*c*).

EXERCISES

1. Find the maximum stress in the aluminum plate shown in figure 5.19. Consider aluminum plate 10 mm thick with hole at the centre of the plate. Assume plane stress condition. Take E = 70 GPa and ν = 0.35. Also calculate the maximum stress by analytical method and compare the results.

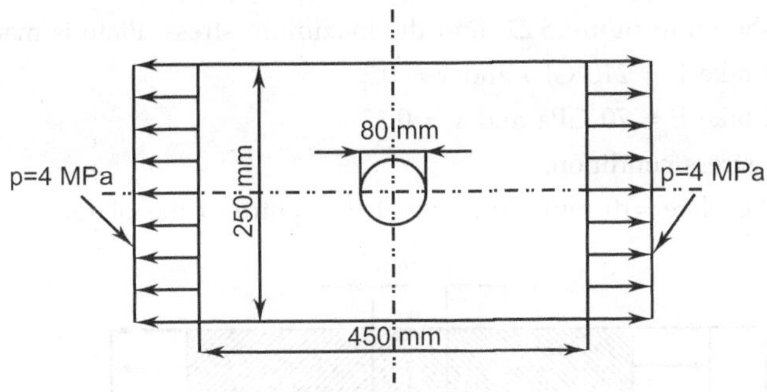

Figure 5.19

2. Find the maximum stress for the plate shown in figure 5.20 if the hole is located halfway between the centre line and the top edge as shown. Take E = 70 GPa and ν = 0.35. Assume plane stress condition.

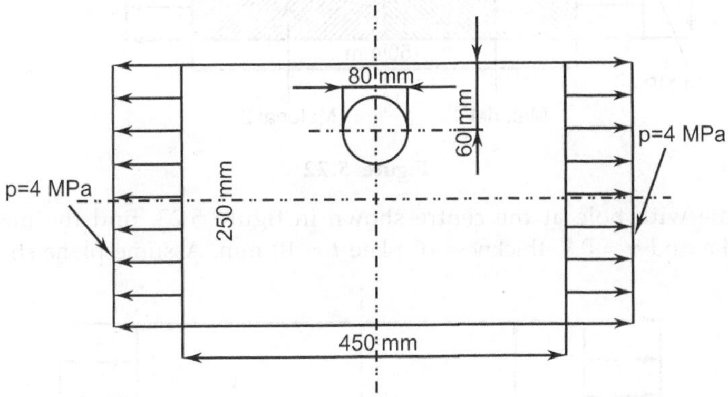

Figure 5.20

[**Hint:** Model half of the plate by taking symmetry about *Y*-axis.]

3. For the plate shown in figure 5.21, find the maximum stress. Take Young's modulus E = 210 GPa, Poisson's ratio = 0.3. Assume plane stress condition. Thickness of the plate = 10 mm with hole at the centre of the plate.

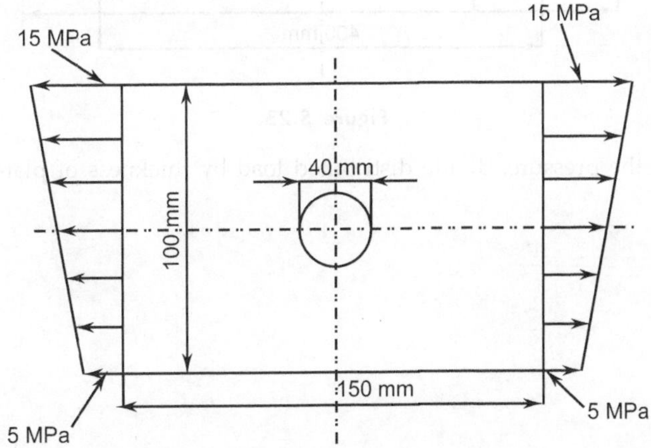

Figure 5.21

4. For the plate shown in figure 5.22, find the maximum stress. Plate is made up of two materials.
For Material 1 take E = 210 GPa and ν = 0.3
For Material 2 take E = 70 GPa and ν = 0.35
Assume plane stress condition.
Thickness of the plate =10 mm with hole at the centre of the plate.

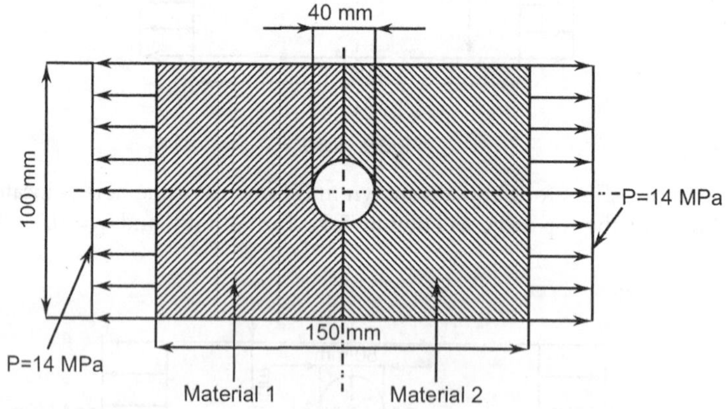

Figure 5.22

5. For the plate with hole at the centre shown in figure 5.23, find the maximum stress. Take E = 210 GPa and ν = 0.3, thickness of plate t = 10 mm. Assume plane stress condition.

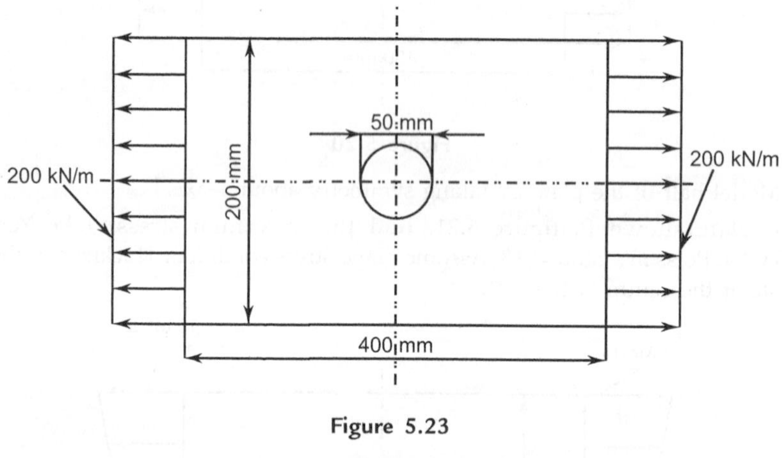

Figure 5.23

[**Hint:** To find the pressure, divide distributed load by thickness of plate.]

Chapter 6

THERMAL ANALYSIS

6.1 INTRODUCTION

Thermal analysis is one of scalar field problems. These problems have only one degree of freedom per node namely temperature. In this chapter, one-dimensional and two-dimensional heat conduction problems are dealt with. In these problems bar element with two end nodes each having temperature (T) as sole degree of freedom is made use of. Nodal heat flow rates (Q) or heat fluxes are analogous quantities to nodal forces, in structural bar element. Governing equation for this element is given by,

$$\frac{Ak}{L}\begin{bmatrix} 1 & -1 \\ -1 & 1 \end{bmatrix}\begin{Bmatrix} T_1 \\ T_2 \end{Bmatrix} = \frac{q}{2}\begin{Bmatrix} L \\ L \end{Bmatrix} + \begin{Bmatrix} Q_1 \\ Q_2 \end{Bmatrix}$$

where, q = heat generation rate per unit length.

$\frac{Ak}{L}\begin{bmatrix} 1 & -1 \\ -1 & 1 \end{bmatrix}$ = element heat conductivity matrix.

6.2 PROCEDURE OF FINITE ELEMENT ANALYSIS (RELATED TO THERMAL PROBLEMS)

Step 1. Select element type.

Step 2. Select temperature distribution function.

Step 3. Define the temperature gradient/temperature and heat flux/temperature gradient relationships.

Step 4. Derive the element conduction matrix and heat flux matrix.

STEP 5. Assemble the element equations to obtain the global equations and introduce boundary conditions.

STEP 6. Solve for the nodal temperatures.

STEP 7. Solve for the element temperature gradients and heat fluxes.

6.3 ONE-DIMENSIONAL HEAT CONDUCTION

PROBLEM 6.1. *A composite wall consists of 3 materials. The outer temperature is $T_0 = 20°C$. Convection heat transfer takes place on the inner surface of the wall with $T_\infty = 800°C$ and $h = 25\ W/m^2°C$. Determine the temperature distribution in the wall. Take, $k_1 = 30\ W/m°C$, $k_2 = 50\ W/m°C$, $k_3 = 20\ W/m°C$.*

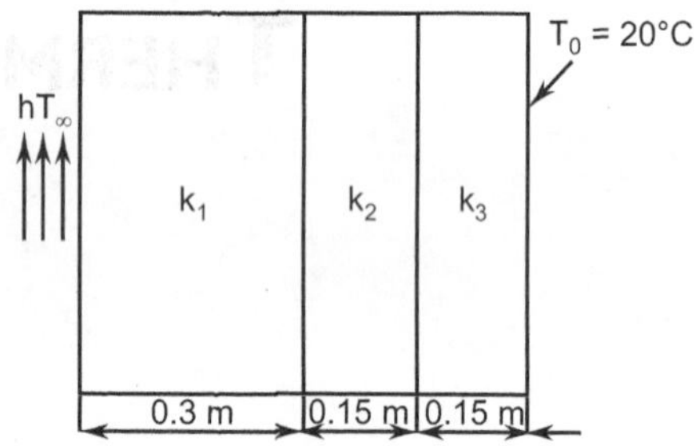

Figure 6.1

SOLUTION. (I) Analytical method

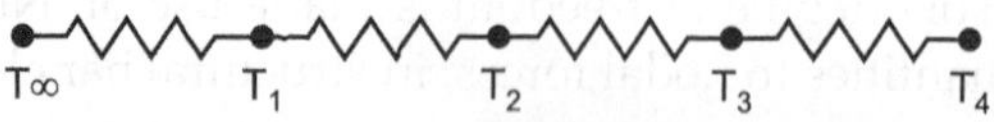

Figure 6.1(a)

Heat flow rate per unit area,

$$Q = \frac{T_\infty - T_4}{\frac{1}{h} + \frac{L_1}{k_1} + \frac{L_2}{k_2} + \frac{L_3}{k_3}} = \frac{800 - 20}{\frac{1}{25} + \frac{0.3}{30} + \frac{0.15}{50} + \frac{0.15}{20}} = 12892.6\ W/m^2$$

Now,

$$Q = h(T_\infty - T_1) = \frac{k_1(T_1 - T_2)}{L_1} = \frac{k_2(T_2 - T_3)}{L_2} = \frac{k_3(T_3 - T_4)}{L_3}$$

$$12892.6 = 25\,(800 - T_1) = \frac{30(T_1 - T_2)}{0.3} = \frac{50(T_2 - T_3)}{0.15} = \frac{20(T_3 - 20)}{0.15}$$

Solving,

$$T_1 = 284.3°C,$$

$$T_2 = 155.37°C,$$

$$T_3 = 116.7°C$$

(II) FEM by hand calculations [Refer figure 6.1(b)]

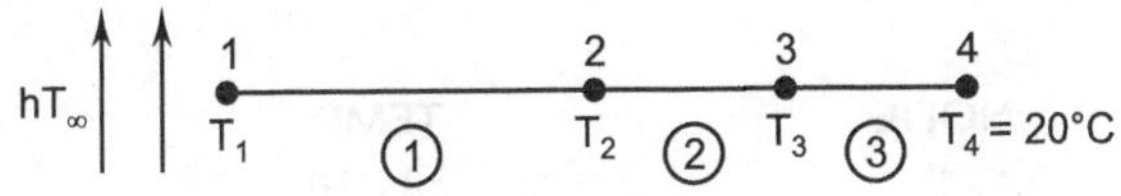

Figure 6.1(b)

Governing equation is,

$$\frac{k}{L}\begin{bmatrix} 1 & -1 \\ -1 & 1 \end{bmatrix}\begin{Bmatrix} T_1 \\ T_2 \end{Bmatrix} = q\begin{Bmatrix} \frac{L}{2} \\ \frac{L}{2} \end{Bmatrix} + \begin{Bmatrix} -Q_1 \\ +Q_2 \end{Bmatrix}$$

Since there is no heat generation specified, $q = 0$.

For element 1,

$$\frac{30}{0.3}\begin{bmatrix} 1 & -1 \\ -1 & 1 \end{bmatrix}\begin{Bmatrix} T_1 \\ T_2 \end{Bmatrix} = 0 + \begin{Bmatrix} -Q_1 \\ +Q_2 \end{Bmatrix}$$

For element 2,

$$\frac{50}{0.15}\begin{bmatrix} 1 & -1 \\ -1 & 1 \end{bmatrix}\begin{Bmatrix} T_2 \\ T_3 \end{Bmatrix} = 0 + \begin{Bmatrix} -Q_2 \\ +Q_3 \end{Bmatrix}$$

For element 3,

$$\frac{20}{0.15}\begin{bmatrix} 1 & -1 \\ -1 & 1 \end{bmatrix}\begin{Bmatrix} T_3 \\ T_4 \end{Bmatrix} = 0 + \begin{Bmatrix} -Q_3 \\ +Q_4 \end{Bmatrix}$$

Global equation after assembly,

$$\begin{bmatrix} 100 & -100 & 0 & 0 \\ -100 & 100 + 333.33 & -333.33 & 0 \\ 0 & -333.33 & 333.33 + 133.33 & -133.33 \\ 0 & 0 & -133.33 & 133.33 \end{bmatrix}\begin{Bmatrix} T_1 \\ T_2 \\ T_3 \\ T_4 \end{Bmatrix} = \begin{Bmatrix} -Q_1 \\ 0 \\ 0 \\ +Q_4 \end{Bmatrix}$$

Boundary conditions are, $T_4 = 20^\circ C$ and

$$Q_1 = -h\,(T_\infty - T_1)$$
$$-Q_1 = 25\,(800 - T_1) = 20000 - 25T_1$$

So modified equation,

$$\begin{bmatrix} 100 + 25 & -100 & 0 \\ -100 & 433.33 & -333.33 \\ 0 & -333.33 & 466.66 \end{bmatrix}\begin{Bmatrix} T_1 \\ T_2 \\ T_3 \end{Bmatrix} = +\begin{Bmatrix} 20000 \\ 0 \\ 0 \end{Bmatrix} + \begin{Bmatrix} 0 \\ 0 \\ 20 \times 133.33 \end{Bmatrix}$$

$$\begin{bmatrix} 125 & -100 & 0 \\ -100 & 433.33 & -333.33 \\ 0 & -333.33 & 466.66 \end{bmatrix}\begin{Bmatrix} T_1 \\ T_2 \\ T_3 \end{Bmatrix} = \begin{Bmatrix} 20000 \\ 0 \\ 0 \end{Bmatrix} + \begin{Bmatrix} 0 \\ 0 \\ 2666.6 \end{Bmatrix}$$

After solving the matrix and simultaneous equations, we get,

$$T_1 = 284.3^\circ C$$
$$T_2 = 155.37^\circ C$$
$$T_3 = 116.7^\circ C.$$

(III) Software results

Temperature values

NODE	TEMP
1	284.30
2	155.37
3	116.69
4	20.000
5	800.00

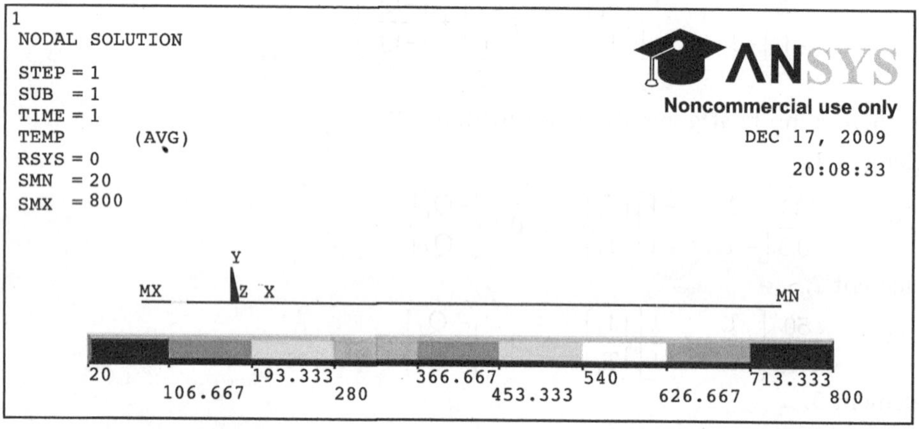

Figure 6.1(c). Temperature distribution in a composite wall (Refer to appendix for colour figures)

ANSWERS OF PROBLEM 6.1

Parameter	Analytical method	FEM-hand calculation	Software results
Temperature			
at node 1	284.3°C	284.3°C	284.3°C
at node 2	155.37°C	155.37°C	155.37°C
at node 3	116.7°C	116.7°C	116.69°C

Procedure for solving the problems using ANSYS®12.0 academic teaching software

FOR PROBLEM 6.1

PREPROCESSING

1. **Main Menu > Preferences, then select Thermal > OK**

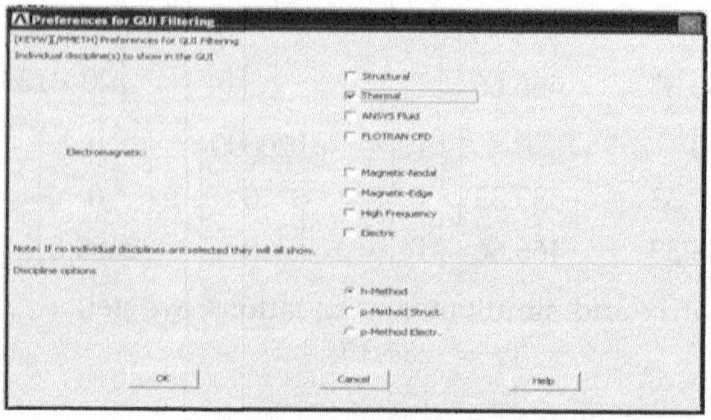

Figure 6.2. Selecting the preferences

2. **Main Menu > Preprocessor > Element Type > Add/Edit/Delete > Add > Click on Link > then on 2D conduction 32 > OK > Add > Click on Link > then on 3D convection 34 > OK > Close**

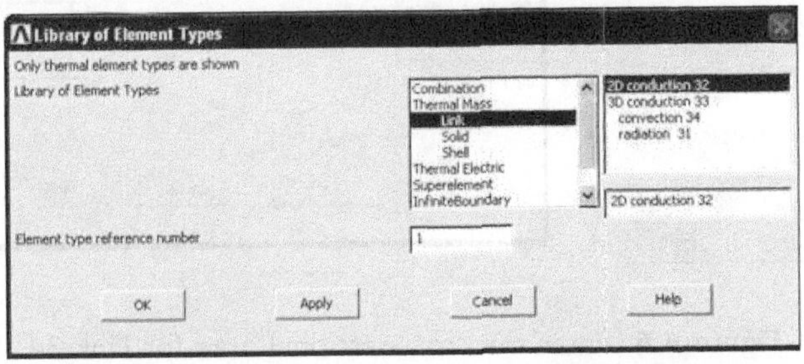

Figure 6.3. Selecting the element for conduction

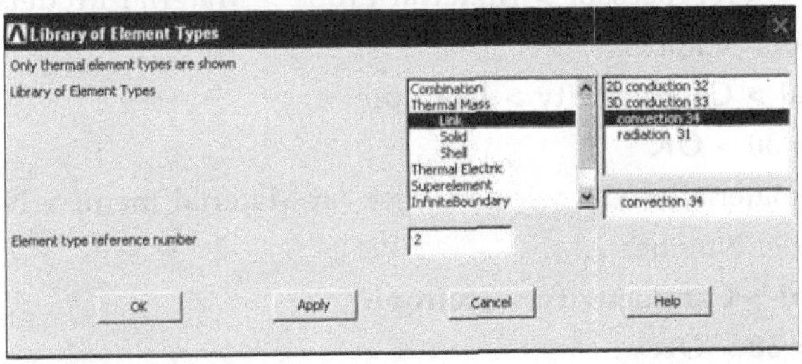

Figure 6.4. Selecting the element for convection

3. **Main Menu > Preprocessor > Real Constants > Add/Edit/Delete > Add > Click on Link 32 > OK**

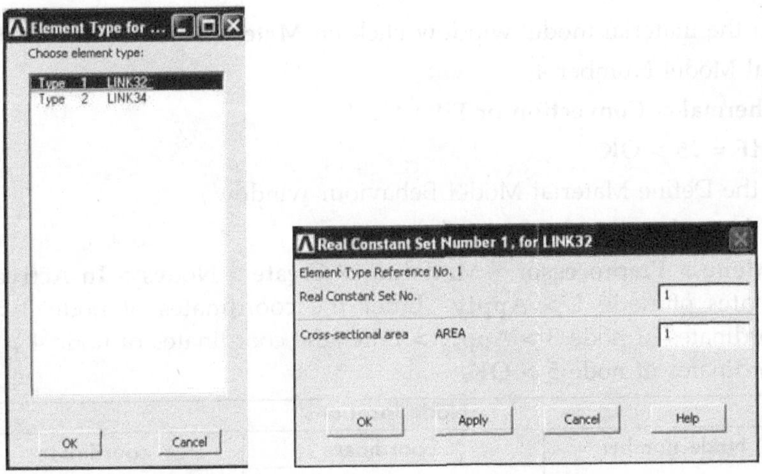

Figure 6.5. Enter the cross-sectional area for Link 32

Enter cross-sectional area AREA > **Enter 1 > OK**

Add > Click on Link 34 > OK

Enter cross-sectional area AREA > **Enter 1 > OK > Close**

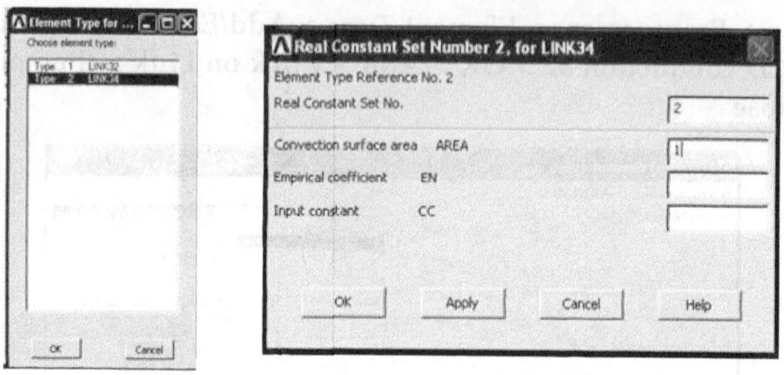

Figure 6.6. Enter the cross-sectional area for Link 34

Enter the material properties.

4. **Main Menu > Preprocessor > Material Props > Material Models**
   Material Model Number 1,
   click **Thermal > Conductivity > Isotropic**
   Enter **KXX** = **30** > **OK**
   Then in the material model window click on **Material menu > New Model > OK**
   Material Model Number 2,
   click **Thermal > Conductivity > Isotropic**
   Enter **KXX** = **50** > **OK**
   Then in the material model window click on **Material menu > New Model > OK**
   Material Model Number 3,
   click **Thermal > Conductivity > Isotropic**
   Enter **KXX** = **20** > **OK**
   Then in the material model window click on **Material menu > New Model > OK**
   Material Model Number 4,
   click **Thermal > Convection or Film Coef.**
   Enter **HF** = **25** > **OK**
   (**Close** the Define Material Model Behaviour window.)
   Create the nodes and elements.
5. **Main Menu > Preprocessor > Modeling > Create > Nodes > In Active CS** Enter the coordinates of node 1 > **Apply** Enter the coordinates of node 2 > **Apply** Enter the coordinates of node 3 > **Apply** > Enter the coordinates of node 4 > **Apply** > Enter the coordinates of node 5 > **OK.**

Node locations		
Node number	**X coordinate**	**Y coordinate**
1	0	0
2	0.3	0
3	0.45	0
4	0.6	0
5	−0.1	0

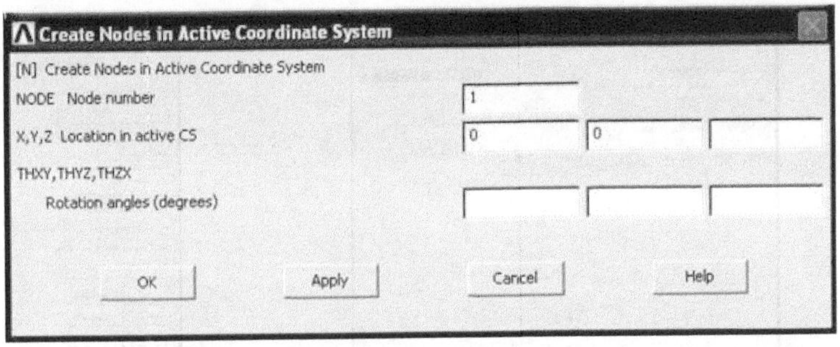

Figure 6.7. Enter the node coordinates

6. **Main Menu > Preprocessor > Modeling > Create > Elements > Elem Attributes > OK > Auto Numbered > Thru nodes** Pick the 1st and 2nd node **> OK**

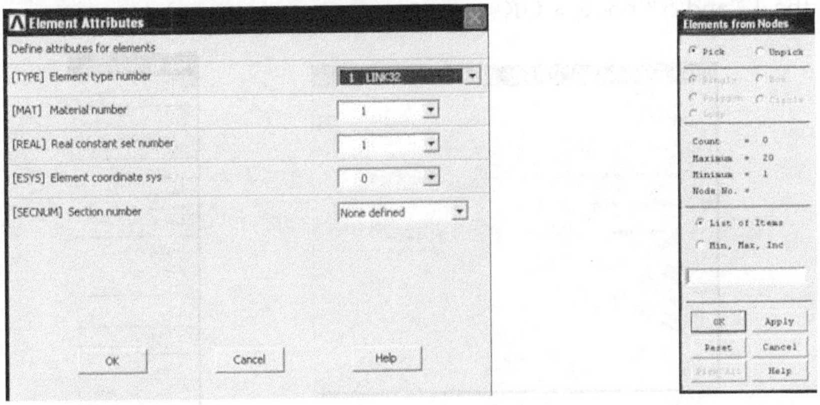

Figure 6.8. Assigning element attributes to element 1 and creating element 1

> Elem Attributes > change the material number to 2 > OK > Auto Numbered > Thru nodes Pick the 2nd and 3rd node **> OK**

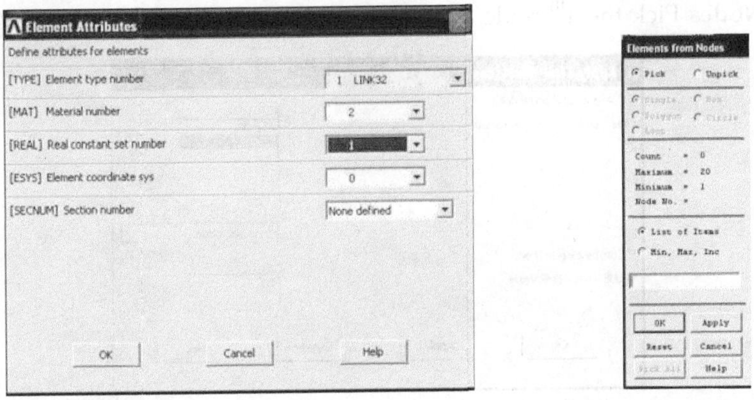

Figure 6.9. Asigning element attributes to element 2 and creating element 2

> Elem Attributes > change the material number to 3 > OK > Auto Numbered > Thru nodes Pick the 3rd and 4th node **> OK**

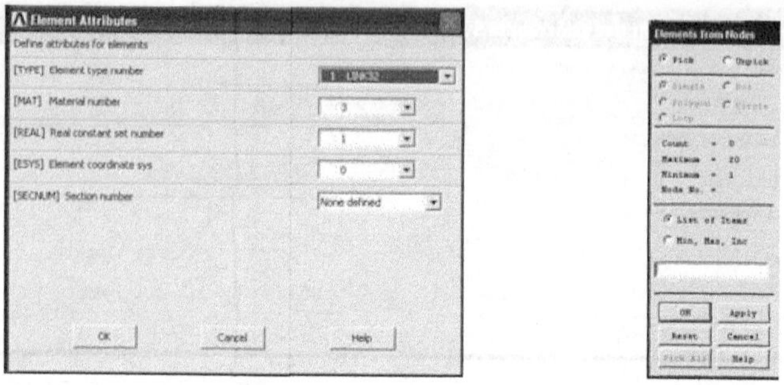

Figure 6.10. Assigning element attributes to element 3 and creating element 3

> Elem Attributes > change the element type to Link 34 > change the material number to 4 > change the Real constant set number to 2 > OK > Auto Numbered >Thru nodes Pick the 1st and 5th node > **OK**

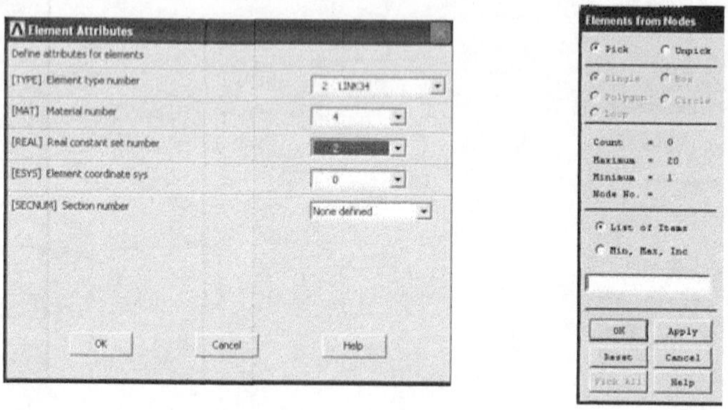

Figure 6.11. Assigning element attributes to element 4 and creating element 4

Apply the boundary conditions and temperature.

7. **Main Menu > Preprocessor > Loads > Define Loads > Apply > Thermal > Temperature > On Nodes** Pick the 4th node > **Apply > Click on TEMP and Enter Value = 20 > OK**

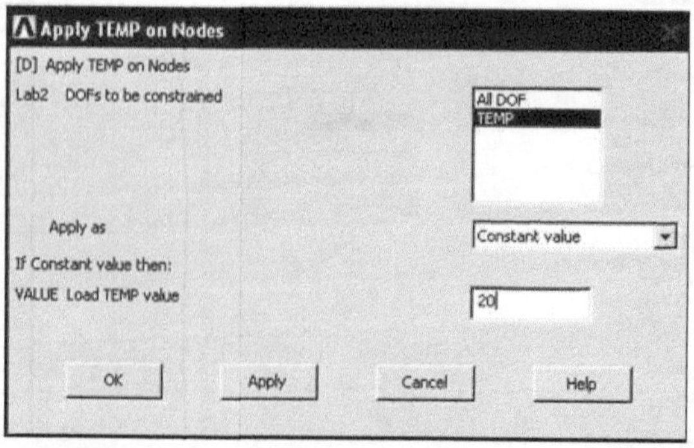

Figure 6.12. Applying temperature on node 4

8. **Main Menu > Preprocessor > Loads > Define Loads > Apply > Thermal > Temperature > On Nodes** Pick the 5th node > **Apply > Click on TEMP and Enter Value = 800 > OK**

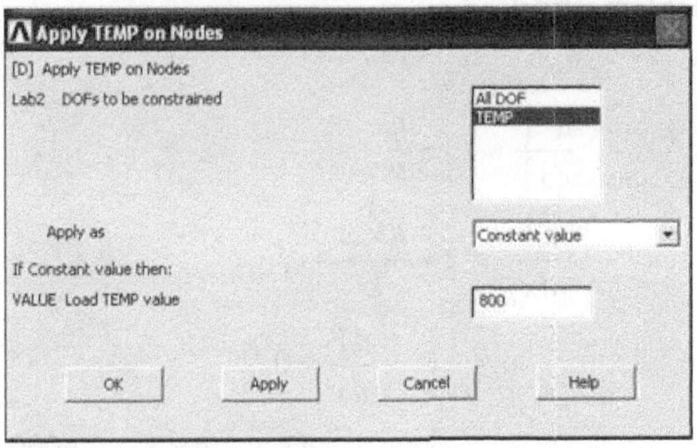

Figure 6.13. Applying temperature on node 5

SOLUTION. The interactive solution proceeds.

9. **Main Menu > Solution > Solve > Current LS > OK**

   The **/STATUS Command** window displays the problem parameters and **the SolveCurrent Load Step** window is shown. Check the solution options in the **/STATUS** window and if all is OK, select **File > Close**

   In the Solve **Current Load Step** window, Select **OK**, and when the solution is complete, **close** the **'Solution is Done!'** window.

POSTPROCESSING

We can now plot the results of this analysis and also list the computed values.

10. **Main Menu > General Postproc > Plot Results > Contour Plot > Nodal Solu > DOF Solution > Temperature > OK**

    This result is shown in figure 6.1(*c*)

11. **Main Menu > General Postproc > List Results > Nodal Solu > Select Temperature > OK**

PROBLEM 6.2. *Heat is generated in a large plate with k = 0.75 W/m°C at the rate of 6000 W/m³. The plate is 40 cm thick. The outside surfaces of plate are exposed to fluid at 35°C with convective heat transfer coefficient of 15 W/m² °C. Determine temperature distribution in wall. Two element model to be used for solution.*

Figure 6.14

SOLUTION. (I) Analytical method

Governing equation is,

$$\frac{d^2T}{dx^2} = -\frac{q}{k}$$

$$\therefore \quad \frac{dT}{dx} = -\frac{q \times x}{k} + c_1$$

At $\quad x = 0, \; \frac{dT}{dx} = 0$

$$\therefore \quad c_1 = 0 \quad \therefore \quad \frac{dT}{dx} = -\frac{q \times x}{k} = \frac{-6000 \times x}{0.75} = -8000x \qquad ...(i)$$

$$T = -\frac{q}{k}\frac{x^2}{2} + c_2 = -4000x^2 + c_2 \qquad ...(ii)$$

Boundary conditions are,

At $x = L$

$$-k\frac{dT}{dx} = h(T_3 - T_\infty)$$

$$-k(8000 \times x) = h(T_3 - T_\infty)$$

$$\therefore \quad -0.75\,[-8000 \times 0.2] = 15(T_3 - 35)$$

$$\therefore \quad T_3 = 115°C$$

We know at $\quad x = 0.2, \; T_3 = 115°C$

Substituting this in equation (*ii*)

$$115 = -4000\,(0.2)^2 + c_2 \rightarrow c_2 = 275$$

Substituting c_2 in equation (*ii*)

$$T = -4000 \times x^2 + 275$$

$$T_2 = T\big|_{x=0.1} = -4000(0.1)^2 + 275 = 235°C$$

$$T_1 = T\big|_{x=0} = 275°C$$

(II) FEM by hand calculations [Refer to figure 6.14(*a*)]

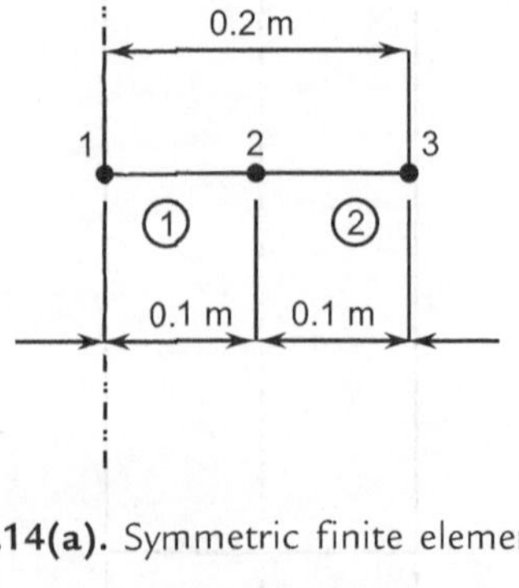

Figure 6.14(a). Symmetric finite element model

Given: $\quad t = 40$ cm $= 0.4$ m, $T_\infty = 35°C$, $h = 15$ W/m^2°C, $k = 0.75$ W/m°C

For element 1,

$$\frac{k_1}{L_1}\begin{bmatrix} 1 & -1 \\ -1 & 1 \end{bmatrix}\begin{Bmatrix} T_1 \\ T_2 \end{Bmatrix} = q\begin{Bmatrix} \frac{L}{2} \\ \frac{L}{2} \end{Bmatrix} + \begin{Bmatrix} -Q_1 \\ +Q_2 \end{Bmatrix}$$

$$\frac{0.75}{0.10}\begin{bmatrix} 1 & -1 \\ -1 & 1 \end{bmatrix}\begin{Bmatrix} T_1 \\ T_2 \end{Bmatrix} = 6000\begin{Bmatrix} 0.05 \\ 0.05 \end{Bmatrix} + \begin{Bmatrix} -Q_1 \\ +Q_2 \end{Bmatrix}$$

$$\begin{matrix} & 1 & 2 & \\ 7.5 & \begin{bmatrix} 1 & -1 \\ -1 & 1 \end{bmatrix} & \begin{matrix} 1 \\ 2 \end{matrix} \end{matrix}\begin{Bmatrix} T_1 \\ T_2 \end{Bmatrix} = 6000\begin{Bmatrix} 0.05 \\ 0.05 \end{Bmatrix} + \begin{Bmatrix} -Q_1 \\ +Q_2 \end{Bmatrix}$$

For element 2,

$$\frac{k_2}{L_2}\begin{bmatrix} 1 & -1 \\ -1 & 1 \end{bmatrix}\begin{Bmatrix} T_2 \\ T_3 \end{Bmatrix} = q\begin{Bmatrix} \frac{L}{2} \\ \frac{L}{2} \end{Bmatrix} + \begin{Bmatrix} -Q_2 \\ +Q_3 \end{Bmatrix}$$

$$\frac{0.75}{0.10}\begin{bmatrix} 1 & -1 \\ -1 & 1 \end{bmatrix}\begin{Bmatrix} T_2 \\ T_3 \end{Bmatrix} = 6000\begin{Bmatrix} 0.05 \\ 0.05 \end{Bmatrix} + \begin{Bmatrix} -Q_2 \\ +Q_3 \end{Bmatrix}$$

$$\begin{matrix} & 2 & 3 & \\ 7.5 & \begin{bmatrix} 1 & -1 \\ -1 & 1 \end{bmatrix} & \begin{matrix} 2 \\ 3 \end{matrix} \end{matrix}\begin{Bmatrix} T_2 \\ T_3 \end{Bmatrix} = 6000\begin{Bmatrix} 0.05 \\ 0.05 \end{Bmatrix} + \begin{Bmatrix} -Q_2 \\ +Q_3 \end{Bmatrix}$$

$$\text{Assembling} \Rightarrow \begin{matrix} & 1 & 2 & 3 & \\ 7.5 & \begin{bmatrix} 1 & -1 & 0 \\ -1 & 1+1 & -1 \\ 0 & -1 & 1 \end{bmatrix} & \begin{matrix} 1 \\ 2 \\ 3 \end{matrix} \end{matrix}\begin{Bmatrix} T_1 \\ T_2 \\ T_3 \end{Bmatrix} = 6000\begin{Bmatrix} 0.05 \\ 0.05+0.05 \\ 0.05 \end{Bmatrix} + \begin{Bmatrix} -Q_1 \\ 0 \\ +Q_3 \end{Bmatrix}$$

Boundary conditions are, $Q_1 = 0$ and $Q_3 = -h(T_3 - T_\infty) => Q_3 = -15\ (T_3 - 35) = -15\ T_3 + 525$

$$7.5\begin{bmatrix} 1 & -1 & 0 \\ -1 & 2 & -1 \\ 0 & -1 & 1 \end{bmatrix}\begin{Bmatrix} T_1 \\ T_2 \\ T_3 \end{Bmatrix} = 6000\begin{Bmatrix} 0.05 \\ 0.1 \\ 0.05 \end{Bmatrix} + \begin{Bmatrix} 0 \\ 0 \\ -15T_3 + 525 \end{Bmatrix}$$

$$\begin{bmatrix} 7.5 & -7.5 & 0 \\ -7.5 & 15 & -7.5 \\ 0 & -7.5 & 7.5 \end{bmatrix}\begin{Bmatrix} T_1 \\ T_2 \\ T_3 \end{Bmatrix} = \begin{Bmatrix} 300+0 \\ 600+0 \\ 300-15T_3+525 \end{Bmatrix}$$

Now,

$$\begin{bmatrix} 7.5 & -7.5 & 0 \\ -7.5 & 15 & -7.5 \\ 0 & -7.5 & 7.5+15 \end{bmatrix}\begin{Bmatrix} T_1 \\ T_2 \\ T_3 \end{Bmatrix} = \begin{Bmatrix} 300 \\ 600 \\ 825 \end{Bmatrix}$$

By solving the above matrix and simultaneous equations we have temperature distribution as,

$$T_1 = 275°C, T_2 = 235°C, T_3 = 115°C$$

$$\therefore \quad \begin{Bmatrix} T_1 \\ T_2 \\ T_3 \end{Bmatrix} = \begin{Bmatrix} 275 \\ 235 \\ 115 \end{Bmatrix}$$

(III) Software results

Due to symmetry of the geometry only half of the finite element model is created for software analysis.

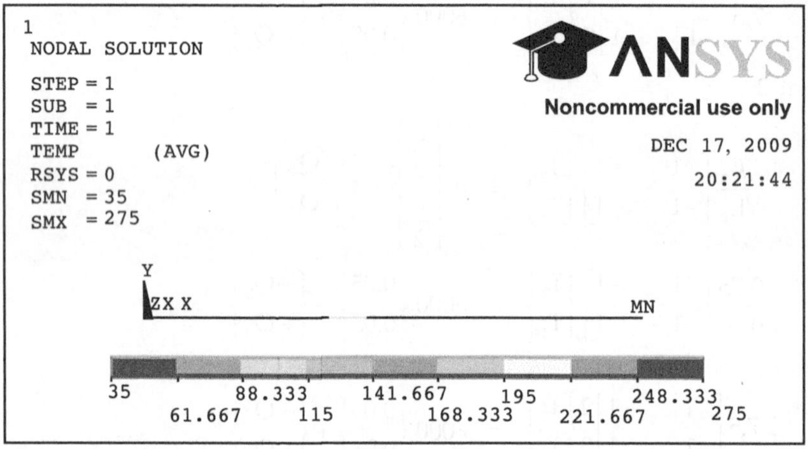

Figure 6.14(b). Temperature distribution in a large plate (Refer to appendix for colour figures)

Temperature values

NODE	TEMP
1	275.00
2	235.00
3	115.00
4	35.000

Answers of Problem 6.2

Parameter	Analytical method	FEM–hand calculation	Software results
Temperature			
at node 1	275°C	275°C	275°C
at node 2	235°C	235°C	235°C
at node 3	115°C	115°C	115°C

Procedure for solving the problems using ANSYS® 12.0 academic teaching software

For Problem 6.2

PREPROCESSING

1. **Main Menu > Preferences, then select Thermal > OK**

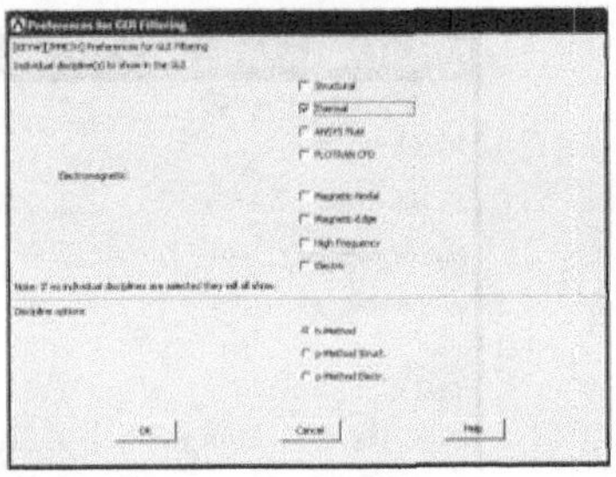

Figure 6.15. Selecting the preferences

2. **Main Menu > Preprocessor > Element Type > Add/Edit/Delete > Add > Click on Link > then on 2D conduction 32 > OK > Add > Click on Link > then on 3D convection 34 > OK > Close**

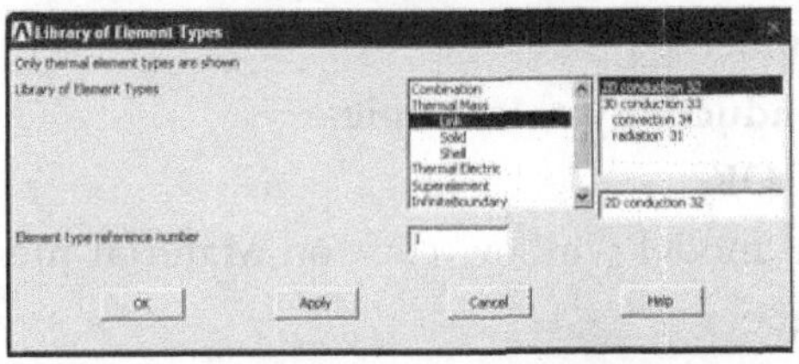

Figure 6.16. Selecting the element for conduction

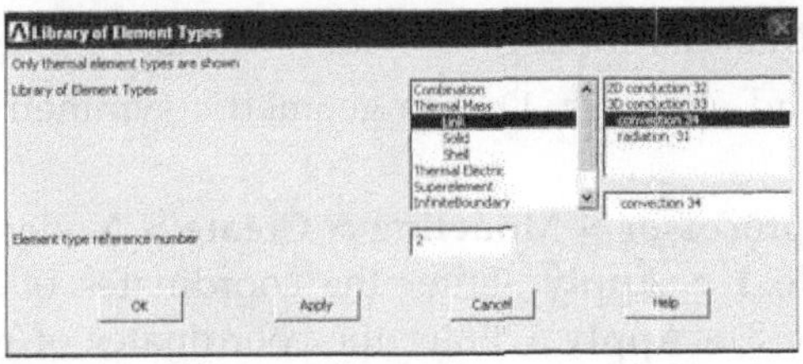

Figure 6.17. Selecting the element for convection

3. **Main Menu > Preprocessor > Real Constants > Add/Edit/Delete > Add > Click on Link 32 > OK**

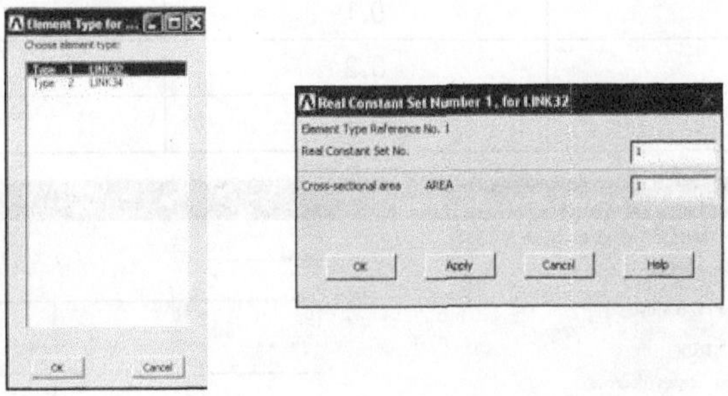

Figure 6.18. Enter the cross-sectional area for Link 32

Enter Cross-sectional area AREA > **Enter 1 > OK**

Add > Click on Link 34 > OK

Enter Cross-sectional area AREA > **Enter 1 > OK**

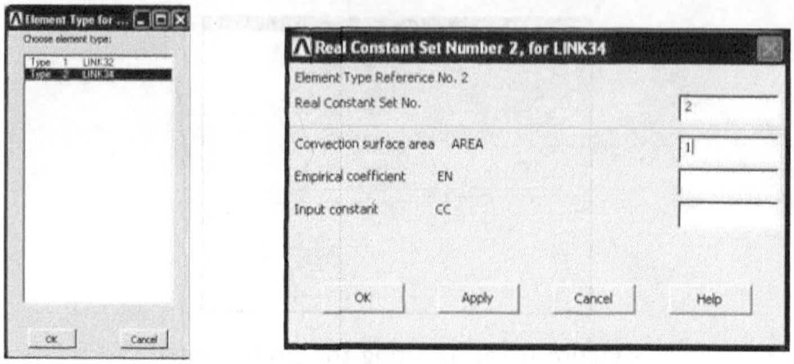

Figure 6.19. Enter the cross-sectional area for Link 34

Enter the material properties.

4. **Main Menu > Preprocessor > Material Props > Material Models**
   Material Model Number 1,
   click **Thermal > Conductivity > Isotropic**
   Enter **KXX = 0.75 > OK**
   Then in the material model window click on **Material menu > New Model > OK**
   Material Model Number 2,
   click **Thermal > Convection or Film Coef.**
   Enter **HF = 15 > OK**
   (**Close** the Define Material Model Behaviour window.)
   Create the nodes and elements. Due to geometric symmetry only half of the model is created.
5. **Main Menu > Preprocessor > Modeling > Create > Nodes > In Active CS** Enter the coordinates of node 1 > **Apply** Enter the coordinates of node 2 > **Apply** Enter the coordinates of node 3 > **Apply** > Enter the coordinates of node 4 > **OK.**

Node locations		
Node number	**X coordinate**	**Y coordinate**
1	0	0
2	0.1	0
3	0.2	0
4	0.3	0

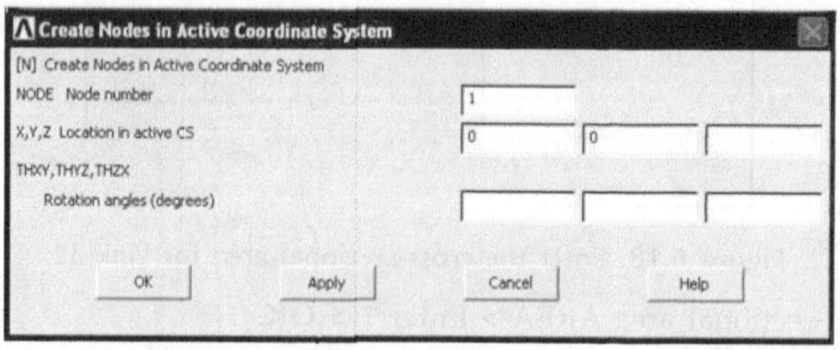

Figure 6.20. Enter the node coordinates

6. **Main Menu > Preprocessor > Modeling > Create > Elements > Elem Attributes > OK > Auto Numbered > Thru nodes** Pick the 1st and 2nd node **> Apply>** then Pick the 2nd and 3rd node > **OK**

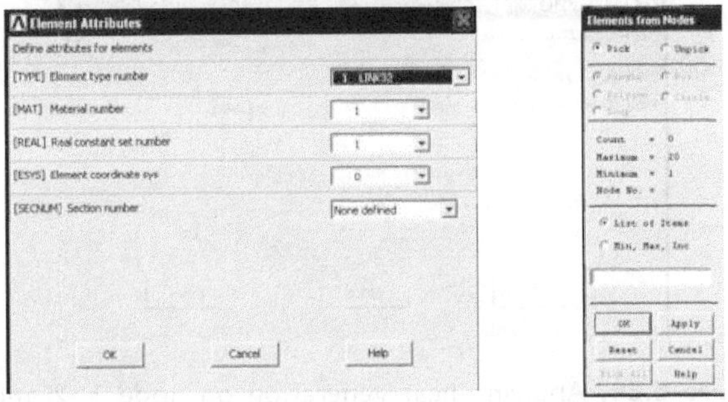

Figure 6.21. Assigning element attributes to element 1 and 2 and creating element 1 and 2

> Elem Attributes >change the element type to Link 34 > change the material number to 2 > change the Real constant set number to 2 > OK > Auto Numbered > Thru nodes Pick the 3rd and 4th node > **OK**

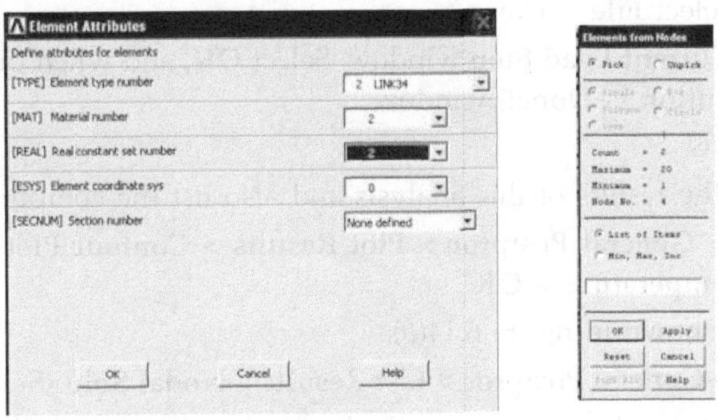

Figure 6.22. Assigning element attributes to element 3 and creating element 3

Apply the boundary conditions and temperature.

7. **Main Menu > Preprocessor > Loads > Define Loads > Apply > Thermal >Temperature > On Nodes** Pick the 4th node > **Apply > Click on TEMP and Enter Value = 35 > OK**

Figure 6.23. Applying temperature on node 4

8. **Main Menu > Preprocessor > Loads > Define Loads > Apply > Thermal >Heat Generat > On Nodes** Pick the 1st, 2nd and 3rd nodes > **Apply >Enter HGEN Value = 6000 > OK**

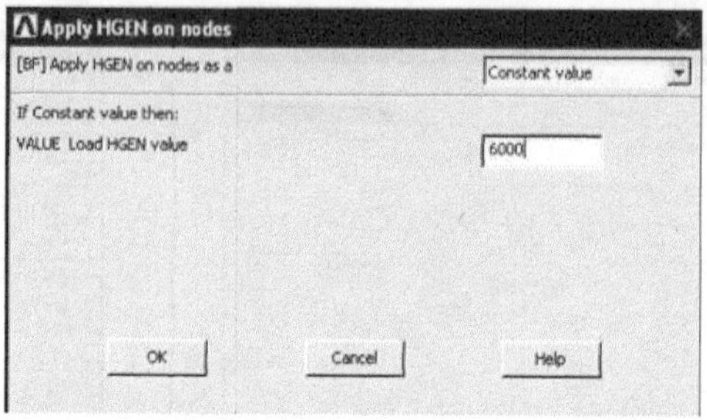

Figure 6.24. Applying heat generation on node 1, 2 and 3

SOLUTION. The interactive solution proceeds.

9. **Main Menu > Solution > Solve > Current LS > OK**

   The **/STATUS Command** window displays the problem parameters and **the SolveCurrent Load Step** window is shown. Check the solution options in the **/STATUS** window and if all is OK, select **File > Close**

   In the Solve **Current Load Step** window, Select **OK**, and when the solution is complete, **close** the **'Solution is Done!'** window.

POSTPROCESSING

We can now plot the results of this analysis and also list the computed values.

10. **Main Menu > General Postproc > Plot Results > Contour Plot > Nodal Solu > DOF Solution > Temperature > OK**

    This result is shown in figure 6.14(*b*).

11. **Main Menu > General Postproc > List Results > Nodal Solu>Select Temperature > OK**

PROBLEM 6.3. *Compute the temperature distribution in a long steel cylinder with inner radius of 125 mm and outer radius of 250 mm. The interior of the cylinder is kept at 300°K and heat lost on the exterior by convection to a fluid whose temperature is 280°K. The convection heat transfer coefficient h is 0.994 W/m^2°K and the thermal conductivity for steel k is 0.031 W/m°K.*

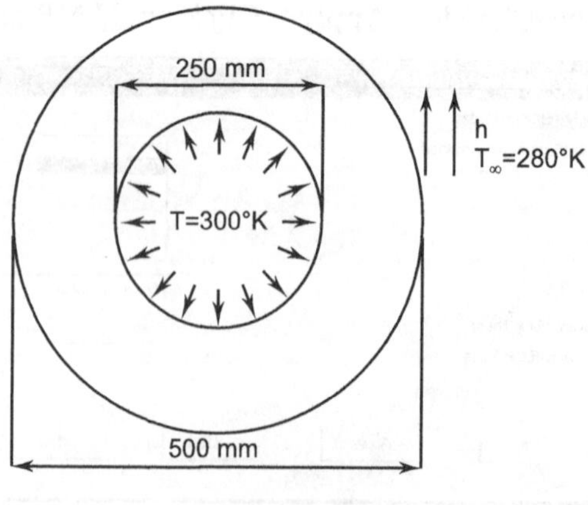

Figure 6.25

SOLUTION. (I) Analytical method

Here problem is solved considering heat flow in radial direction.

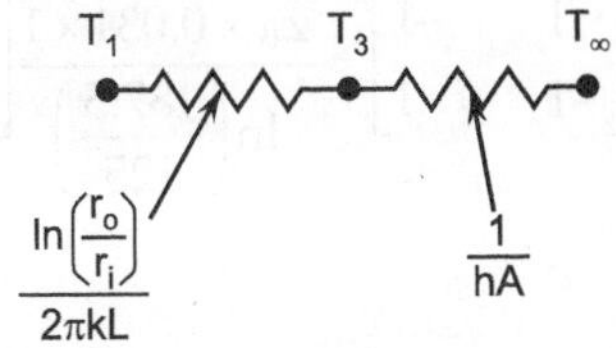

Figure 6.25(a)

$$r_o = 250 \text{ mm} = 0.25 \text{ m and } r_i = 125 \text{ mm} = 0.125 \text{ m}$$

Assume unit length of the cylinder

$$Q = \frac{T_1 - T_\infty}{\frac{\ln\left(\frac{r_o}{r_i}\right)}{2\pi k L} + \frac{1}{hA}} = \frac{300 - 280}{\frac{\ln\left(\frac{250}{125}\right)}{2\pi \times 0.031 \times 1} + \frac{1}{0.994(2\pi \times 0.25)}} = 4.762934 \text{ W}$$

Now, $$Q = \frac{T_1 - T_3}{\frac{\ln\left(\frac{r_o}{r_i}\right)}{2\pi k L}} \Rightarrow 4.762934 = \frac{300 - T_3}{\frac{\ln\left(\frac{250}{125}\right)}{2\pi \times 0.031 \times 1}}$$

$$T_3 = 280.51°\text{C}$$

Let $$T = T_2 \text{ at } r = 187.5 \text{ mm then,}$$

$$4.762934 = \frac{300 - T_2}{\frac{\ln\left(\frac{250}{187.5}\right)}{2\pi \times 0.031 \times 1}}$$

$$\therefore \quad T_2 = 283.1°\text{C}$$

(II) FEM by hand calculations

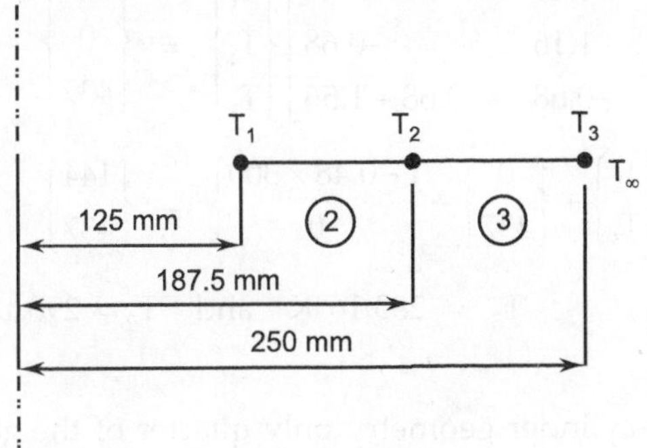

Figure 6.25(b)

$$r_1 = 125 \text{ mm}, r_2 = 187.5 \text{ mm and } r_3 = 250 \text{ mm}$$

Element matrices are,

$$k_{1C} = \frac{2\pi kL}{\ln\left(\frac{r_2}{r_1}\right)}\begin{bmatrix}1 & -1\\ -1 & 1\end{bmatrix} = \frac{2\pi \times 0.031 \times 1}{\ln\left(\frac{187.5}{125}\right)}\begin{bmatrix}1 & -1\\ -1 & 1\end{bmatrix} = 0.48\begin{bmatrix}1 & -1\\ -1 & 1\end{bmatrix}\begin{matrix}T_1\\ T_2\end{matrix} \quad (\text{columns } T_1,\ T_2)$$

$$k_{2C} = \frac{2\pi kL}{\ln\left(\frac{r_3}{r_2}\right)}\begin{bmatrix}1 & -1\\ -1 & 1\end{bmatrix} = \frac{2\pi \times 0.031 \times 1}{\ln\left(\frac{250}{187.5}\right)}\begin{bmatrix}1 & -1\\ -1 & 1\end{bmatrix} = 0.68\begin{bmatrix}1 & -1\\ -1 & 1\end{bmatrix}\begin{matrix}T_2\\ T_3\end{matrix} \quad (\text{columns } T_2,\ T_3)$$

Global conduction matrix is,

$$[K_C] = \begin{bmatrix}0.48 & -0.48 & 0\\ -0.48 & 0.48+0.68 & -0.68\\ 0 & -0.68 & 0.68\end{bmatrix} = \begin{bmatrix}0.48 & -0.48 & 0\\ -0.48 & 1.16 & -0.68\\ 0 & -0.68 & 0.68\end{bmatrix}$$

Global equation is,

$$\begin{bmatrix}0.48 & -0.48 & 0\\ -0.48 & 1.16 & -0.68\\ 0 & -0.6 & 0.68\end{bmatrix}\begin{Bmatrix}T_1\\ T_2\\ T_3\end{Bmatrix} = \begin{Bmatrix}Q_1\\ Q_2\\ Q_3\end{Bmatrix}$$

Applying boundary conditions, $T_1 = 300°K$ and $Q_3 = -hA_0(T_3 - T_\infty)$

$$A_0 = 2\pi r_3 = 2\pi \times 0.25 = 1.57 \text{ m}^2$$

Therefore, $Q_3 = -hA_0(T_3 - T_\infty) = -0.994 \times 1.57(T_3 - 280) = -(1.56\,T_3 - 437)$

$$\begin{bmatrix}0.48 & -0.48 & 0\\ -0.48 & 1.16 & -0.68\\ 0 & -0.68 & 0.68\end{bmatrix}\begin{Bmatrix}T_1\\ T_2\\ T_3\end{Bmatrix} = \begin{Bmatrix}0\\ 0\\ -(1.56T_3 - 437)\end{Bmatrix}$$

$$\begin{bmatrix}\cancel{0.48} & \cancel{-0.48} & \cancel{0}\\ -0.48 & 1.16 & -0.68\\ 0 & -0.68 & 0.68+1.56\end{bmatrix}\begin{Bmatrix}\cancel{T_1}\\ T_2\\ T_3\end{Bmatrix} = \begin{Bmatrix}\cancel{0}\\ 0\\ 437\end{Bmatrix}$$

$$\begin{bmatrix}1.16 & -0.68\\ -0.68 & 2.24\end{bmatrix}\begin{Bmatrix}T_2\\ T_3\end{Bmatrix} + \begin{Bmatrix}0\\ 437\end{Bmatrix} - \begin{Bmatrix}-0.48 \times 300\\ 0\end{Bmatrix} = \begin{Bmatrix}144\\ 437\end{Bmatrix}$$

Solving, we get $T_3 = 283.16°K$ and $T_2 = 290.13°K$

(III) Software results

Due to symmetry of the cylinder geometry only quarter of the geometry is drawn for finite element analysis.

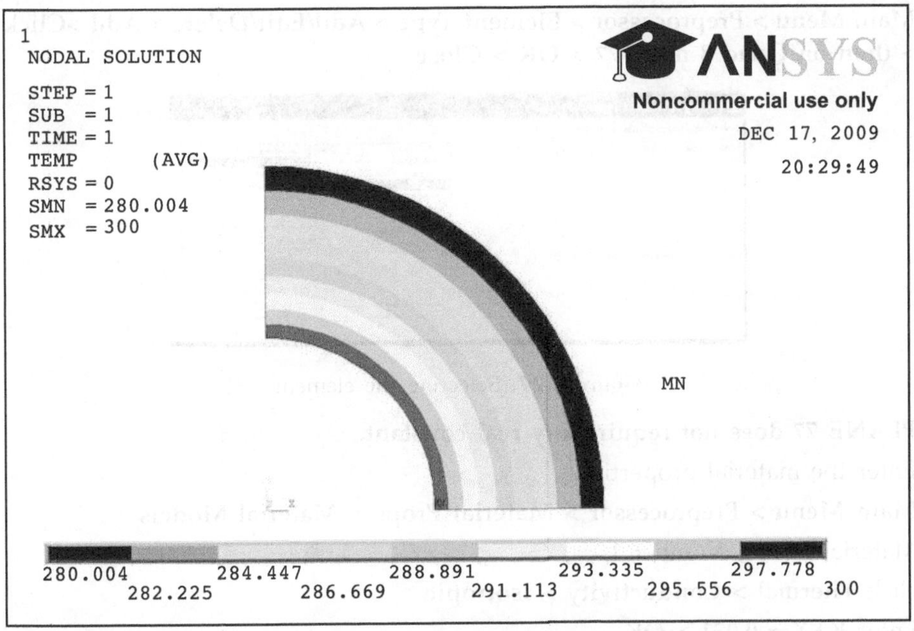

Figure 6.25(c). Temperature distribution in a long cylinder (Refer to appendix for colour figures)

The temperature on the interior is 300°K and on the outside wall it is found to be 280.004°K.

ANSWERS OF PROBLEM 6.3

Parameter	Analytical method	FEM – hand calculation	Software results
Temperature on the interior surface	300°K	300°K	300°K
Temperature at radius 187.5 mm	283.1°K	290.13°K	288.891°K
Temperature on the outside wall	280.51°K	283.16°K	280.004°K

Procedure for solving the problems using ANSYS®12.0 academic teaching software

FOR PROBLEM 6.3

PREPROCESSING

1. **Main Menu > Preferences, then select Thermal > OK**

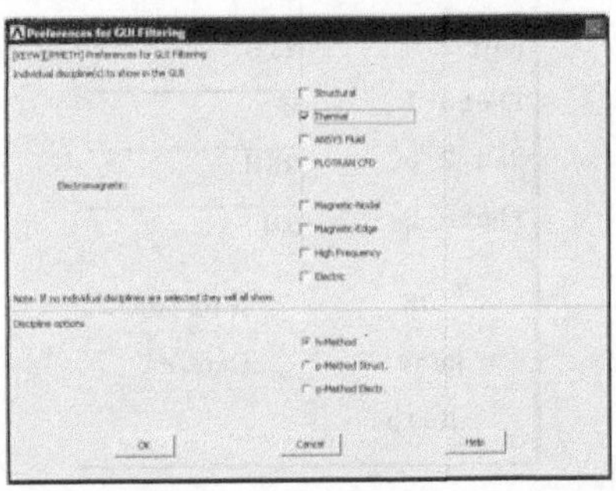

Figure 6.26. Selecting the preferences

2. **Main Menu > Preprocessor > Element Type > Add/Edit/Delete > Add >Click on Solid > then on Quad 8 node 77 > OK > Close**

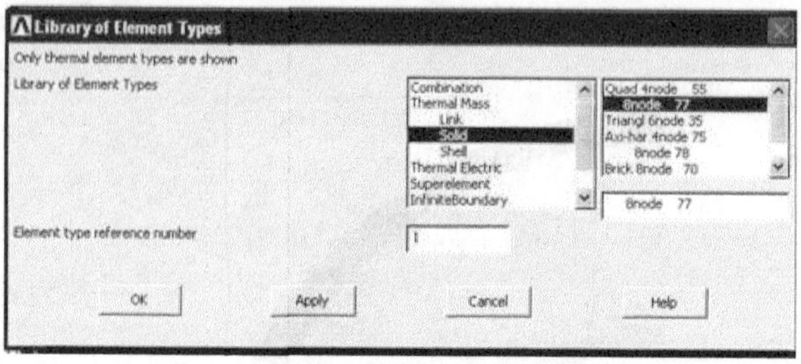

Figure 6.27. Selecting the element

3. **PLANE 77 does not require any real constant.**

   Enter the material properties.

4. **Main Menu > Preprocessor > Material Props > Material Models**

   Material Model Number 1,

   click **Thermal > Conductivity > Isotropic**

   Enter **KXX = 0.031 > OK**

   (**Close** the Define Material Model Behaviour window.)

   Recognize symmetry of the problem, and a quadrant of a section through the cylinder is created.

5. **Main Menu > Preprocessor > Modeling > Create > Areas > Circle > Partial Annulus.** Enter the data as shown below.

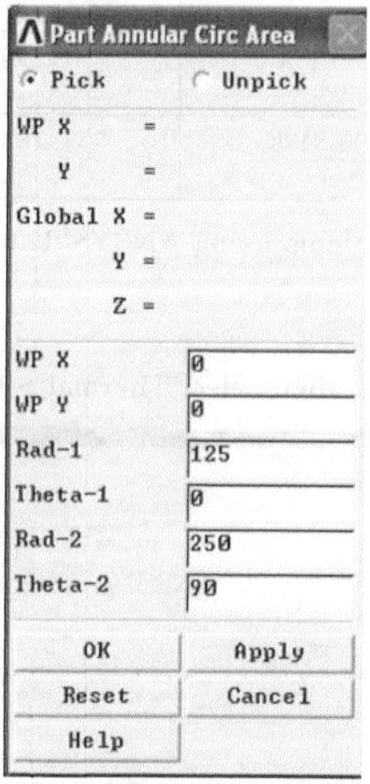

Figure 6.28. Create partial annular area

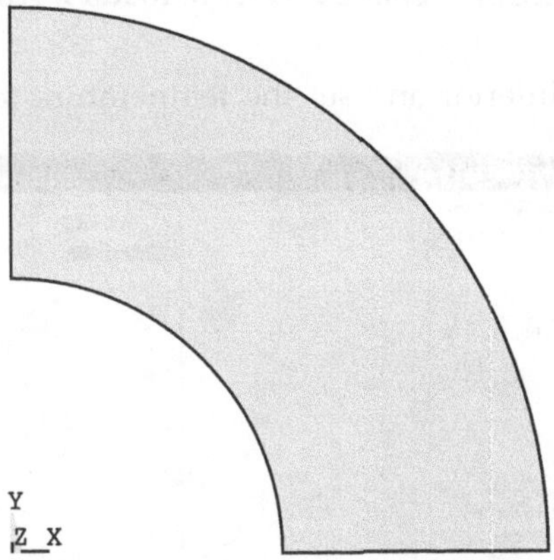

Figure 6.29. Quadrant of a cylinder

6. **Main Menu > Preprocessor > Meshing > Mesh Tool**

   The **Mesh Tool** dialog box appears. In that dialog box click on the **Smart Size** and move the slider available below the **Smart Size to 2** (*i.e.*, towards **Fine** side). Then close the Mesh Tool box.

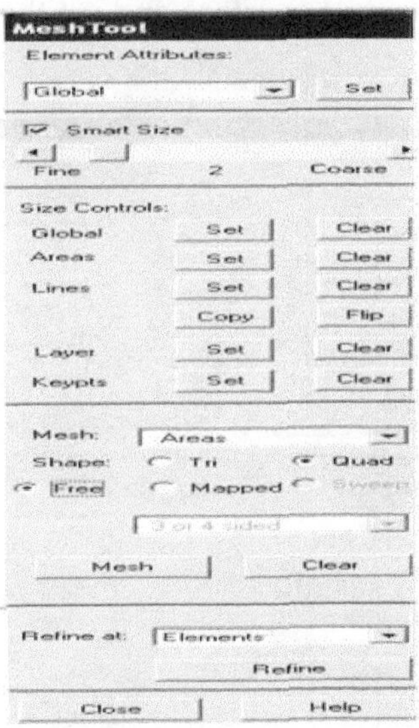

Figure 6.30. Mesh tool box

7. **Main Menu > Preprocessor > Meshing > Mesh > Areas > Free,** Pick the quadrant **>OK**

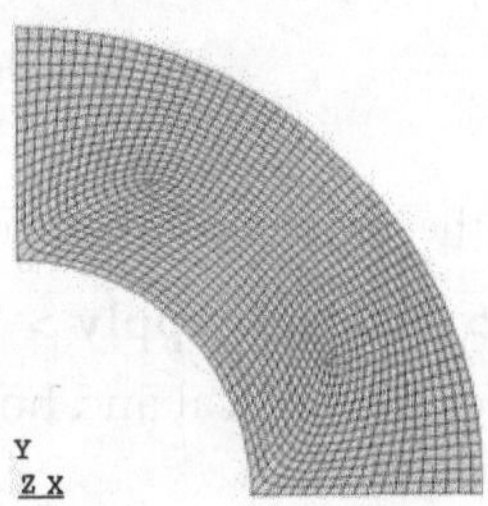

Figure 6.31. Quad element mesh

8. **Main Menu > Preprocessor > Loads > Define loads > Apply > Thermal > Temperatures > On Lines**

   Select the line on the interior and set the temperature to 300.

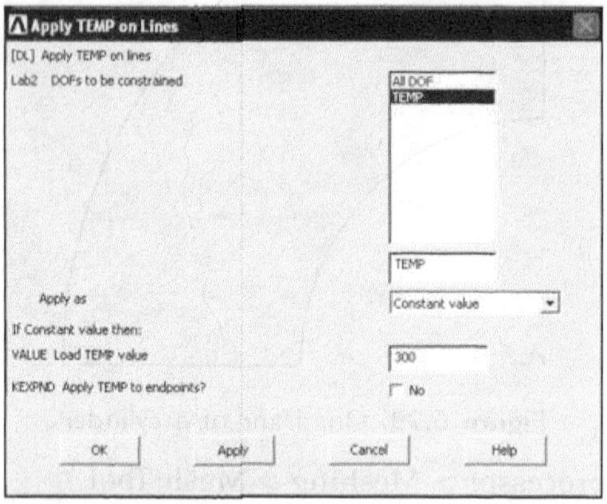

Figure 6.32. Setting the temperature on the interior of the cylinder

9. **Main Menu > Preprocessor > Loads > Apply > Convection > On Lines**

   Select the lines defining the outer surface and set the convection coefficient to 0.994 and the fluid temp to 280.

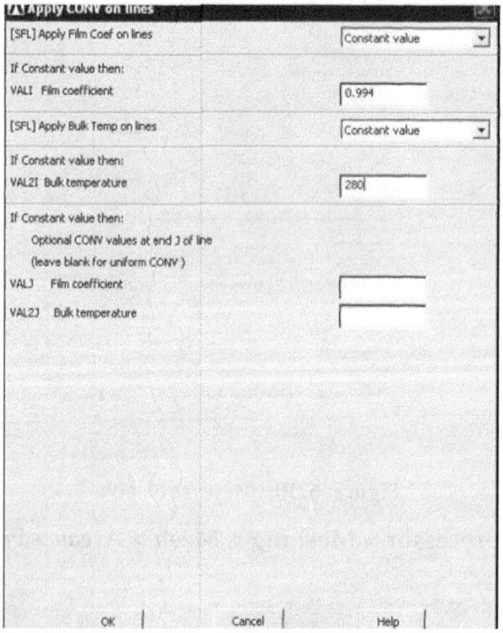

Figure 6.33. Setting the convection coefficient on outer surface

10. **Main Menu > Preprocessor > Loads > Apply > Heat Flux > On Lines**

    To account for symmetry, select the vertical and horizontal lines of symmetry and set the heat flux to zero.

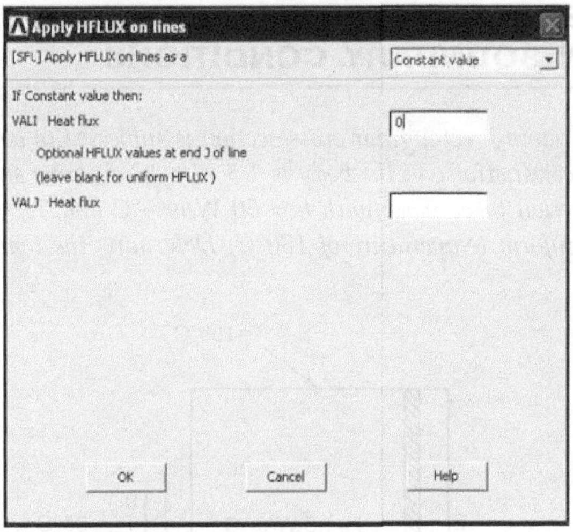

Figure 6.34. Setting the heat flux

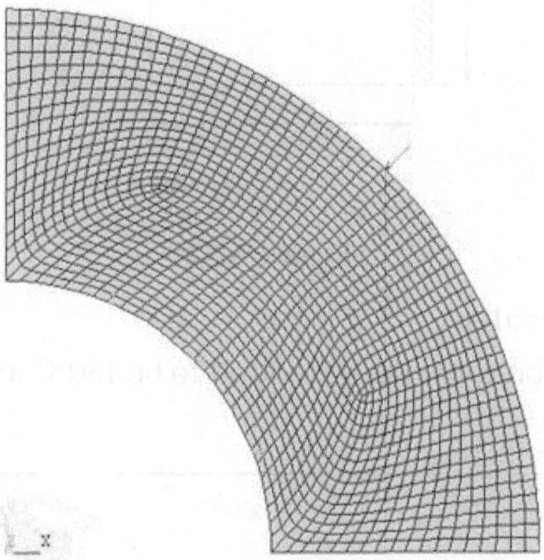

Figure 6.35. Model with boundary conditions

SOLUTION. The interactive solution proceeds.

11. **Main Menu > Solution > Solve > Current LS > OK**

    The **/STATUS Command** window displays the problem parameters and **the SolveCurrent Load Step** window is shown. Check the solution options in the **/STATUS** window and if all is OK, select **File > Close**

    In the Solve **Current Load Step** window, Select **OK**, and when the solution is complete, **close** the **'Solution is Done!'** window.

POSTPROCESSING

We can now plot the results of this analysis and also list the computed values.

12. **Main Menu > General Postproc > Plot Results > Contour Plot > Nodal Solu > DOF Solution > Temperature > OK**

    This result is shown in figure 6.25(*b*)

13. **Main Menu > General Postproc > List Results > Nodal Solu > Select Temperature > OK**

6.4 TWO-DIMENSIONAL PROBLEM WITH CONDUCTION AND WITH CONVECTION BOUNDARY CONDITIONS

PROBLEM 6.4. *A body having rectangular cross-section is subjected to boundary conditions as shown in figure 6.36. The thermal conductivity of the body is 1.5 W/m°C. On one side of the body it is insulated and on the other side convection takes place with h = 50 W/m^2–°C and T_∞ = 35°C. The top and bottom sides are maintained at a uniform temperature of 180°C. Determine the temperature distribution in the body.*

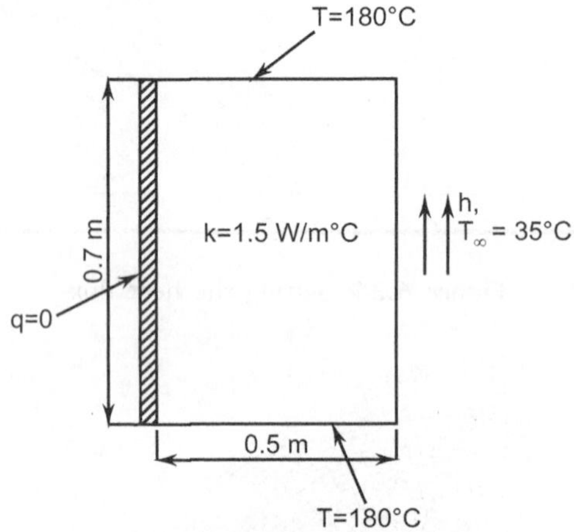

Figure 6.36

SOLUTION. (I) Software results

The temperature at the top and bottom edges is found to be 180°C and at the right edge temperature is found to be 46.802°C.

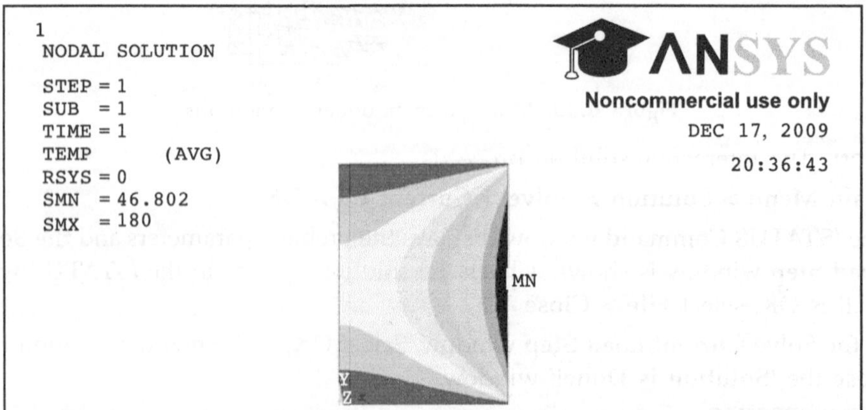

Figure 6.36(a). Temperature distribution in a body of rectangular cross-section (Refer to appendix for colour figures)

EXERCISES

1. Determine the temperature distribution for the two-dimensional body shown in figure 6.37, subjected to boundary conditions as shown in the figure. The top and bottom edges are insulated. The left side of the body is maintained at a temperature of 45°C. On the right side convection process takes place with heat transfer coefficient h = 100 W/m^2°C and T_∞ = 20°C. The thermal conductivity of the body is k = 45 W/m°C.

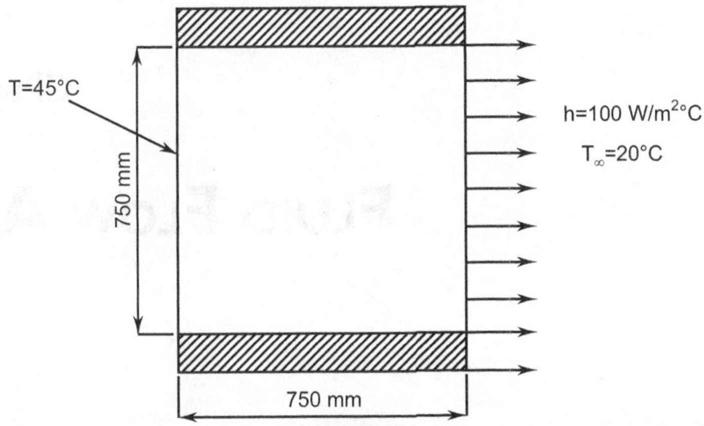

Figure 6.37

2. Determine the temperature distribution for the two dimensional body shown in figure 6.38. The temperature of 200°C is maintained at top and bottom edges. Left and right edges are insulated. Heat is generated at the rate of q = 2000 W/m^3 in a body as shown in figure. Let k = 35 W/m°C.

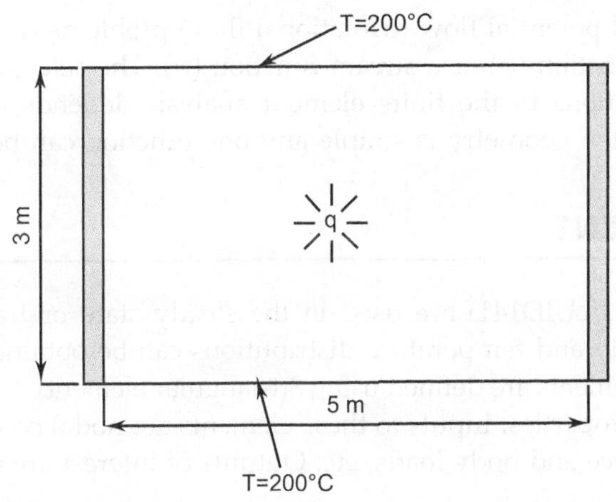

Figure 6.38

Chapter 7

FLUID FLOW ANALYSIS

7.1 POTENTIAL DISTRIBUTION IN THE 2-D BODIES

This chapter covers the finite element solution of ideal or potential flow (inviscid, incompressible flow) problems. Typical examples of potential flow are flow over a cylinder, flow around an airfoil and flow out of an orifice.

The two-dimensional potential flow (irrotational flow) problems can be formulated in terms of a velocity potential function (φ) or a stream function (ψ). The selection between velocity and stream function formulations in the finite element analysis depends on the ease of applying boundary conditions. If the geometry is simple any one function can be used.

7.2 FLUID ELEMENT

Fluid elements (*e.g.*, FLUID141) are used in the steady state or transient analysis of fluid systems. Pressure, velocity and temperature distributions can be obtained using these elements. Two dimensional fluid elements are defined using 3(triangular element) or 4 (quadrilateral element) nodes add by isotropic properties. Inputs to these elements are nodal co-ordinates, real constants, material properties, surface and body loads, etc. Outputs of interest are nodal values of pressure and velocity.

7.3 PROCEDURE OF FINITE ELEMENT ANALYSIS (RELATED TO FLUID FLOW PROBLEMS)

Step 1. Select element type—the basic 3 node triangular element can be used.

Step 2. Choose a potential function.

Step 3. Define the gradient/potential and velocity/gradient relationships.

Step 4. Derive the element stiffness matrix and equations.

Step 5. Assemble the element equations to obtain the global equations and introduce boundary conditions.

Step 6. Solve for the nodal potentials.

Step 7. Solve for the element velocities and volumetric rates.

The finite element solution using software of potential flow problems is illustrated below. Only potential function formulation is considered. Two cases are considered in this chapter.

Case 1. Potential flow over a cylinder

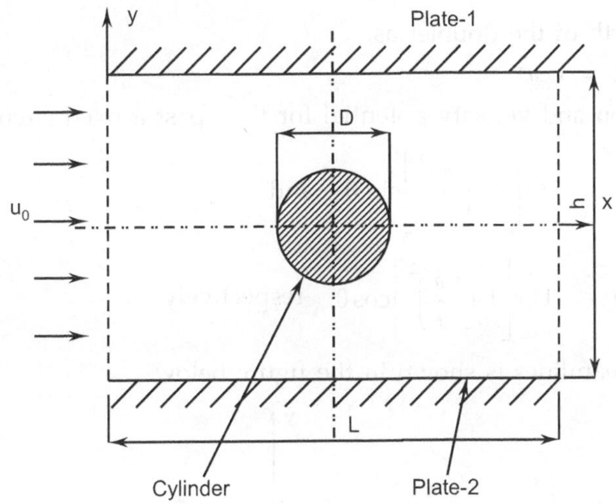

Figure 7.1. Potential flow over a cylinder

Above figure depicts the steady-state irrotational flow of an ideal fluid over a cylinder, confined between two parallel plates. We assume that, at the inlet, velocity is uniform, say u_0. Here we have to determine the flow velocities near the cylinder.

Flow past a fixed circular cylinder can be obtained by combining uniform flow with a doublet.

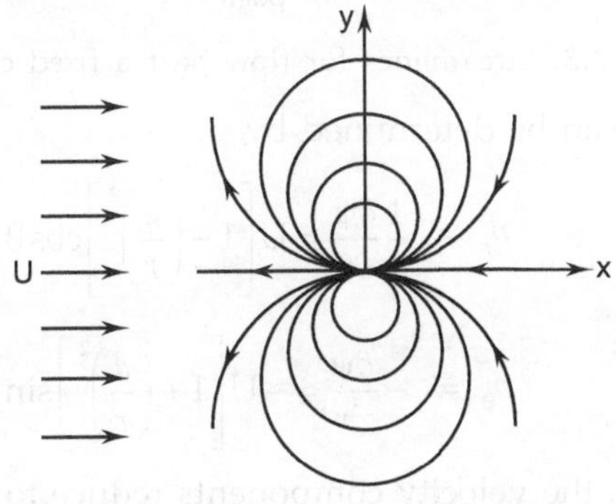

Figure 7.2. Superposition of a uniform flow and a doublet

The superimposed stream function and velocity potential are given by,

$$\psi = \psi_{\text{uniform flow}} + \psi_{\text{doublet}} = U.r.\sin\theta - K.\frac{\sin\theta}{r} \quad \text{and}$$

$$\phi = \phi_{\text{uniform flow}} + \phi_{\text{doublet}} = U.r.\cos\theta - K.\frac{\cos\theta}{r} \quad \text{respectively.}$$

Since the streamline that passes through the stagnation point has a value of zero, the stream function on the surface of the cylinder of radius 'a' is then given by,

$$\psi = U.a.\sin\theta - K.\frac{\sin\theta}{a} = 0$$

which gives the strength of the doublet as,

$$K = U.a^2$$

The stream function and velocity potential for flow past a fixed circular cylinder become

$$\psi = U.r\left[1-\left(\frac{a}{r}\right)^2\right]\sin\theta \quad \text{and}$$

$$\phi = U.r\left[1+\left(\frac{a}{r}\right)^2\right]\cos\theta \quad \text{respectively.}$$

The plot of the streamlines is shown in the figure below.

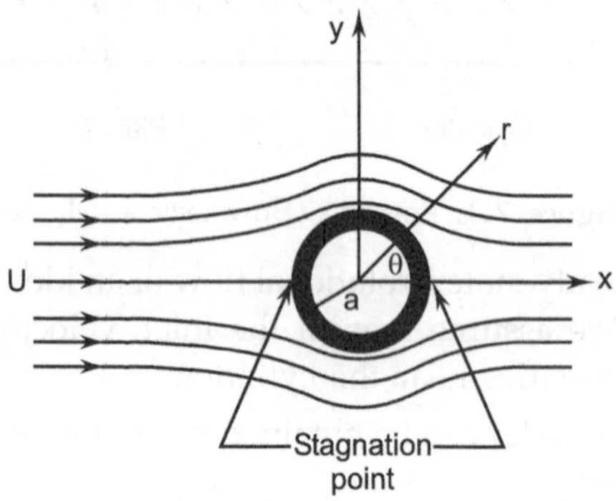

Figure 7.3. Streamlines for flow past a fixed cylinder

The velocity components can be determined by,

$$v_r = \frac{1}{r}\frac{\partial\psi}{\partial\theta} = U\left[1-\left(\frac{a}{r}\right)^2\right]\cos\theta$$

$$v_\theta = -\frac{\partial\psi}{\partial r} = -U\left[1+\left(\frac{a}{r}\right)^2\right]\sin\theta$$

Along the cylinder ($r = a$), the velocity components reduce to

$$v_r = 0 \quad \text{and} \quad v_\theta = -2U\sin\theta$$

The radial velocity component is always zero along the cylinder while the tangential velocity component varies from 0 at the stagnation point ($\theta = \pi$) to a maximum velocity of 2U at the top and bottom of the cylinder ($\theta = \pi/2$ or $-\pi/2$).

Case 2. Potential flow around an airfoil

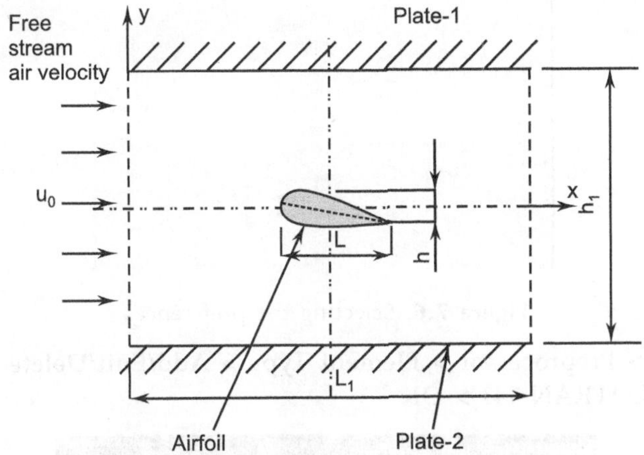

Figure 7.4. Potential flow around an airfoil

PROBLEM 7.1. *Flow over a circular cylinder between two parallel plates is as shown in figure 7.5. Assume unit thickness. Find the velocity distribution for the flow over a circular cylinder. Consider the flow of a liquid over a circular cylinder. Take liquid as water. Water density = 1000 kg/m^3 and viscosity = 0.001 N-s/m^2.*

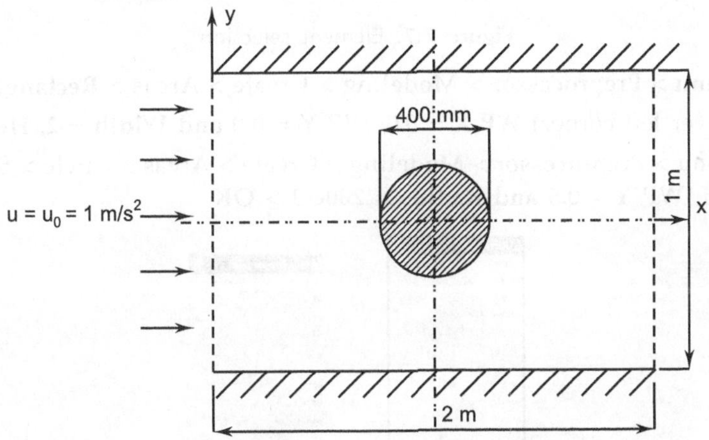

Figure 7.5. Flow over a cylinder

SOLUTION. (I) Software results

Procedure for solving the problem using ANSYS®12.0 academic teaching software

PREPROCESSING

1. **Main Menu >Preferences, then select FLOTRAN CFD > OK**

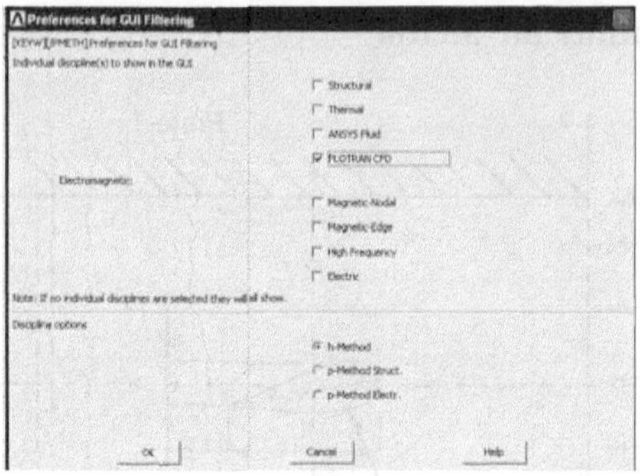

Figure 7.6. Selecting the preferences

2. **Main Menu > Preprocessor > Element Type > Add/Edit/Delete > Add > FLOTRAN CFD > 2D FLOTRAN 141 > OK**

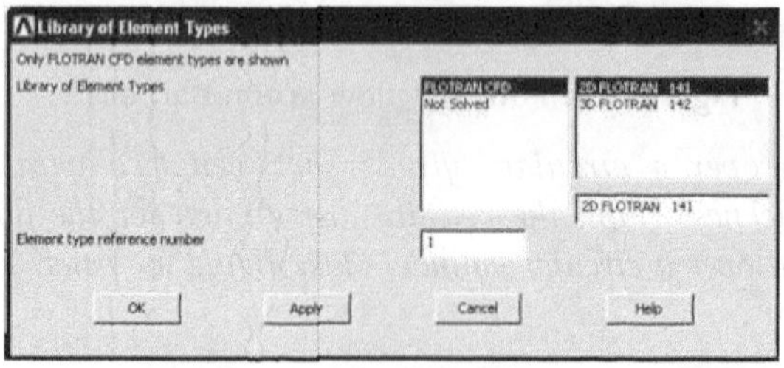

Figure 7.7. Element selection

3. **Main Menu > Preprocessor > Modeling > Create > Areas > Rectangle > By 2 Corners** Enter (lower left corner) **WP X = 0.0, WP Y = 0.0 and Width = 2, Height = 1 > OK**
4. **Main Menu > Preprocessor > Modeling > Create > Areas > Circle > Solid Circle.** Enter **WP X = 1, WP Y = 0.5 and Radius = 200e-3 > OK**

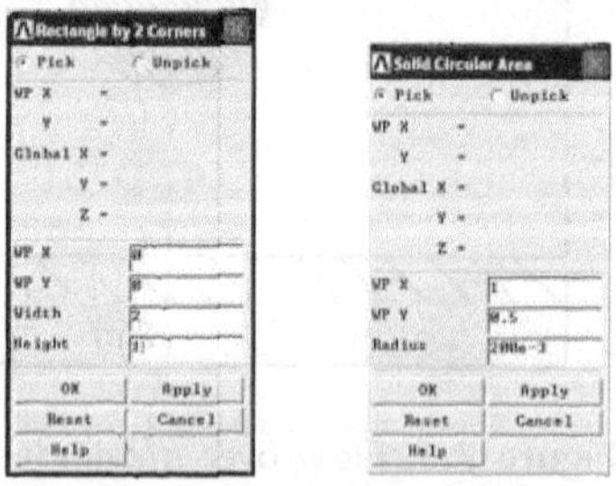

Figure 7.8. Create areas

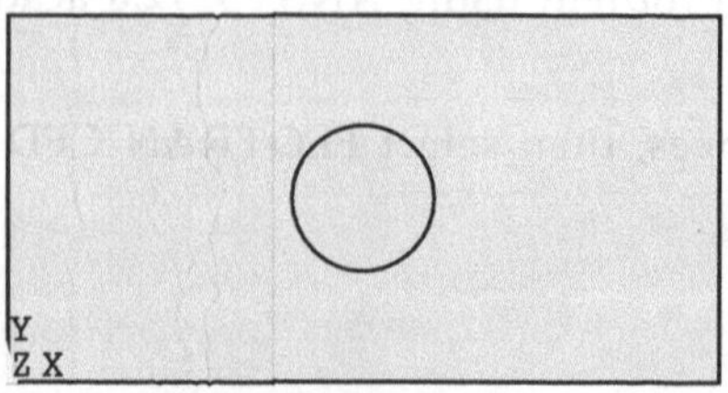

Figure 7.9. Rectangle and circle.

Now **subtract** the circle from the rectangle. (Read the messages in the window at the bottom of the screen as necessary)

5. **Main Menu > Preprocessor > Modeling > Operate > Booleans > Subtract > Areas >** Pick the rectangle **> OK,** then pick the circle **> OK**

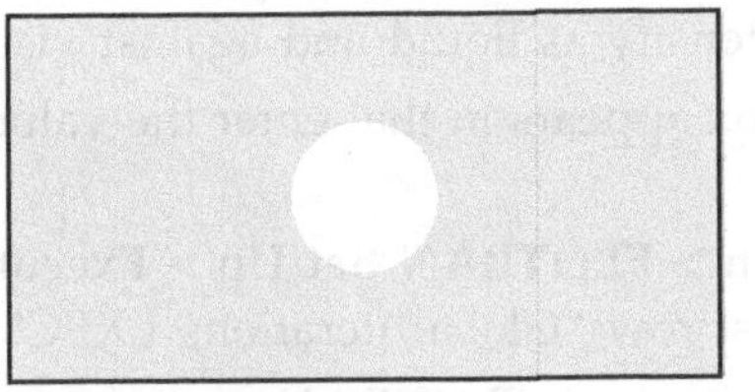

Figure 7.10. Geometry for the flow over a cylinder

Create a mesh of quadrilateral elements over the area.

6. **Main Menu > Preprocessor > Meshing > Mesh Tool**

   The **Mesh Tool** dialog box appears. Close the **Mesh Tool** box.

7. **Main Menu > Preprocessor > Meshing > Mesh > Areas > Free** Pick the area **> OK**

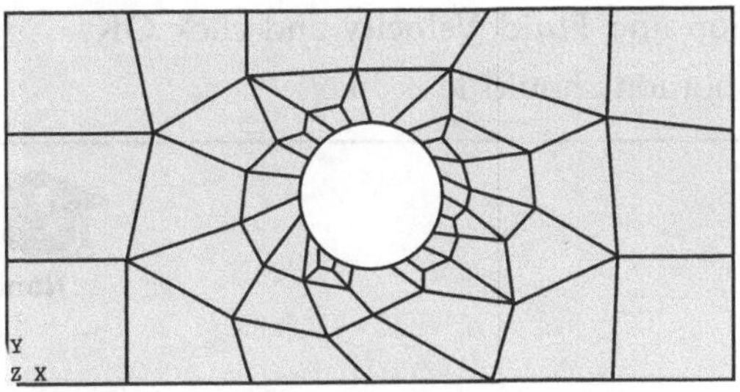

Figure 7.11. Quadrilateral element mesh

Apply the velocity boundary conditions and pressure.

8. **Main Menu > Preprocessor > Loads > Define Loads > Apply > Fluid/CFD > Velocity > On Lines** Pick the left edge of the plate **> OK > Enter VX = 1 > OK**

   (VX = 1 means an initial velocity of 1 m/s^2)

9. **Main Menu > Preprocessor > Loads > Define Loads > Apply > Fluid/CFD > Velocity > On Lines** Pick the edges around the cylinder **> OK > Enter VX = 0 and VY = 0 > OK**

10. **Main Menu > Preprocessor > Loads > Define Loads > Apply > Fluid/CFD > Pressure DOF > On Lines** Pick the top, bottom and right edges of the plate **> OK > OK**

    Once all the boundary conditions are applied the cylinder with plate look like as below.

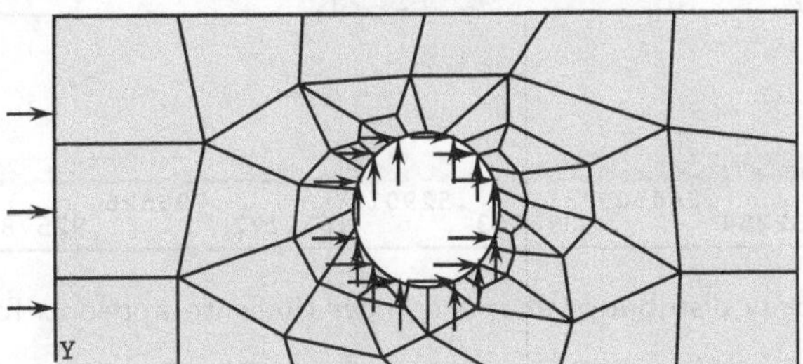

Figure 7.12. Model with boundary conditions.

The model-building step is now complete, and we can proceed to the solution. First to be safe, save the model.

SOLUTION. The interactive solution proceeds.

11. **Main Menu > Solution > FLOTRAN Set Up > Fluid Properties >** A dialog box appears in that select against density as liquid and against viscosity as liquid. > **OK**

    Then another dialog box appears in that enter the value of density = 1000 and viscosity value = 0.001 > **OK**

12. **Main Menu > Solution > FLOTRAN Set Up > Execution Ctrl >** a dialog box appears in that Enter in the first row "Global iterations EXEC" = 200

13. **Main Menu > Solution > Run FLOTRAN**

    When the solution is complete, **close** the **'Solution is Done!'** window.

POSTPROCESSING

We can now plot the results of this analysis and also list the computed values.

14. **Main Menu > General Postproc > Read Results > Last Set**

15. **General Postproc > Plot Results > Contour Plot > Nodal Solu**

    Select **DOF Solution** and **Fluid Velocity** and click **OK**

    This is what the solution should look like:

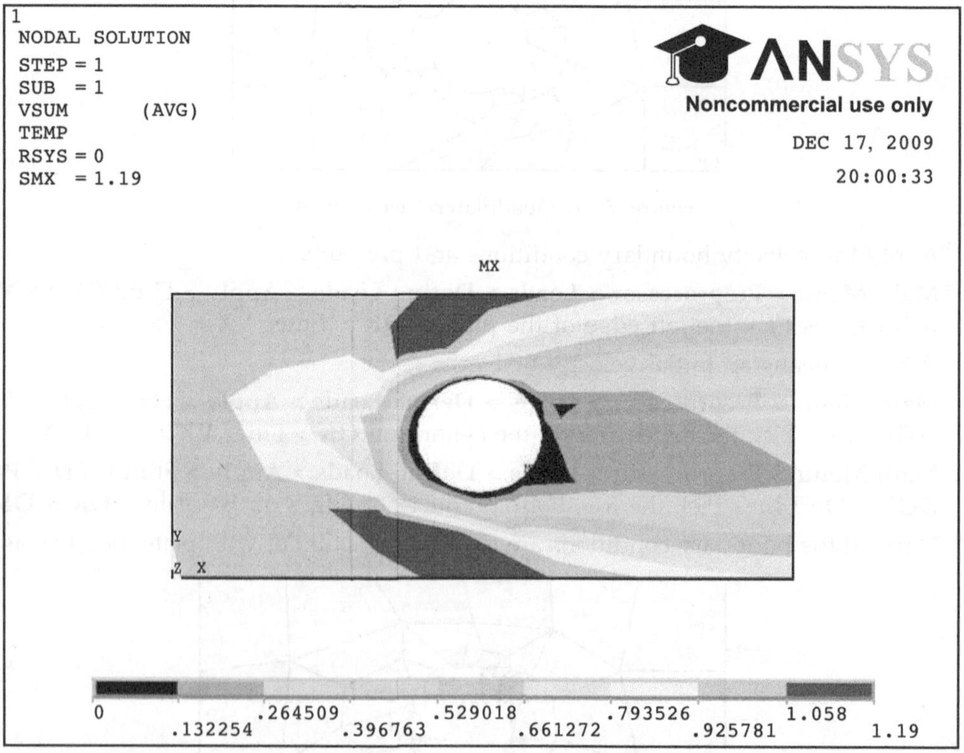

Figure 7.13. Velocity distribution over a cylinder (Refer to appendix for colour figures)

16. Next, go to **Main Menu > General Postproc > Plot Results > Vector Plot > Predefined.** One window will appear then click **OK**

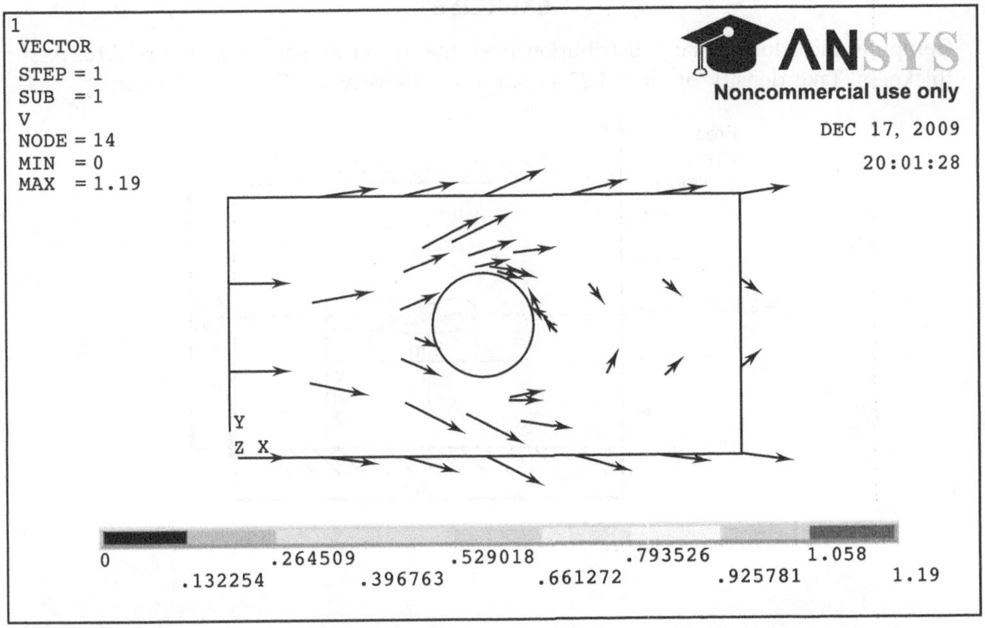

Figure 7.14. Vector plot of the fluid velocity (Refer to appendix for colour figures)

17. **General Postproc > Plot Results > Contour Plot > Nodal Solu**
    Select **DOF Solution** and **Pressure** and Click **OK**

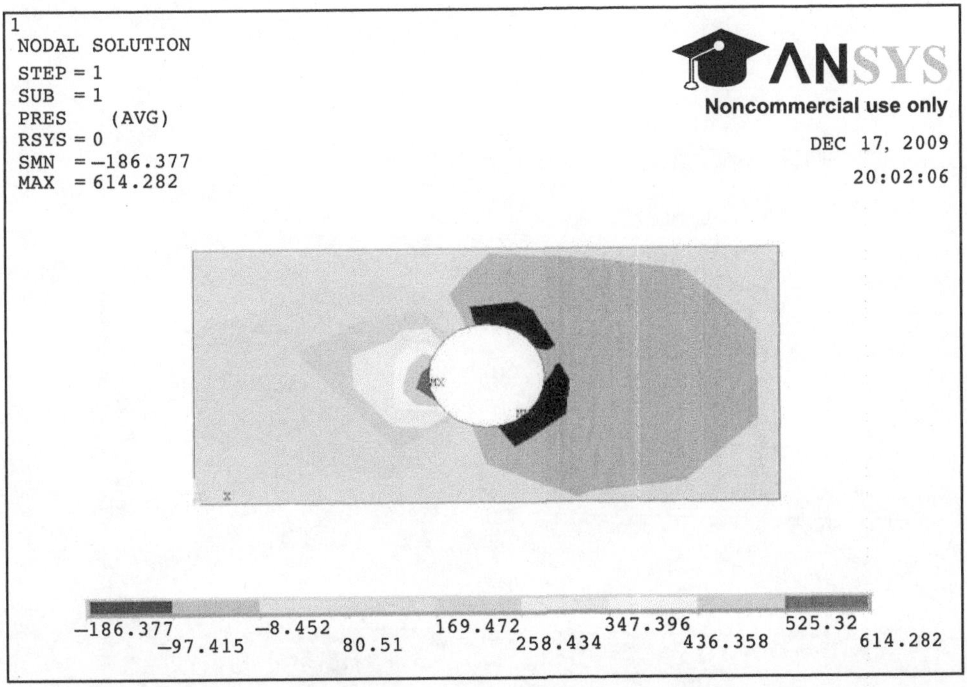

Figure 7.15. Pressure distribution over a cylinder (Refer to appendix for colour figures)

EXERCISE

1. Compute and plot velocity distribution over the airfoil as shown in figure 7.16. Assume unit thickness. Take density of air = 1.23 kg/m^3 and viscosity = 1.79 × 10^{-5} N-s/m^2.

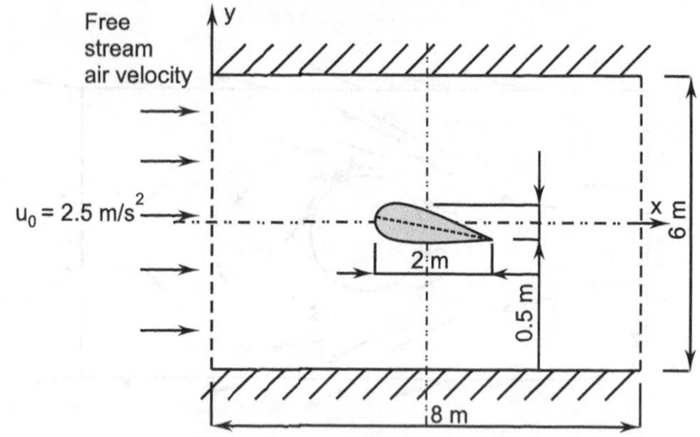

Figure 7.16. Flow over an airfoil

Chapter 8

DYNAMIC ANALYSIS

8.1 INTRODUCTION

Structural dynamics encompasses modal analysis, harmonic response analysis and transient response analysis. Modal analysis consists of finding natural frequencies and corresponding modal shapes of structures. Finding amplitude of vibration when the loads vary sinusoidal with time is known as harmonic response analysis. Finding the structural response to arbitrary time dependent loading is referred to as transient response analysis.

In this chapter one-dimensional problems relating to these topics are dealt with. In vibration analysis mass matrix and damping matrix will also come into picture in addition to stiffness matrix.

Governing equation of undamped free vibration assumes the form,

$$([k] - \omega^2[m])\{q\} = 0$$

For non trivial solution of this equation is

$$([k] - \omega^2[m]) = 0$$

Solution of this equation gives natural frequencies (ω). Substituting value of w back into the governing equation gives modal shapes defined by $\{q\}$.

Mass matrices for bar element and beam elements are given by

$$[m]_{\text{bar}} = \rho AL\begin{bmatrix} \frac{1}{3} & \frac{1}{6} \\ \frac{1}{6} & \frac{1}{3} \end{bmatrix}$$

$$[m]_{\text{beam}} = \frac{\rho AL}{420}\begin{bmatrix} 156 & 22L & 54 & -13L \\ 22L & 4L^2 & 13L & -3L^2 \\ 54 & 13L & 156 & -22L \\ -13L & -3L^2 & -22L & 4L^2 \end{bmatrix}$$

8.2 PROCEDURE OF FINITE ELEMENT ANALYSIS (RELATED TO DYNAMIC PROBLEMS)

STEP 1. Select element type.

STEP 2. Select a displacement function.

STEP 3. Define the strain/displacement and stress/strain relationships.

STEP 4. Derive the element stiffness and mass matrices and equations.

STEP 5. Assemble the element equations to obtain the global equations and introduce boundary conditions.

STEP 6. Solve for the natural frequencies and mode shapes.

8.3 FIXED-FIXED BEAM FOR NATURAL FREQUENCY DETERMINATION

PROBLEM 8.1. *Determine the first two natural frequencies for the fixed-fixed beam shown in figure 8.1. The beam is made of steel with modulus of elasticity E = 209 GPa, Poisson's ratio = 0.3, length L = 0.75 m, cross-section area A = 625 mm², mass density ρ = 7800 kg/m³, moment of inertia I = 34700 mm⁴.*

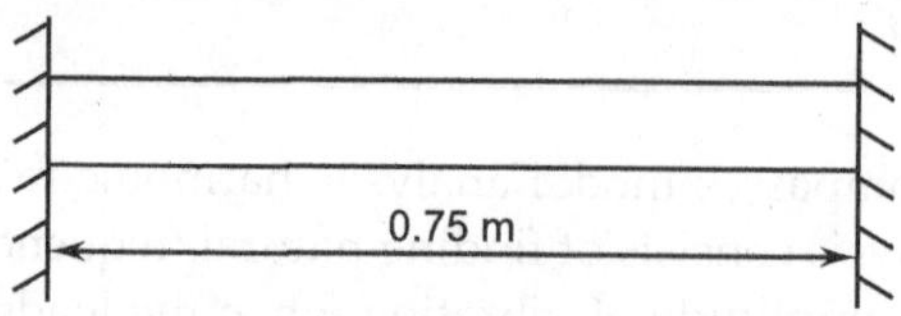

Figure 8.1

SOLUTION. (I) Analytical method

$$\therefore \quad \omega_1 = \frac{22.4}{L^2}\sqrt{\frac{EI}{\rho A}} = \frac{22.4}{(0.75)^2}\sqrt{\frac{209\times10^9\times34700\times10^{-12}}{7800\times625\times10^{-6}}} = 1535.95 \text{ rad/s}$$

$$\text{Frequency} \quad f_1 = \frac{\omega_1}{2\pi} = \frac{1535.95}{2\pi} = 244.45 \text{ Hz}$$

$$\therefore \quad \omega_2 = \frac{61.7}{L^2}\sqrt{\frac{EI}{\rho A}} = \frac{61.7}{(0.75)^2}\sqrt{\frac{209\times10^9\times34700\times10^{-12}}{7800\times625\times10^{-6}}} = 4230.71 \text{ rad/s}$$

$$\text{Frequency} \quad f_2 = \frac{\omega_2}{2\pi} = \frac{4230.71}{2\pi} = 673.34 \text{ Hz}$$

(II) FEM by hand calculations

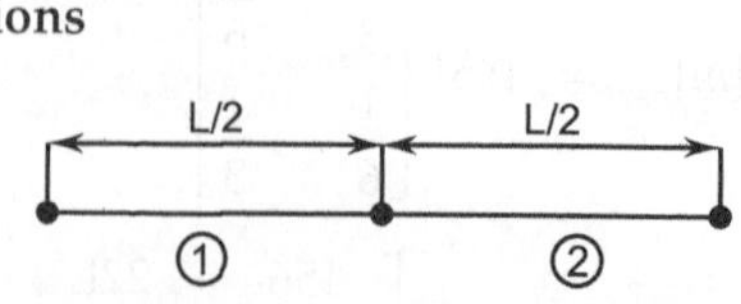

Figure 8.1(a). Finite element model

Mass matrices,

$$[M_1] = [M_2] = \frac{\rho \times A \times \frac{L}{2}}{420}\begin{bmatrix} 156 & 22\left(\frac{L}{2}\right) & 54 & -13\left(\frac{L}{2}\right) \\ 22\left(\frac{L}{2}\right) & 4\left(\frac{L}{2}\right)^2 & 13\left(\frac{L}{2}\right) & -3\left(\frac{L}{2}\right)^2 \\ 54 & 13\left(\frac{L}{2}\right) & 156 & -22\left(\frac{L}{2}\right) \\ 13\left(\frac{L}{2}\right) & -3\left(\frac{L}{2}\right)^2 & 22\left(\frac{L}{2}\right) & 4\left(\frac{L}{2}\right)^2 \end{bmatrix}$$

$$= \frac{\rho AL}{840}\begin{bmatrix} 156 & 11L & 54 & -6.5L \\ 11L & L^2 & 6.5L & -3\left(\frac{L^2}{4}\right) \\ 54 & 13\left(\frac{L}{2}\right) & 156 & -11L \\ -6.5L & -3\left(\frac{L^2}{4}\right) & -11L & L^2 \end{bmatrix}$$

$$[k_1] = [k_2] = \frac{EI}{\left(\frac{L}{2}\right)^3}\begin{bmatrix} 12 & 6\left(\frac{L}{2}\right) & -12 & \left(\frac{L}{2}\right) \\ & 4\left(\frac{L}{2}\right)^2 & -6\left(\frac{L}{2}\right) & 2\left(\frac{L}{2}\right)^2 \\ & & 12 & -6\left(\frac{L}{2}\right) \\ \text{Symmetric} & & & 4\left(\frac{L}{2}\right)^2 \end{bmatrix} = \frac{8EI}{L^3}\begin{bmatrix} 12 & 3L & -12 & 3L \\ & L^2 & -3L & \frac{L^2}{2} \\ & & 12 & -3L \\ \text{Symmetric} & & & L^2 \end{bmatrix}$$

Global $[M] = \frac{\rho AL}{840}\begin{bmatrix} 156+156 & -11L+11L \\ -11L+11L & L^2+L^2 \end{bmatrix} = \frac{\rho AL}{840}\begin{bmatrix} 312 & 0 \\ 0 & 2L^2 \end{bmatrix}$

Global $[K] = \frac{8EI}{L^3}\begin{bmatrix} 12+12 & -3L+3L \\ -3L+3L & L^2+L^2 \end{bmatrix} = \frac{8EI}{L^3}\begin{bmatrix} 24 & 0 \\ 0 & 2L^2 \end{bmatrix}$

Governing Equation is $([K] - \omega^2[M])\{q\} = 0$

$$\left(\frac{8EI}{L^3}\begin{bmatrix} 24 & 0 \\ 0 & 2L^2 \end{bmatrix} - \omega^2\frac{\rho AL}{840}\begin{bmatrix} 312 & 0 \\ 0 & 2L^2 \end{bmatrix}\right)\{q\} = 0$$

$$\left(\begin{bmatrix} 24 & 0 \\ 0 & 2L^2 \end{bmatrix} - \frac{\omega^2 \rho AL L^3}{840 \times 8EI}\left(\begin{bmatrix} 312 & 0 \\ 0 & 2L^2 \end{bmatrix}\right)\right)\{q\} = 0$$

$$\left(\begin{bmatrix} 24 - 312a & 0 \\ 0 & 2L^2 - 2L^2 a \end{bmatrix}\right)\{q\} = 0, \quad \text{where } a = \frac{\omega^2 \rho AL^4}{6720EI}$$

For non-trivial solution,

$$\begin{bmatrix} 24-312a & 0 \\ 0 & 2L^2-2L^2a \end{bmatrix} = 0$$

Solving $$a = 1 \quad \text{or} \quad a = 0.076923$$

$$\frac{\omega^2\rho AL^4}{6720EI} = 1 \quad \text{or} \quad \frac{\omega^2\rho AL^4}{6720EI} = 0.076923$$

$$\therefore \quad \omega_2 = \frac{81.98}{L^2}\sqrt{\frac{EI}{\rho A}} \qquad \therefore \quad \omega_1 = \frac{22.74}{L^2}\sqrt{\frac{EI}{\rho A}}$$

Given:

$$E = 209 \text{ GPa},$$
$$A = 625 \times 10^{-6} \text{ m}^2$$
$$I = 34700 \times 10^{-12} \text{ m}^4$$
$$\rho = 7800 \text{ kg/m}^3$$
$$L = 0.75 \text{ m}$$

$$\therefore \quad \omega_1 = \frac{22.74}{(0.75)^2}\sqrt{\frac{209\times10^9\times34700\times10^{-12}}{7800\times625\times10^{-6}}} = 1559.26 \text{ rad/s}$$

Frequency $$f_1 = \frac{\omega_1}{2\pi} = \frac{1559.26}{2\pi} = 248.164 \text{ Hz}$$

$$\therefore \quad \omega_2 = \frac{81.98}{(0.75)^2}\sqrt{\frac{209\times10^9\times34700\times10^{-12}}{7800\times625\times10^{-6}}} = 5621.29 \text{ rad/s}$$

Frequency $$f_2 = \frac{\omega_2}{2\pi} = \frac{5621.29}{2\pi} = 894.66 \text{ Hz}$$

(III) Software results

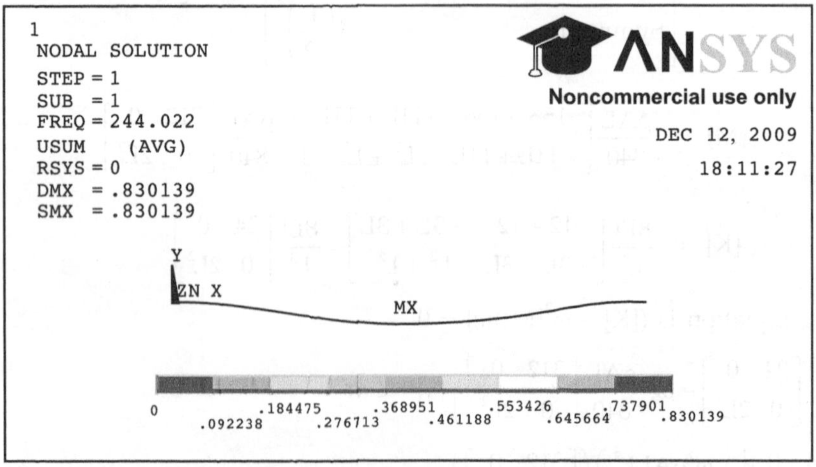

Figure 8.1(b). Deflection pattern for a fixed-fixed beam for mode 1 (Refer to appendix for colour figures)

Frequency values (in Hz)

SET	TIME/FREQ	LOAD STEP	SUBSTEP	CUMULATIVE
1	244.02	1	1	1
2	671.69	1	2	2

Following are the mode shapes,

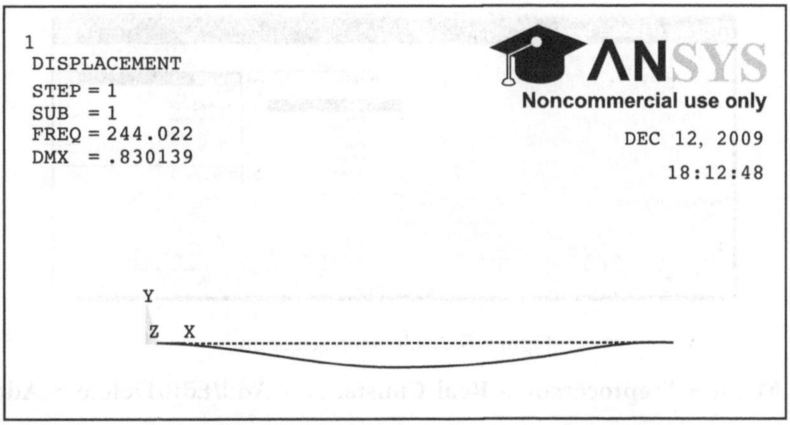

Figure 8.1(c). Mode 1 for fixed-fixed beam (Refer to appendix for colour figures)

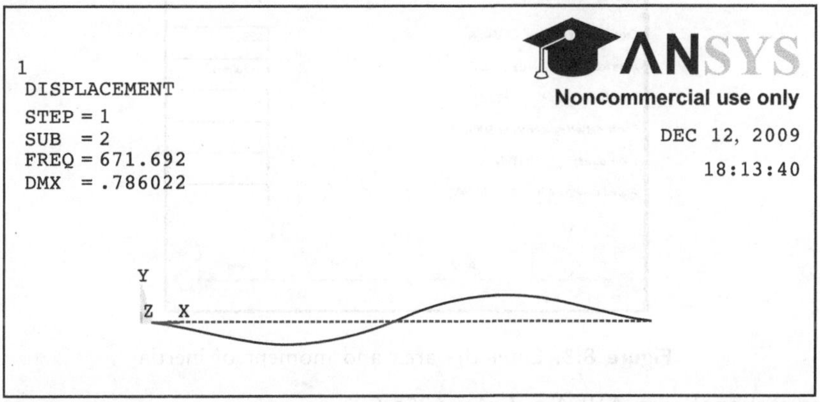

Figure 8.1(d). Mode 2 for fixed-fixed beam (Refer to appendix for colour figures)

Answers of Problem 8.1

Parameter	Analytical method	FEM–hand calculation (with 2 elements)	Software results (with 10 elements)
Natural frequency			
f_1	244.45 Hz	248.16 Hz	244.02 Hz
f_2	673.34 Hz	894.66 Hz	671.69 Hz

Procedure for solving the problem using ANSYS® 12.0 academic teaching software

For problem 8.1

PREPROCESSING

1. **Main Menu > Preprocessor > Element Type > Add/Edit/Delete > Add > Beam > 2D elastic 3 > OK > Close**

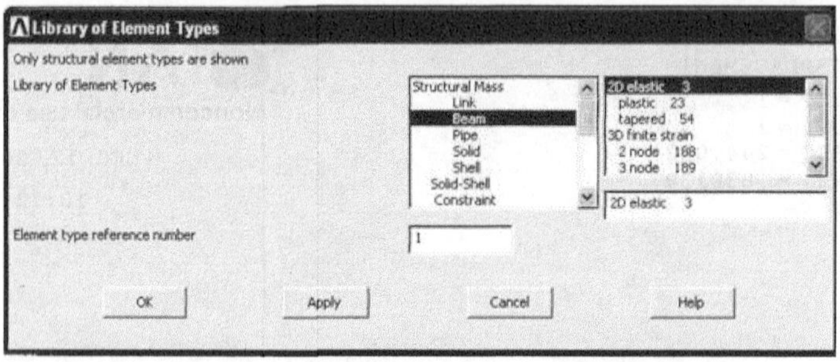

Figure 8.2. Element selection

2. **Main Menu > Preprocessor > Real Constants > Add/Edit/Delete > Add > OK**

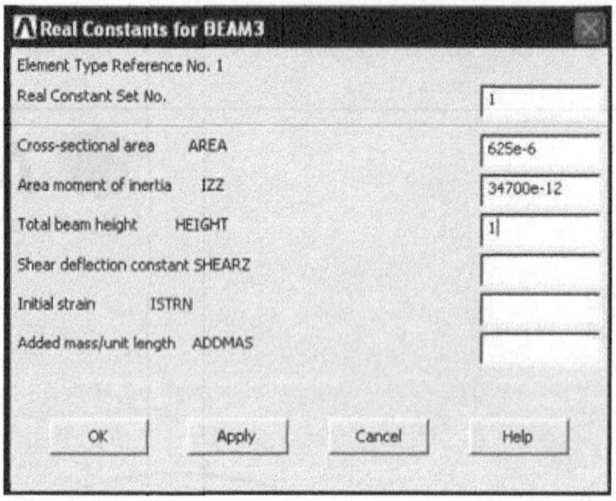

Figure 8.3. Enter the area and moment of inertia

Cross-sectional area AREA > **Enter 625e-6**

Area moment of inertia IZZ > **Enter 34700e-12**

Total beam height HEIGHT > **Enter 1 > OK > Close**

Enter the material properties.

3. **Main Menu > Preprocessor > Material Props > Material Models**

   Material Model Number 1, click **Structural > Linear > Elastic > Isotropic**

   Enter **EX = 209E9 and PRXY = 0.3** > **OK**

   click **Structural > Linear > Density**

   Enter DENS = 7800 > **OK**

   (**Close** the Define Material Model Behaviour window.)

   Create the keypoints and lines as shown in the figure.

4. **Main Menu > Preprocessor > Modeling > Create > Keypoints > In Active CS** Enter the coordinates of keypoint 1 > **Apply** Enter the coordinates of keypoint 2 > **OK.**

Keypoint locations		
Keypoint number	**X coordinate**	**Y coordinate**
1	0	0
2	0.75	0

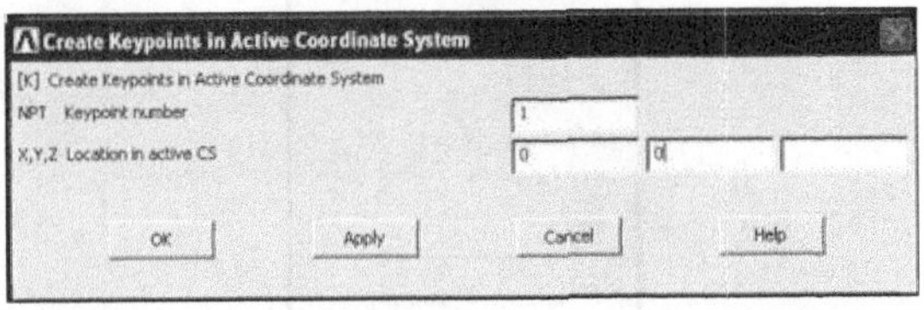

Figure 8.4. Enter the keypoint coordinates

5. **Main Menu > Preprocessor > Modeling > Create > Lines > Lines > Straight Line** Pick the 1st and 2nd keypoint > **OK**

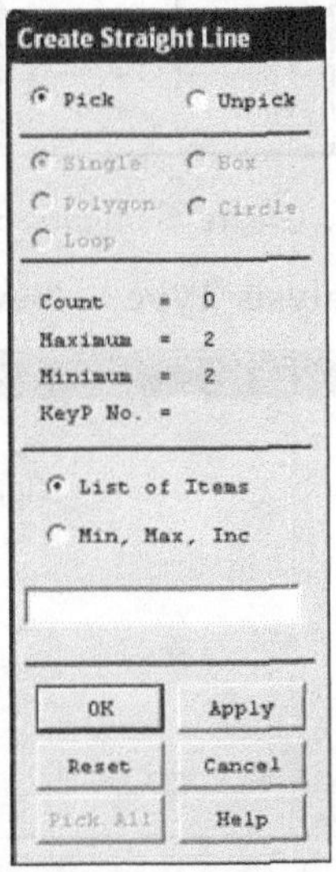

Figure 8.5. Pick the keypoints to create lines

6. **Main Menu > Preprocessor > Meshing > Size Cntrls > ManualSize > Lines > All Lines >** Enter **NDIV No. of element divisions = 10**

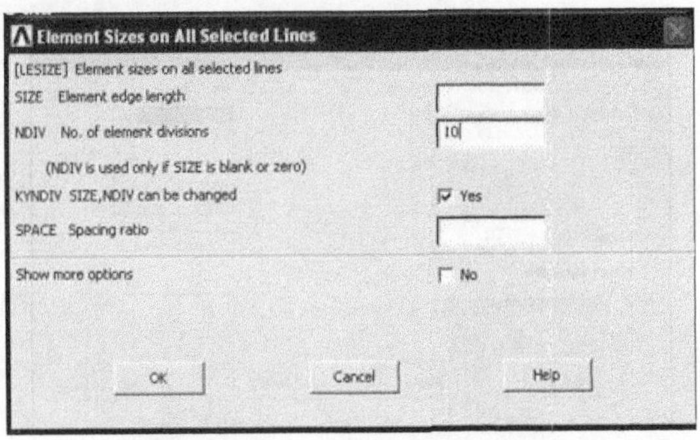

Figure 8.6. Specify element length

7. **Main Menu > Preprocessor > Meshing > Mesh > Lines> Click Pick All.**

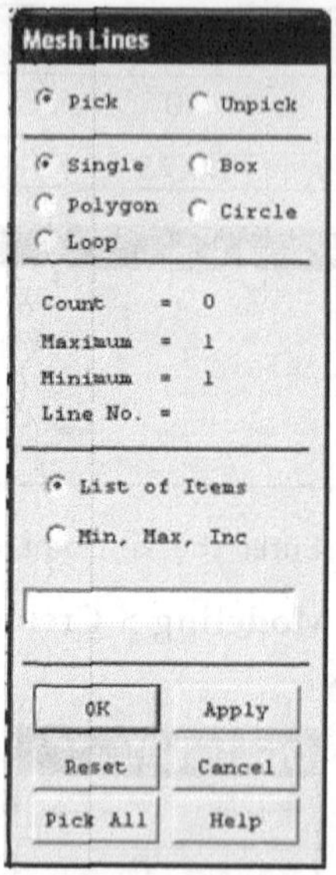

Figure 8.7. Create elements by meshing

8. **Main Menu > Solution > Analysis Type > New Analysis > Select Modal > OK**

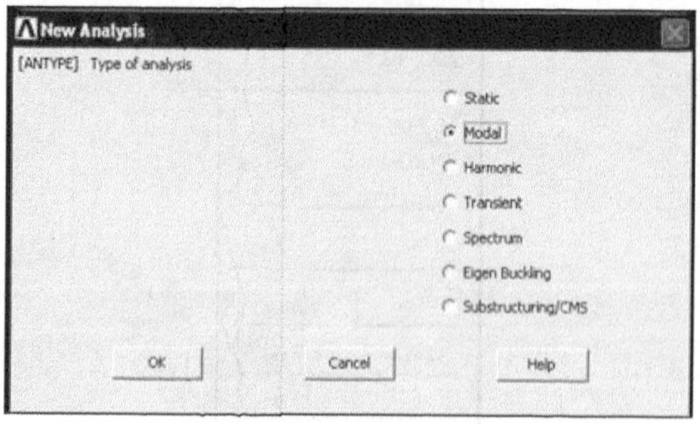

Figure 8.8. Define the type of analysis

9. **Main Menu > Preprocessor > Loads > Define Loads > Apply > Structural > Displacement > On Nodes** Pick the left most node and right most node **> Apply > Select All DOF > OK**

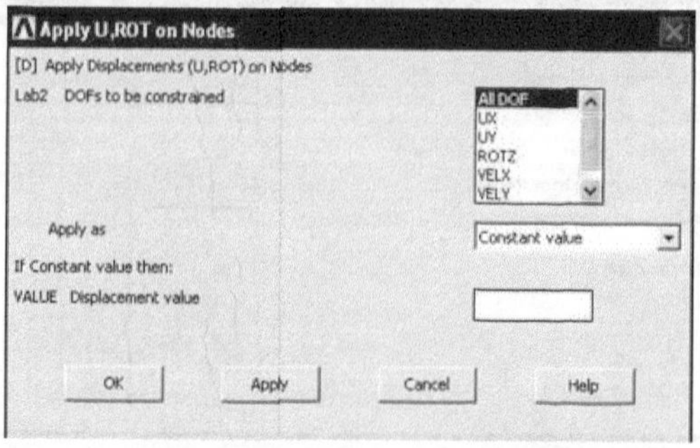

Figure 8.9. Apply the displacement constraint

10. **Main Menu > Solution> Analysis Type> Analysis Options > Select PCG Lanczos option**

    Enter No. of modes to extract = 2

    NMODE No. of modes to expand = 2 > OK

    After OK one more window will appear, for that also click OK.

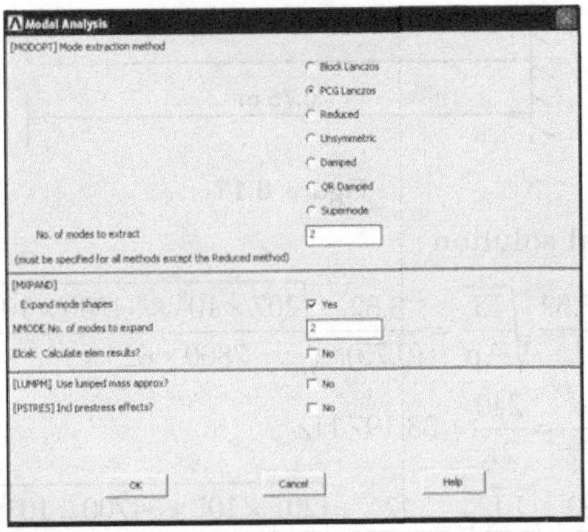

Figure 8.10. Select the number of modes to extract

SOLUTION. The interactive solution proceeds.

11. **Main Menu > Solution > Solve > Current LS > OK**

    The **/STATUS Command** window displays the problem parameters and **the SolveCurrent Load Step** window is shown. Check the solution options in the **/STATUS**window and if all is OK, select **File > Close**

    In the Solve **Current Load Step** window, Select **OK**, and when the solution is complete, **close** the **'Solution is Done!'** window.

POSTPROCESSING

12. **Main Menu > General Postproc > Results Summary**

    This result is shown as frequency values in Hz.

13. **Main Menu > General Postproc > Read Results > First Set**
14. **Main Menu > General Postproc > Plot Results > Deformed Shape > Click Def + undeformed > OK**

    This result is the first mode shown in figure 8.1(*c*).

15. **Main Menu > General Postproc > Read Results > Next Set**
16. **Main Menu > General Postproc > Plot Results > Deformed Shape > Click Def + undeformed > OK**

    This result is the second mode shown in figure 8.1(*d*).

8.4 TRANSVERSE VIBRATIONS OF A CANTILEVER BEAM

PROBLEM 8.2. *Determine the first four natural frequencies for the cantilever beam shown in figure 8.11. The beam is made of steel with modulus of elasticity, E = 207 GPa, Poisson's ratio = 0.3, length L = 0.75 m, cross-section area A = 625 mm^2, mass density ρ = 7800 kg/m^3, moment of inertia I = 34700 mm^4.*

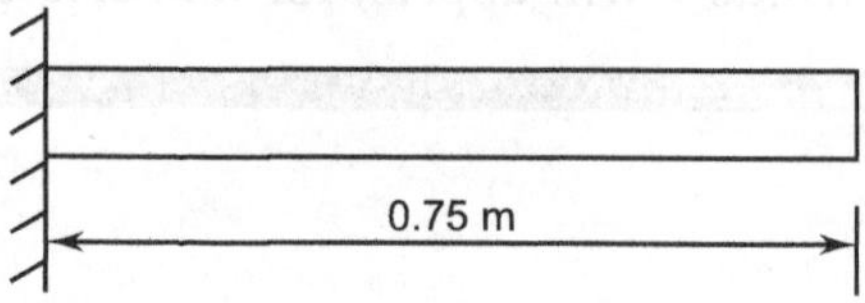

Figure 8.11

SOLUTION. (I) Analytical solution

$$\omega_1 = \frac{3.52}{L^2}\sqrt{\frac{EI}{A\rho}} = \frac{3.52}{(0.75)^2}\sqrt{\frac{207\times10^9\times34700\times10^{-12}}{7800\times625\times10^{-6}}} = 240 \text{ rad/s}$$

$$f_1 = \frac{\omega_1}{2\pi} = \frac{240}{2\pi} = 38.197 \text{ Hz}$$

$$\omega_2 = \frac{22}{L^2}\sqrt{\frac{EI}{A\rho}} = \frac{22}{(0.75)^2}\sqrt{\frac{207\times10^9\times34700\times10^{-12}}{7800\times625\times10^{-6}}} = 1501 \text{ rad/s}$$

$$f_2 = \frac{\omega_2}{2\pi} = \frac{1501}{2\pi} = 238.89 \text{ Hz}$$

$$\omega_3 = \frac{61.7}{L^2}\sqrt{\frac{EI}{A\rho}} = \frac{61.7}{(0.75)^2}\sqrt{\frac{207\times10^9\times34700\times10^{-12}}{7800\times625\times10^{-6}}} = 4210 \text{ rad/s}$$

$$f_3 = \frac{\omega_3}{2\pi} = \frac{4210}{2\pi} = 670.04 \text{ Hz}$$

$$\omega_4 = \frac{121}{L^2}\sqrt{\frac{EI}{A\rho}} = \frac{121}{(0.75)^2}\sqrt{\frac{207\times10^9\times34700\times10^{-12}}{7800\times625\times10^{-6}}} = 8257 \text{ rad/s}$$

$$f_4 = \frac{\omega_4}{2\pi} = \frac{8257}{2\pi} = 1314.14 \text{ Hz}$$

(II) FEM by hand calculation

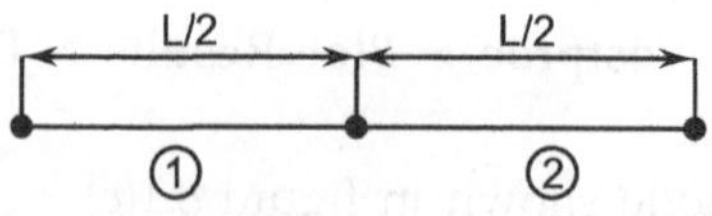

Figure 8.11(a). Finite element model

Stiffness matrices are,

$$\text{Stiffness matrix } [k_1] = [k_2] = \frac{EI}{\left(\frac{L}{2}\right)^3}\begin{bmatrix} 12 & 6\left(\frac{L}{2}\right) & -12 & 6\left(\frac{L}{2}\right) \\ 6\left(\frac{L}{2}\right) & 4\left(\frac{L}{2}\right)^2 & -6\left(\frac{L}{2}\right) & 2\left(\frac{L}{2}\right)^2 \\ -12 & -6\left(\frac{L}{2}\right) & 12 & -6\left(\frac{L}{2}\right) \\ 6\left(\frac{L}{2}\right) & 2\left(\frac{L}{2}\right)^2 & -6\left(\frac{L}{2}\right) & 4\left(\frac{L}{2}\right)^2 \end{bmatrix}$$

Global Stiffness matrix

$$[K]=\frac{8EI}{L^3}\begin{bmatrix}24 & 0 & -12 & 3L\\ 0 & 2L^2 & -3L & \frac{L^2}{2}\\ -12 & -3L & 12 & -3L\\ 3L & \frac{L^2}{2} & -3L & L^2\end{bmatrix} = 136209.07\begin{bmatrix}24 & 0 & -12 & 2.25\\ 0 & 1.125 & -2.25 & 0.28125\\ -12 & -2.25 & 12 & -2.25\\ 2.25 & 0.28125 & -2.25 & 0.5625\end{bmatrix}$$

Mass matrices

$$[M_1] = [M_2]=\frac{\rho A\left(\frac{L}{2}\right)}{420}\begin{bmatrix}156 & 22\left(\frac{L}{2}\right) & 54 & -13\left(\frac{L}{2}\right)\\ 22\left(\frac{L}{2}\right) & 4\left(\frac{L}{2}\right)^2 & 13\left(\frac{L}{2}\right) & -3\left(\frac{L}{2}\right)^2\\ 54 & 13\left(\frac{L}{2}\right) & 156 & -22\left(\frac{L}{2}\right)\\ -13\left(\frac{L}{2}\right) & -3\left(\frac{L}{2}\right)^2 & -22\left(\frac{L}{2}\right) & 4\left(\frac{L}{2}\right)^2\end{bmatrix}$$

Global mass matrix

$$[M] = \frac{\rho AL}{840}\begin{bmatrix}312 & 0 & 54 & -13\left(\frac{L}{2}\right)\\ 0 & 2L^2 & 13\left(\frac{L}{2}\right) & -3\left(\frac{L}{2}\right)^2\\ 54 & 13\left(\frac{L}{2}\right) & 156 & -22\left(\frac{L}{2}\right)\\ -13\left(\frac{L}{2}\right) & -3\left(\frac{L}{2}\right)^2 & -22\left(\frac{L}{2}\right) & 4\left(\frac{L}{2}\right)^2\end{bmatrix}$$

$$= 4.352676\times 10^{-3}\begin{bmatrix}312 & 0 & 54 & -4.875\\ 0 & 1.125 & 4.875 & -0.421875\\ 54 & 4.875 & 156 & -8.25\\ -4.875 & -0.421875 & -8.25 & 0.5625\end{bmatrix}$$

Governing Equation is $([K] - \omega^2 [M]) = \begin{Bmatrix}\omega_2\\ \theta_2\\ \omega_3\\ \theta_3\end{Bmatrix} = 0$

For non-trivial solution

Det $([K] - \omega^2 [M]) = 0$

Substituting and Solving, we get

$$\omega^2 = \begin{Bmatrix} \omega_1^2 \\ \omega_2^2 \\ \omega_3^2 \\ \omega_4^2 \end{Bmatrix} = 10^8 \begin{Bmatrix} 0.0006 \\ 0.0230 \\ 0.2630 \\ 2.2159 \end{Bmatrix}$$

$$\omega_1 = 245 \text{ rad/s} \Rightarrow f_1 = \frac{\omega_1}{2\pi} = \frac{245}{2\pi} = 38.993 \text{ Hz}$$

$$\omega_2 = 1517 \text{ rad/s} \Rightarrow f_2 = \frac{\omega_2}{2\pi} = \frac{1517}{2\pi} = 241.44 \text{ Hz}$$

$$\omega_3 = 5128 \text{ rad/s} \Rightarrow f_3 = \frac{\omega_3}{2\pi} = \frac{5128}{2\pi} = 816.147 \text{ Hz}$$

$$\omega_4 = 14885.9 \text{ rad/s} \Rightarrow f_4 = \frac{\omega_4}{2\pi} = \frac{14885.9}{2\pi} = 2369.16 \text{ Hz}$$

(III) Software results

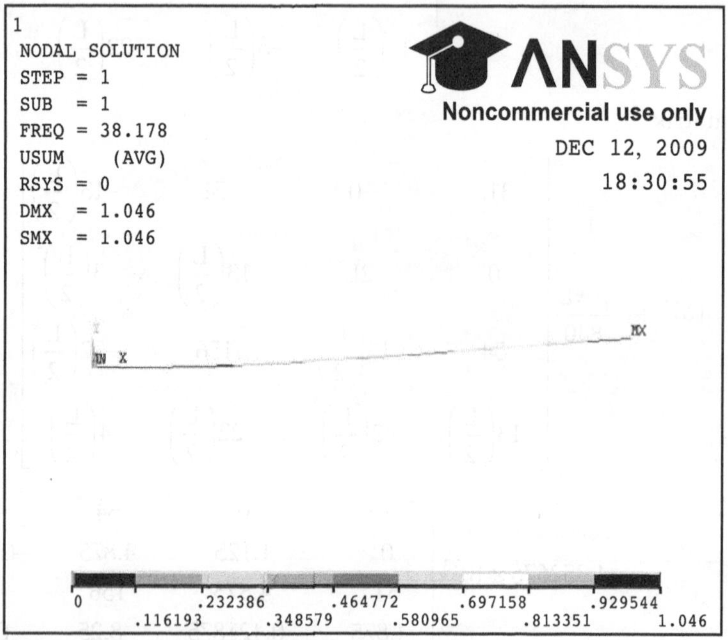

Figure 8.11(b). Deflection pattern for a fixed-fixed beam for mode 1 (Refer to appendix for colour figures)

Frequency values (in Hz)

SET	TIME/FREQ	LOAD STEP	SUBSTEP	CUMULATIVE
1	38.178	1	1	1
2	238.94	1	2	2
3	667.71	1	3	3
4	1305.2	1	4	4

Following are the mode shapes,

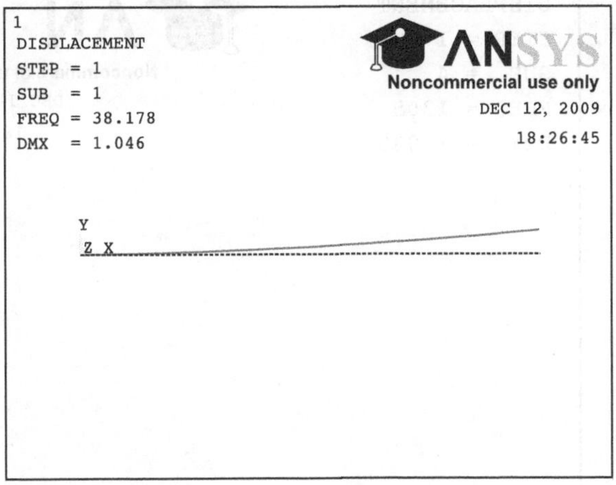

Figure 8.11(c). Mode 1 for cantilever beam (Refer to appendix for colour figures)

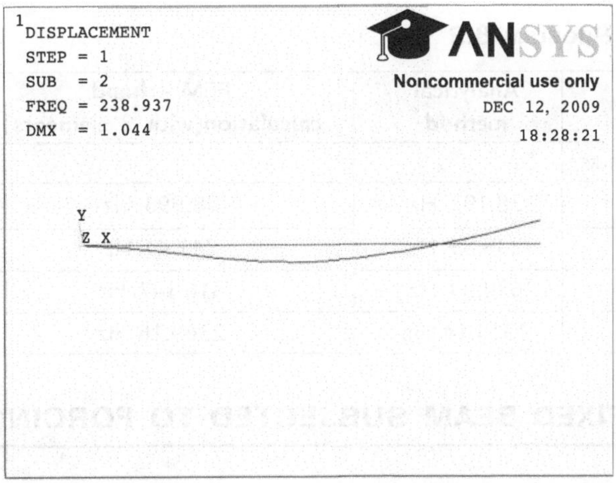

Figure 8.11(d). Mode 2 for cantilever beam (Refer to appendix for colour figures)

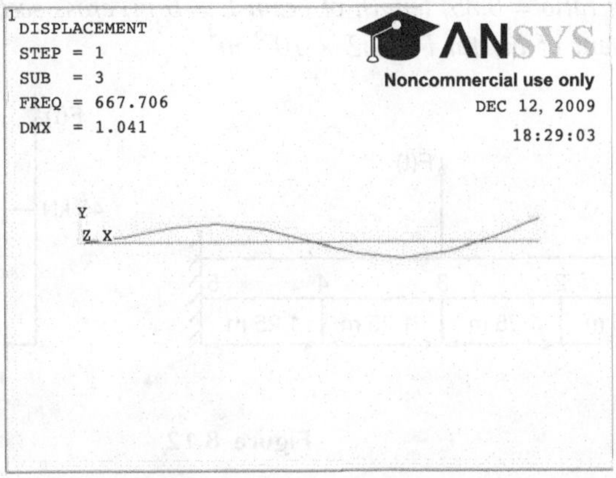

Figure 8.11(e). Mode 3 for cantilever beam (Refer to appendix for colour figures)

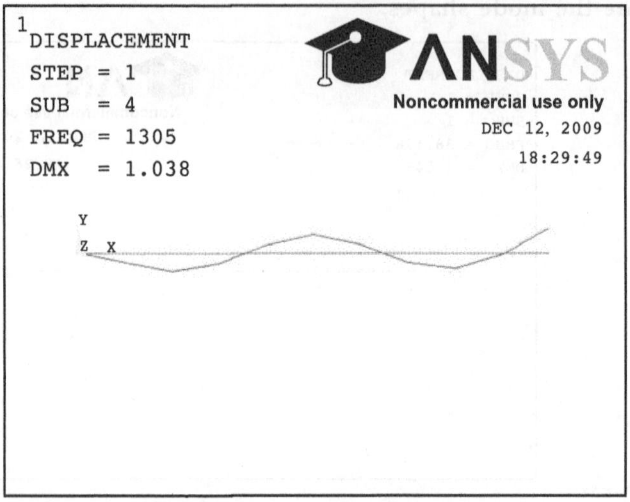

Figure 8.11(f). Mode 4 for cantilever beam (Refer to appendix for colour figures)

ANSWERS OF PROBLEM 8.2

Parameter	Analytical method	FEM – hand calculation(with 2 elements)	Software results (with 10 elements)
Natural frequency			
f_1	38.197 Hz	38.993 Hz	38.178 Hz
f_2	238.89 Hz	241.44 Hz	238.94 Hz
f_3	670.04 Hz	816.147 Hz	667.71 Hz
f_4	1314.14 Hz	2369.16 Hz	1305.2 Hz

8.5 FIXED-FIXED BEAM SUBJECTED TO FORCING FUNCTION

PROBLEM 8.3. *For the fixed-fixed beam subjected to the time dependent forcing function shown in figure 8.12. Determine the displacement response for 0.2 seconds. Use time step integration of 0.01 sec. Let E = 46 GPa, Poisson's ratio = 0.35, length of beam L = 5 m, cross-section area A= 1* m^2*, mass density,* ρ *= 1750* kg/m^3*, moment of inertia* $I = 4.2 \times 10^{-5}\ m^4$*.*

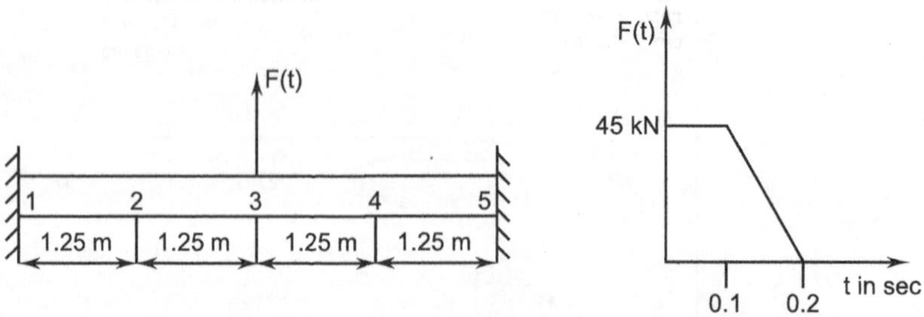

Figure 8.12

SOLUTION. (I) Software results

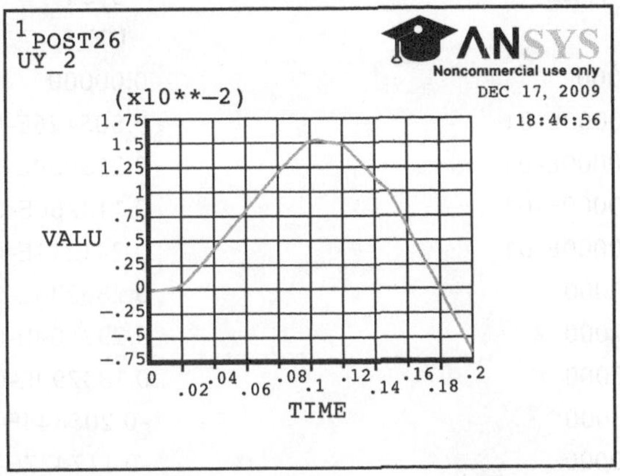

Figure 8.12(a). Displacement response for 0.2 sec for node 2 (Refer to appendix for colour figures)

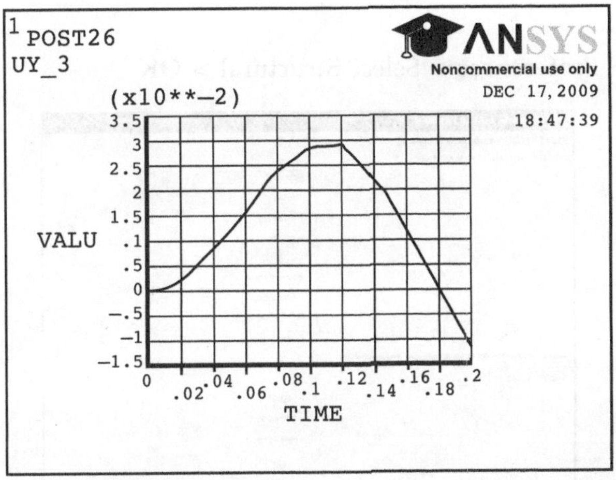

Figure 8.12(b). Displacement response for 0.2 sec for node 3 (Refer to appendix for colour figures)

Displacement values (in metres) for node 2

TIME	2 UY UY_2
0.0000	0.00000
0.10000E-01	0.421220E-05
0.20000E-01	0.284618E-03
0.50000E-01	0.602161E-02
0.80000E-01	0.121677E-01
0.10000	0.153042E-01
0.12000	0.148820E-01
0.15000	0.979873E-02
0.18000	0.868368E-04
0.20000	-0.649350E-02

Displacement values (in metres) for node 3

TIME	3 UY UY_3
0.0000	0.00000
0.10000E-01	0.505126E-03
0.20000E-01	0.218959E-02
0.50000E-01	0.113766E-01
0.80000E-01	0.241211E-01
0.10000	0.286233E-01
0.12000	0.292504E-01
0.15000	0.183799E-01
0.18000	-0.205644E-03
0.20000	-0.117477E-01

Procedure for solving the problem using ANSYS®12.0 academic teaching software

FOR PROBLEM 8.3

PREPROCESSING

1. **Main Menu > Preferences >** Select **Structural > OK**

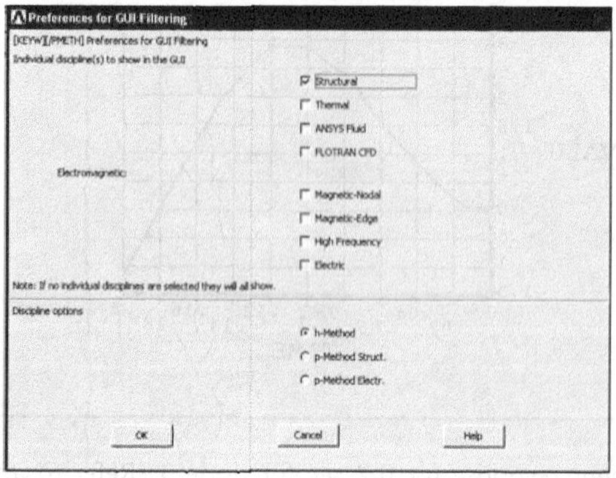

Figure 8.13. Selecting the preferences

2. **Main Menu > Preprocessor > Element Type > Add/Edit/Delete > Add >Beam > 2D elastic 3 > OK > Close**

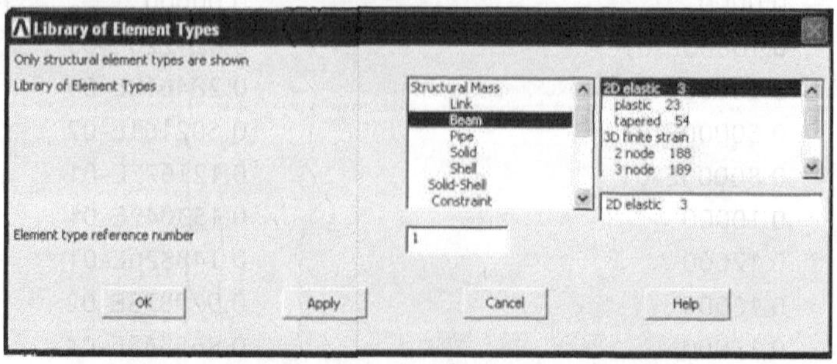

Figure 8.14. Element selection

3. **Main Menu > Preprocessor > Real Constants > Add/Edit/Delete > Add > OK**

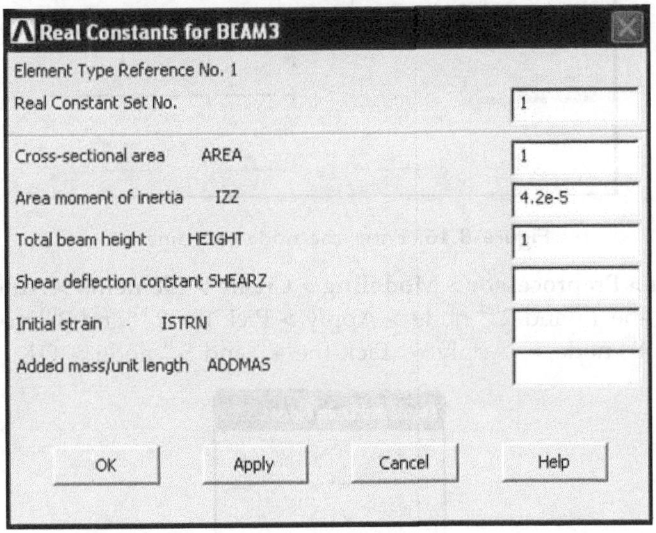

Figure 8.15. Enter the area and moment of inertia

Cross-sectional area AREA >**Enter 1**

Area moment of inertia IZZ >**Enter 4.2e–5**

Total beam height HEIGHT>**Enter 1> OK > Close**

Enter the material properties.

4. **Main Menu > Preprocessor > Material Props > Material Models**

   Material Model Number 1, click **Structural > Linear > Elastic > Isotropic**

   Enter **EX = 46E9 and PRXY = 0.35 > OK**

   click **Structural > Linear > Density**

   Enter DENS = 1750 > OK

   (**Close** the Define Material Model Behaviour window.)

   Create the nodes and elements as shown in the figure.

5. **Main Menu > Preprocessor > Modeling > Create > Nodes > In Active CS** Enter the coordinates of node 1 > **Apply** > Enter the coordinates of node 2 > **Apply** > Enter the coordinates of node 3 > **Apply** > Enter the coordinates of node 4 > **Apply** Enter the coordinates of node 5 > **OK.**

Node locations		
Node number	X coordinate	Y coordinate
1	0	0
2	1.25	0
3	2.5	0
4	3.75	0
5	5	0

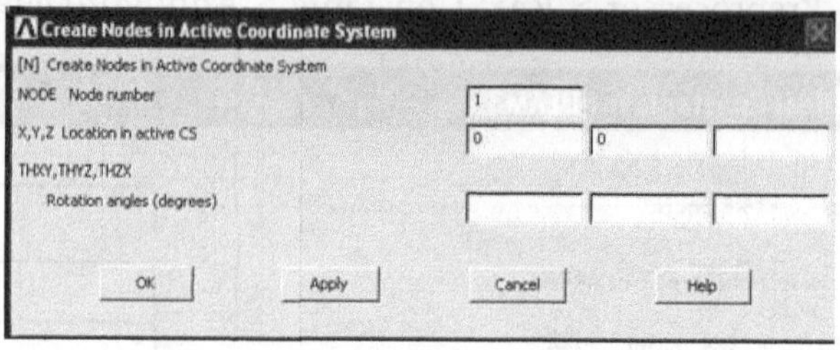

Figure 8.16. Enter the node coordinates.

6. **Main Menu > Preprocessor > Modeling > Create > Elements > Auto Numbered > Thru Nodes** Pick the 1st and 2nd node > Apply > Pick the 2nd and 3rd node > Apply > Pick the 3rd and 4th node > Apply > Pick the 4th and 5th node > **OK**

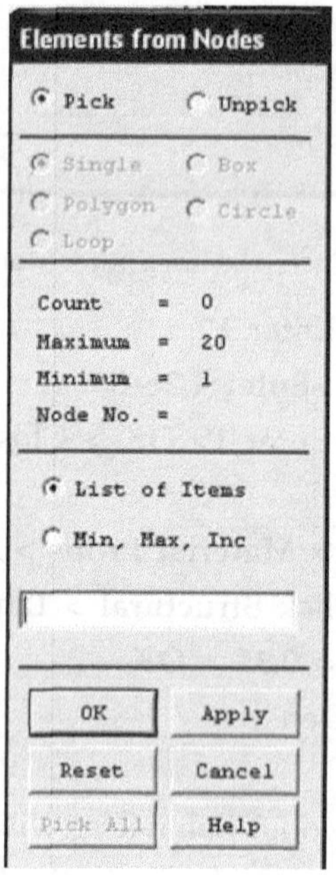

Figure 8.17. Pick the nodes to create elements

7. **Main Menu > Solution > Analysis Type > New Analysis > Select Transient > OK**

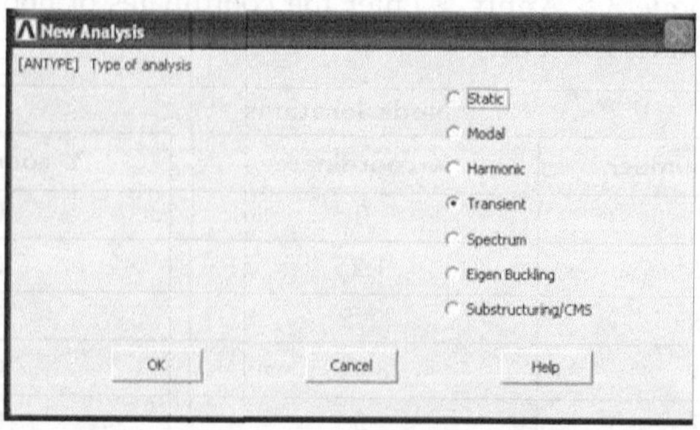

Figure 8.18. Define the type of analysis

then select > **Reduced > OK**

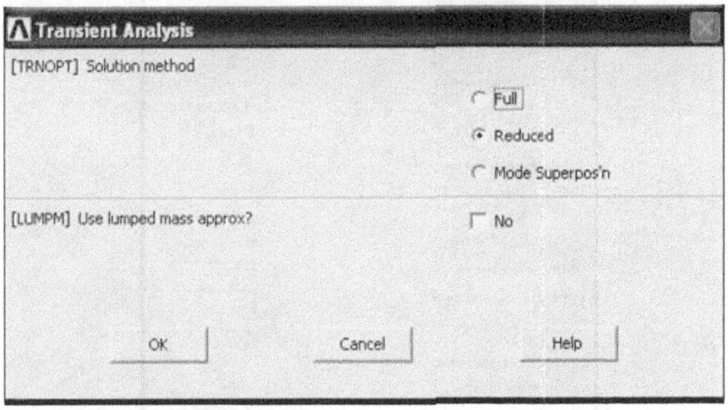

Figure 8.19. Define the type of transient analysis

8. **Main Menu > Preprocessor > Loads > Define Loads > Apply > Structural >Displacement > On Nodes >** Pick the left most node and right most node > **Apply > Select All DOF > OK**

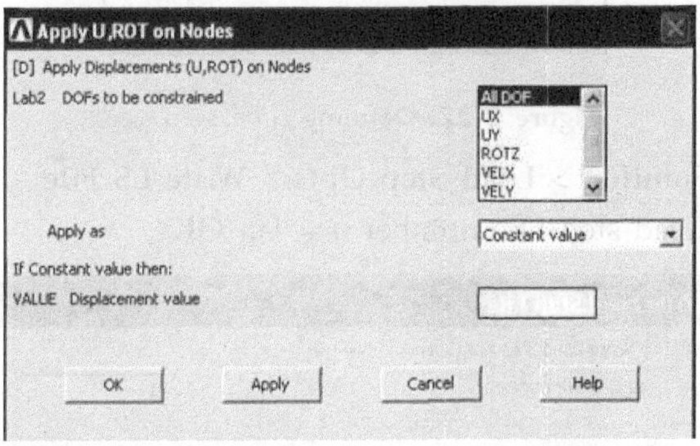

Figure 8.20. Apply the displacement constraint

9. **Main Menu > Solution > Master DOFs > User Selected > Define >** Pick 2nd, 3rd and 4th node > **Apply > Select UY from Lab 1 1st degree of freedom > OK**

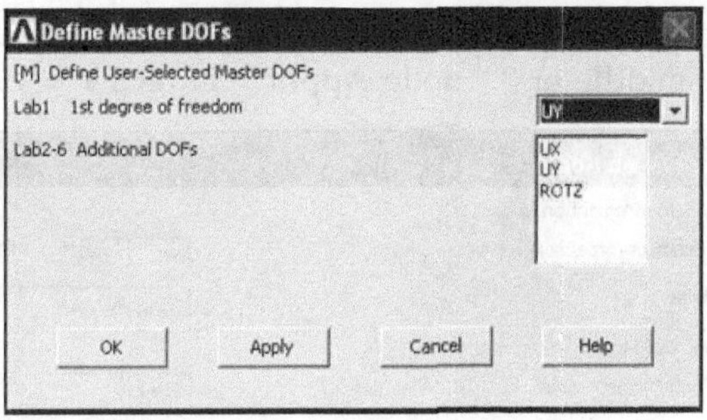

Figure 8.21. Defining master DOF

10. **Main Menu > Solution > Load Step Opts > Time / Frequenc > Time-Time Step**
    Enter [TIME] Time at end of load step – 0
    Enter [DELTIM] Time step size – 0.01 **< OK**

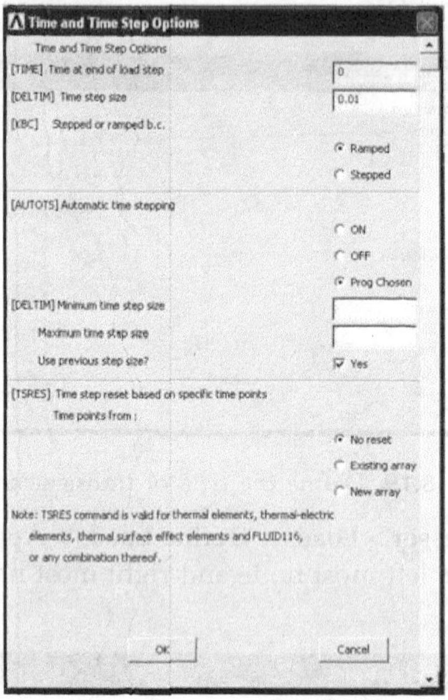

Figure 8.22. Defining time step size

11. **Main Menu > Solution > Load Step Opts > Write LS File**
    Enter LSNUM Load step fine number n = 1 **> OK**

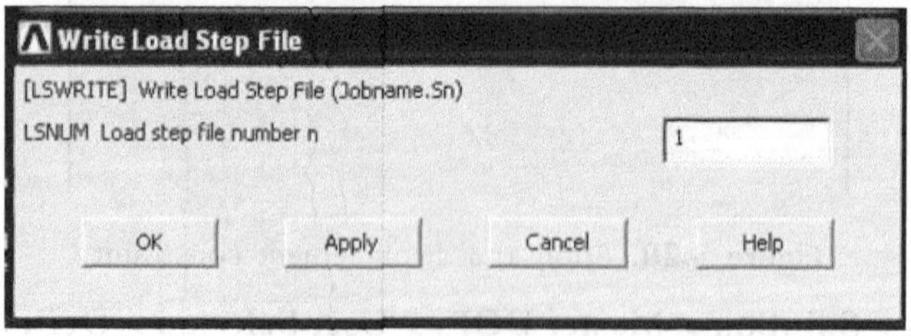

Figure 8.23. Creating LS file

12. **Main Menu > Solution > Define Loads > Apply > Structural > Force/Moment > On Nodes >** Pick the middle or 3rd node **Apply > Enter FY = 45e3 > OK**

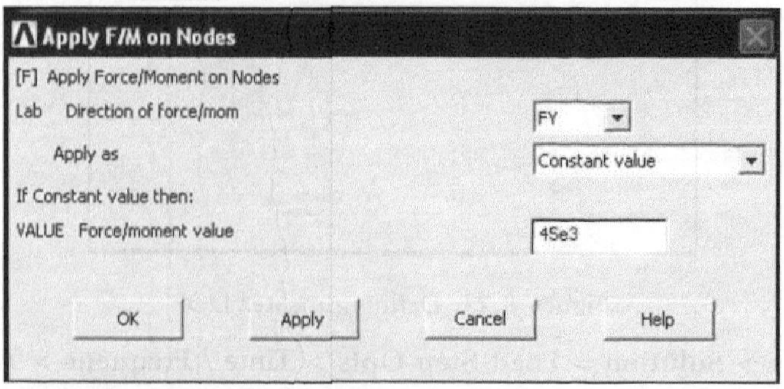

Figure 8.24. Applying force on node

13. **Main Menu > Solution > Load Step Opts > Time/Frequenc > Time-Time Step**
Enter [TIME] Time at end of load step – 0.01 > **OK**

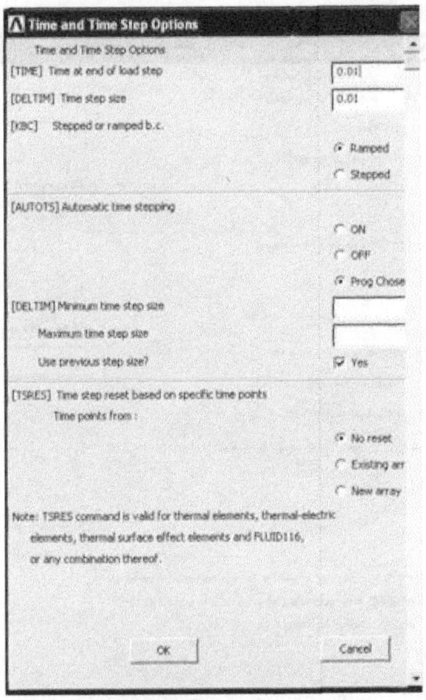

Figure 8.25. Defining time at the end of 1st load step

14. **Main Menu > Solution> Load Step Opts > Write LS File**
Enter LSNUM Load step fine number n = 2 > **OK**

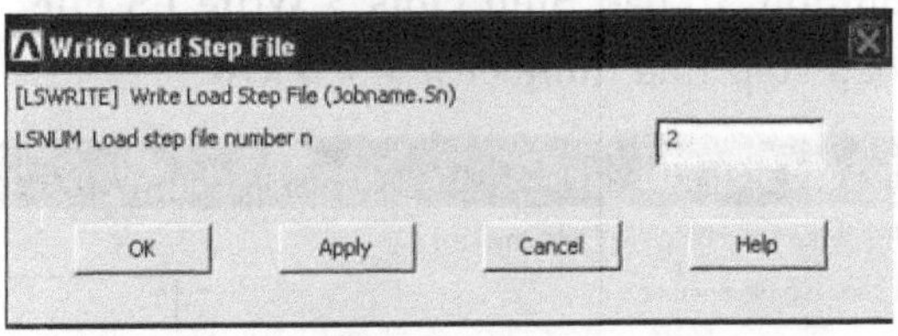

Figure 8.26. Creating LS file for 1st load step

Similarly repeat steps 13 and 14 for Time at end of load step of 0.02, 0.05, 0.08 and 0.1 and each time create a LS file with next numbers (n) *i.e.*, 3, 4, 5 and 6.

15. **Main Menu > Solution > Define Loads > Delete > Structural > Force/Moment > On Nodes >** Pick the middle or 3rd node **Apply > OK**

16. **Main Menu > Solution > Define Loads > Apply > Structural > Force/Moment > On Nodes >** Pick the middle or 3rd node **Apply > Enter FY = 36e3 > OK**

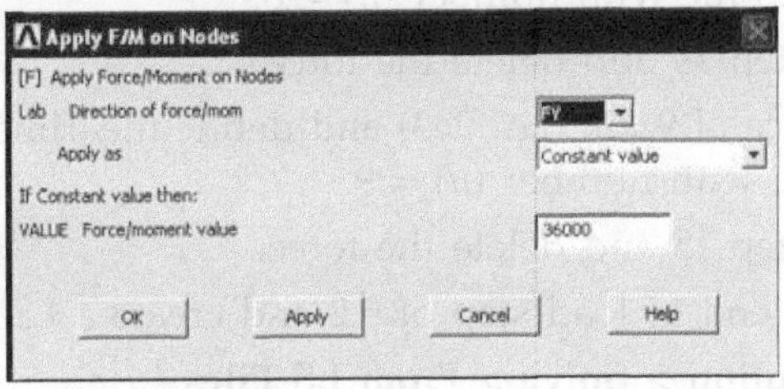

Figure 8.27. Applying force on node

17. **Main Menu > Solution > Load Step Opts > Time/Frequenc > Time-Time Step**
    Enter [TIME] Time at end of load step – 0.12 > **OK**

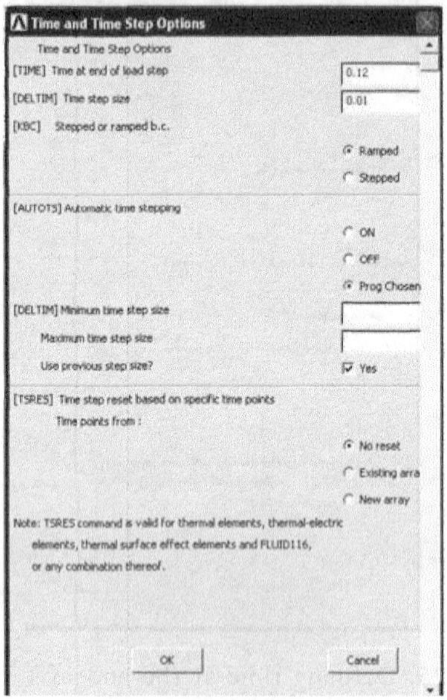

Figure 8.28. Defining time at the end of 6th load step

18. **Main Menu > Solution > Load Step Opts > Write LS File**
    Enter LSNUM Load step fine number n = 7 > **OK**

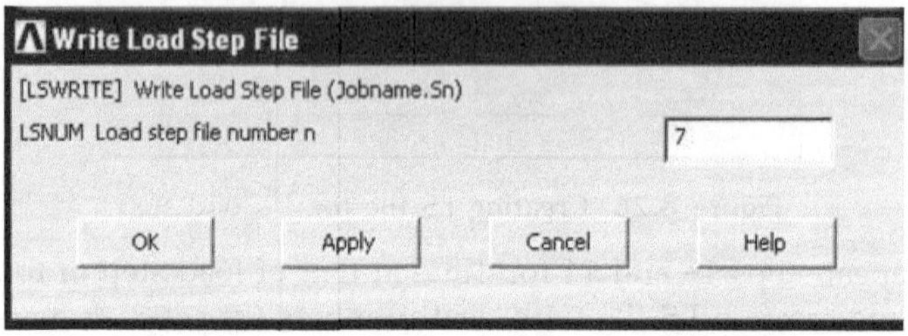

Figure 8.29. Creating LS file for 6th load step

Similarly repeat step 15 and delete the force.

Then apply the force of 22.5 kN (*i.e.*, 22.5e3) and define the Time at end of load step of 0.15 and create a LS file with number (n) = 8.

Again repeat the step 15 and delete the force.

Then apply the force of 9 kN (*i.e.*, 9e3) and define the Time at end of load step of 0.18 and create a LS file with number (n) = 9.

Again repeat the step 15 and delete the force.

Define the Time at end of load step of 0.2 and create a LS file with number (n) = 10.

19. **Main Menu > Solution > Solve > From LS Files**
    Enter LSMIN Starting LS file number = 1

Enter LSMAX Ending LS file number = 10

LSINC File number increment = 1

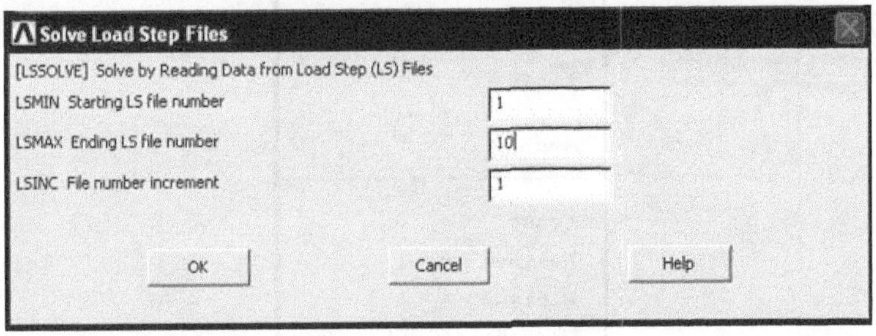

Figure 8.30. Solving from LS files

20. **Main Menu > TimeHist Postpro**

The following dialog box will appear

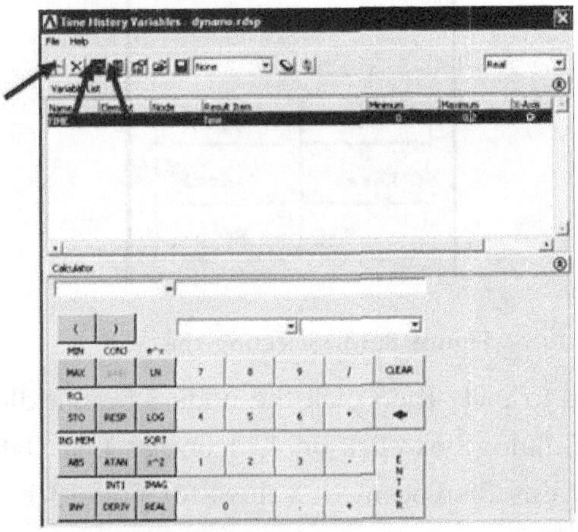

Figure 8.31. Timehist dialog box

In that dialog box click on first icon *i.e.*, on Add data, one more dialog box will appear as shown below. Then click on **DOF Solution > y-Component of displacement > OK.**

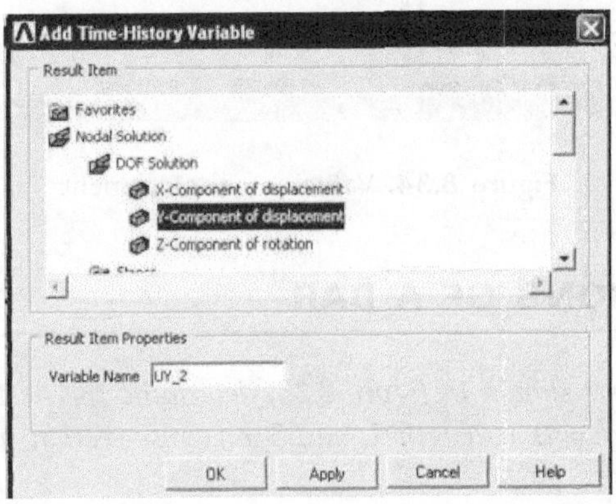

Figure 8.32. Selecting the displacement in *y*-direction

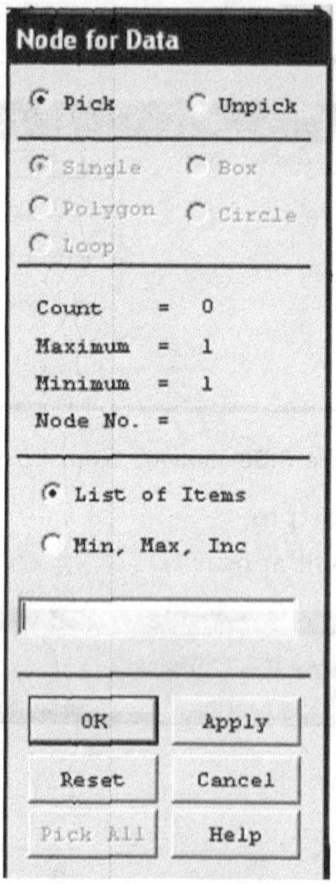

Figure 8.33. Selecting the node

It asks for the node to pick, then pick the node 3 or middle node < OK.

Then in a Timehist dialog box click on 4th icon *i.e.,* List Data (Refer figure 8.31)

This result is shown as displacement values for node 3 in the software results of the problem.

Then in a Timehist dialog box click on 3rd icon *i.e.,* Graph Data (Refer figure 8.31)

This result is shown in figure 8.12(*b*) for node 3 in the software results of the problem.

Maximum displacement values (in metres)

Name	Element	Node	Result Item	Minimum	Maximum	X-Axis
TIME			Time	0	0.2	(•)
UY_2		2	Y-Component of displacement	-0.0064935	0.0153042	( )
UY_3		3	Y-Component of displacement	-0.0117477	0.0292504	( )

Figure 8.34. Values of displacement

8.6 AXIAL VIBRATIONS OF A BAR

Problem 8.4. *For the bar shown in figure 8.35, determine the first two natural frequencies. Let* E *= 207 GPa, Poisson's ratio = 0.3, length* L *= 2.5 m, cross-section area* A *= 1* m^2*, mass density* $\rho = 7800\ kg/m^3$.

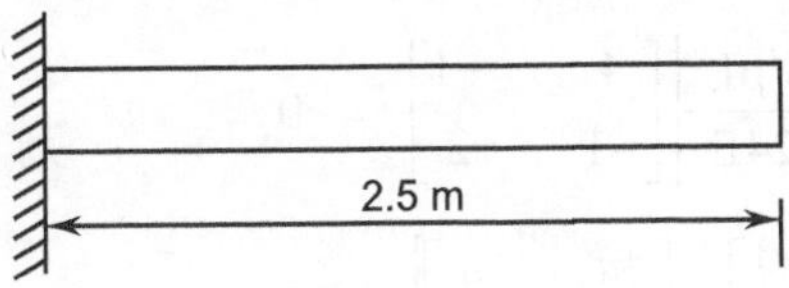

Figure 8.35

SOLUTION. (I) Analytical method

$$\therefore \quad \omega_1 = \frac{1.57}{L}\sqrt{\frac{E}{\rho}} = \frac{1.57}{2.5}\sqrt{\frac{207\times10^9}{7800}} = 3235.17 \text{ rad/s} \Rightarrow f_1 = \frac{\omega_1}{2\pi} = \frac{3235.17}{2\pi} = 514.89 \text{ Hz}$$

$$\therefore \quad \omega_2 = \frac{4.71}{L}\sqrt{\frac{E}{\rho}} = \frac{4.71}{2.5}\sqrt{\frac{207\times10^9}{7800}} = 9705.52 \text{ rad/s} \Rightarrow f_2 = \frac{\omega_2}{2\pi} = \frac{9705.52}{2\pi} = 1544.68 \text{ Hz}$$

(II) FEM by hand calculations

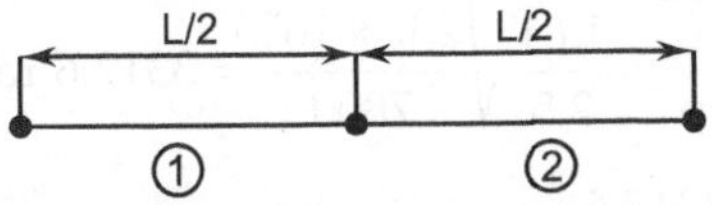

Figure 8.35(a). Finite element model

$$[M_1] = [M_2] = \frac{\rho A\left(\frac{L}{2}\right)}{6}\begin{bmatrix}2 & 1\\1 & 2\end{bmatrix} = \frac{\rho AL}{12}\begin{bmatrix}2 & 1\\1 & 2\end{bmatrix}$$

$$[K_1] = [K_2] = \frac{EA}{\left(\frac{L}{2}\right)}\begin{bmatrix}1 & -1\\-1 & 1\end{bmatrix} = \frac{2EA}{L}\begin{bmatrix}1 & -1\\-1 & 1\end{bmatrix}$$

Global matrices,

$$[M] = \frac{\rho AL}{12}\begin{bmatrix}2 & 1 & 0\\1 & 4 & 1\\0 & 1 & 2\end{bmatrix}$$

$$[K] = \frac{2EA}{L}\begin{bmatrix}1 & -1 & 0\\-1 & 2 & -1\\0 & -1 & 1\end{bmatrix}$$

Governing Equation: $([K]-\omega^2\,[M]) = \begin{Bmatrix}u_1\\u_2\\u_3\end{Bmatrix} = [0]$

$$\left(\frac{2EA}{L}\begin{bmatrix}1 & -1 & 0\\-1 & 2 & -1\\0 & -1 & 1\end{bmatrix}\right) - \omega^2\frac{\rho AL}{12}\begin{bmatrix}2 & 1 & 0\\1 & 4 & 1\\0 & 1 & 2\end{bmatrix}\begin{Bmatrix}u_1\\u_2\\u_3\end{Bmatrix} = 0$$

Boundary conditions are, $u_1 = 0$

Applying Boundary Conditions and for nontrivial solution,

$$\left|\begin{bmatrix} 2 & -1 \\ -1 & 1 \end{bmatrix} - \frac{\omega^2 \rho L^2}{24E}\begin{bmatrix} 4 & -1 \\ -1 & 2 \end{bmatrix}\right| = 0$$

i.e.,

$$\left|\begin{bmatrix} 2 & -1 \\ -1 & 1 \end{bmatrix} - \begin{bmatrix} 4a & a \\ a & 2a \end{bmatrix}\right| = 0$$

where $a = \dfrac{\rho\omega^2 L^2}{24E}$

$$\left|\begin{bmatrix} 2-4a & -1-a \\ -1-a & 1-2a \end{bmatrix}\right| = 0, \text{ Solving}$$

$$a = 0.1081941 = \frac{\omega^2 \rho L^2}{24E} \quad \text{or} \quad a = 1.3203772 = \frac{\omega^2 \rho L^2}{24E}$$

$$\therefore \quad \omega_1 = \frac{1.61}{L}\sqrt{\frac{E}{\rho}} = \frac{1.61}{2.5}\sqrt{\frac{207 \times 10^9}{7800}} = 3317.6 \text{ rad/s}$$

$$f_1 = \frac{\omega_1}{2\pi} = \frac{3317.6}{2\pi} = 528.01 \text{ Hz}$$

$$\therefore \quad \omega_2 = \frac{5.63}{L}\sqrt{\frac{E}{\rho}} = \frac{5.63}{2.5}\sqrt{\frac{207 \times 10^9}{7800}} = 11601.3 \text{ rad/s}$$

$$f_2 = \frac{\omega}{2\pi} = \frac{11601.3}{2\pi} = 1846.4 \text{ Hz}$$

(III) Software results

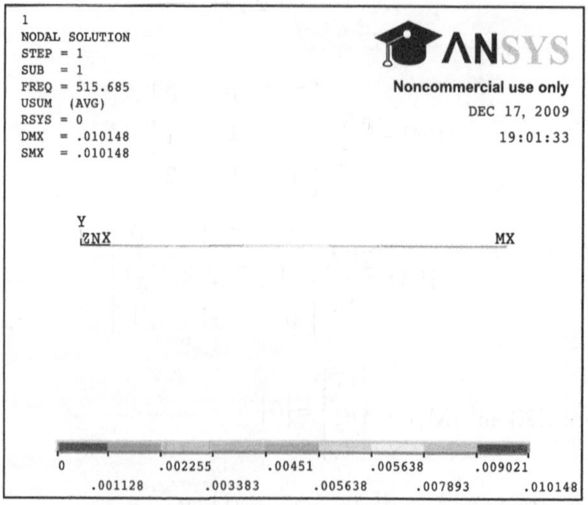

Figure 8.35(b). Deflection pattern for a bar (Refer to appendix for colour figures)

Frequency values (in Hz)

SET	TIME/FREQ	LOAD STEP	SUBSTEP	CUMULATIVE
1	515.68	1	1	1
2	1559.8	1	2	2

ANSWERS OF PROBLEM 8.4

Parameter	Analytical method	FEM – hand calculation (with 2 elements)	Software results (with 10 elements)
Natural frequency			
f_1	514.89 Hz	528.01 Hz	515.68 Hz
f_2	1544.68 Hz	1846.4 Hz	1559.8 Hz

Procedure for solving the problem using ANSYS®12.0 academic teaching software

FOR PROBLEM 8.4

PREPROCESSING

1. **Main Menu > Preprocessor > Element Type > Add/Edit/Delete > Add >Link > 2D spar 1 > OK > Close**

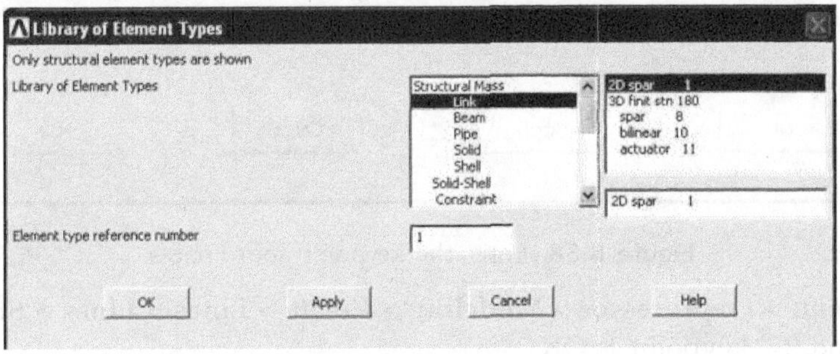

Figure 8.36. Element selection

2. **Main Menu > Preprocessor > Real Constants > Add/Edit/Delete > Add > OK**

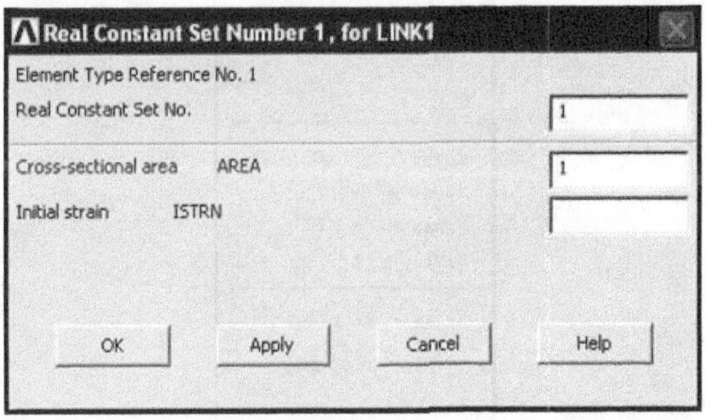

Figure 8.37. Enter the cross-sectional area

Cross-sectional area AREA >**Enter 1>OK>Close**

Enter the material properties.

3. **Main Menu > Preprocessor > Material Props > Material Models**

   Material Model Number 1, click **Structural > Linear > Elastic > Isotropic**

   Enter **EX = 207E9 and PRXY = 0.3 > OK**

   click **Structural > Linear > Density**

   Enter DENS = 7800

(**Close** the Define Material Model Behaviour window.)

Create the keypoints and lines as shown in the figure.

4. **Main Menu > Preprocessor > Modeling > Create > Keypoints > In Active CS** Enter the coordinates of keypoint 1 > **Apply** Enter the coordinates of keypoint 2 > **OK.**

Keypoint locations		
Keypoint number	**X coordinate**	**Y coordinate**
1	0	0
2	2.5	0

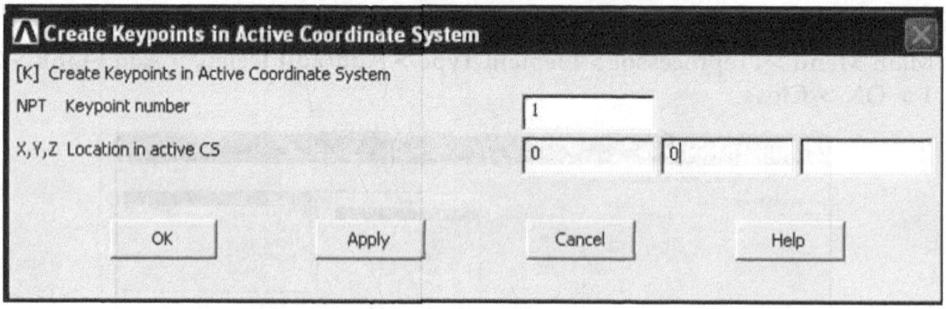

Figure 8.38. Enter the keypoint coordinates

5. **Main Menu > Preprocessor > Modeling > Create > Lines > Lines > Straight Line** Pick the 1st and 2nd keypoint > **OK**

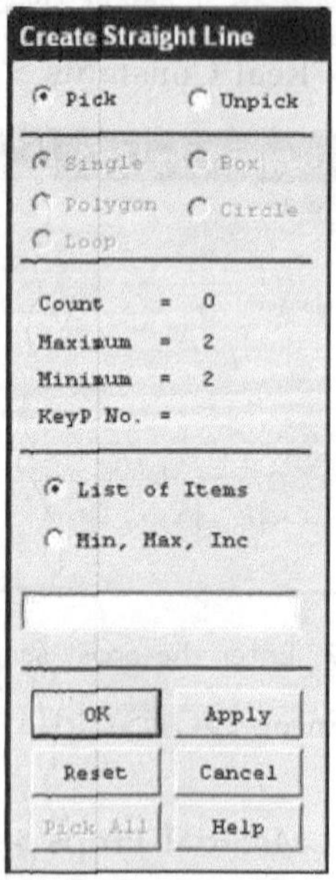

Figure 8.39. Pick the keypoints to create lines

6. **Main Menu > Preprocessor > Meshing > Size Cntrls > ManualSize > Lines > All Lines** > Enter **NDIV No. of element divisions = 10**

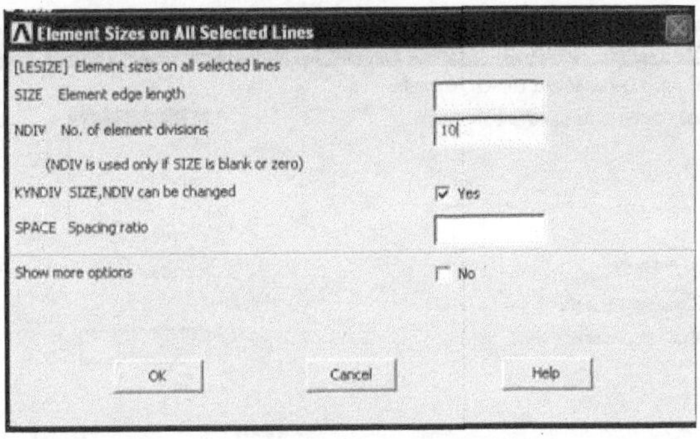

Figure 8.40. Specify element length

7. **Main Menu > Preprocessor > Meshing > Mesh > Lines > Click Pick All.**

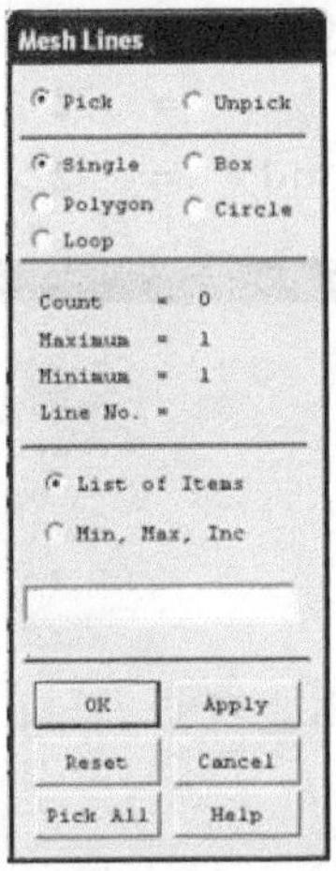

Figure 8.41. Create elements by meshing

8. **Main Menu > Solution > Analysis Type > New Analysis > Select Modal > OK**

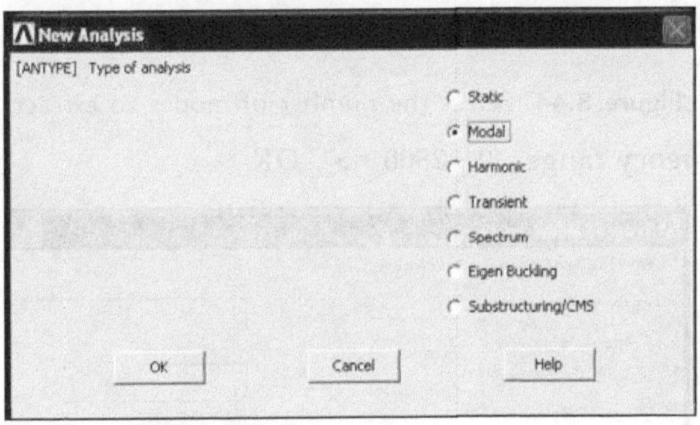

Figure 8.42. Define the type of analysis

9. **Main Menu > Preprocessor > Loads > Define Loads > Apply > Structural > Displacement > On Nodes** Pick the left most node **> Apply > Select All DOF > OK**

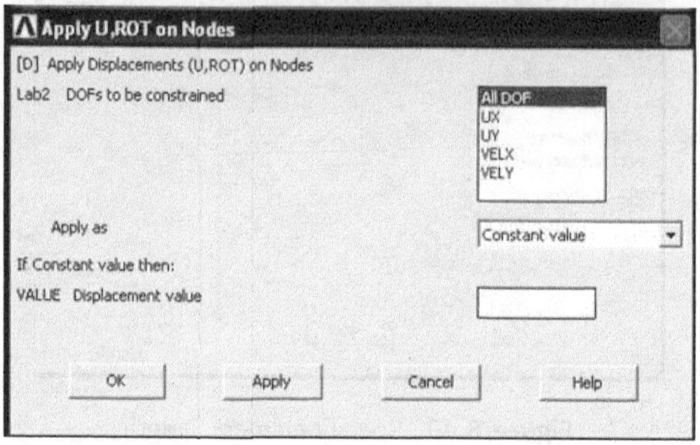

Figure 8.43. Apply the displacement constraint

10. **Main Menu > Solution > Analysis Type > Analysis Options > Select Reduced option**

    Enter No. of modes to extract = 2

    NMODE No. of modes to expand = 2 > OK

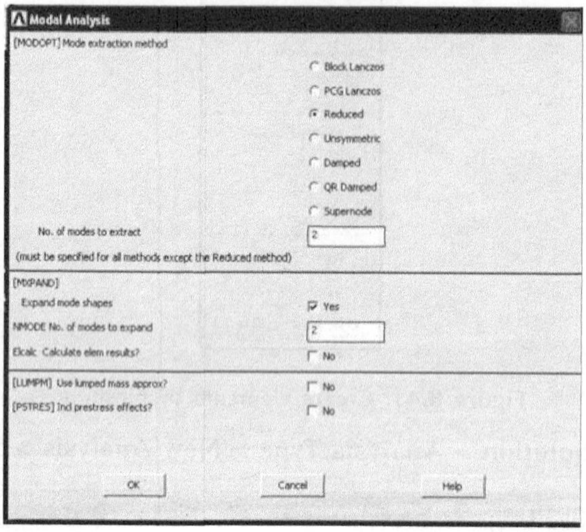

Figure 8.44. Select the number of modes to extract

Enter FREQE Frequency range 0 2500 > OK

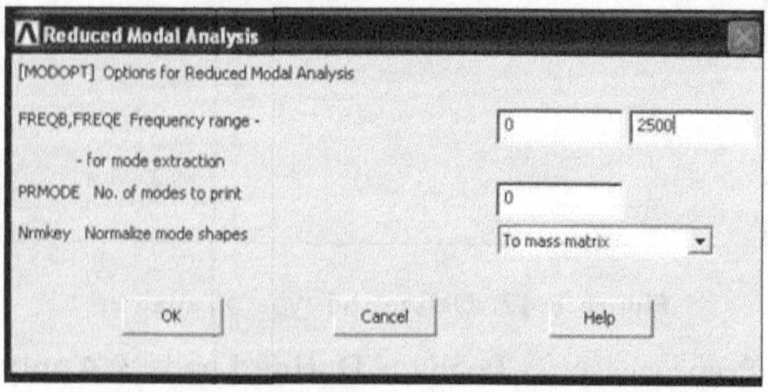

Figure 8.45. Enter the frequency range

11. **Main Menu > Solution > Master DOFs > User Selected > Define > Pick all nodes except left most node > OK > Select UX from Lab 1 1st degree of freedom > OK**

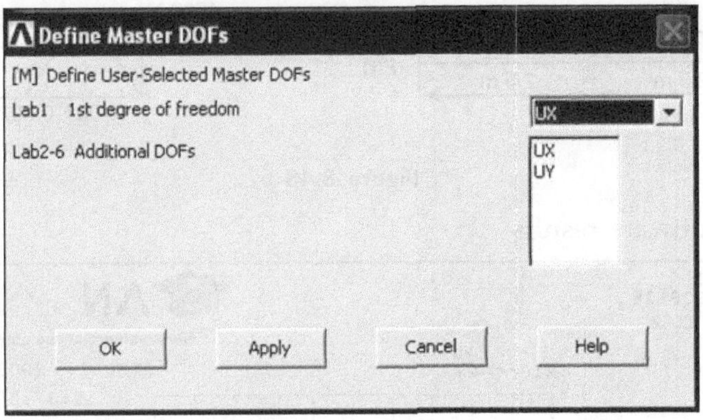

Figure 8.46 Defining the master degree of freedom

Figure 8.47. Model with Master DOF applied

SOLUTION. The interactive solution proceeds.

12. **Main Menu > Solution > Solve > Current LS > OK**

    The **/STATUS Command** window displays the problem parameters and **the SolveCurrent Load Step** window is shown. Check the solution options in the **/STATUS**window and if all is OK, select **File > Close**

    In the Solve **Current Load Step** window, Select **OK**, and when the solution is complete, **close** the **'Solution is Done!'** window.

POSTPROCESSING

13. **Main Menu > General Postproc > Results Summary**

    This result is shown as frequency values in Hz.

14. **Main Menu > General Postproc > Read Results > First Set**

15. **Main Menu > General Postproc > Plot Results > Contour Plot > Nodal Solu > DOF Solution, click on Displacement vector sum > OK**

    This result is shown in figure 8.35(*b*).

8.7 BAR SUBJECTED TO FORCING FUNCTION

PROBLEM 8.5. *For the bar shown in figure 8.48 subjected to time dependent forcing function as shown, determine the nodal displacements for five time steps using 2 finite elements. Let E = 207 GPa, Poisson's ratio = 0.3, length of beam L = 5 m, cross-section area A= 625 × 10^{-6} m^2, mass density ρ = 7800 kg/m^3. Use time step of integration 0.00025 seconds.*

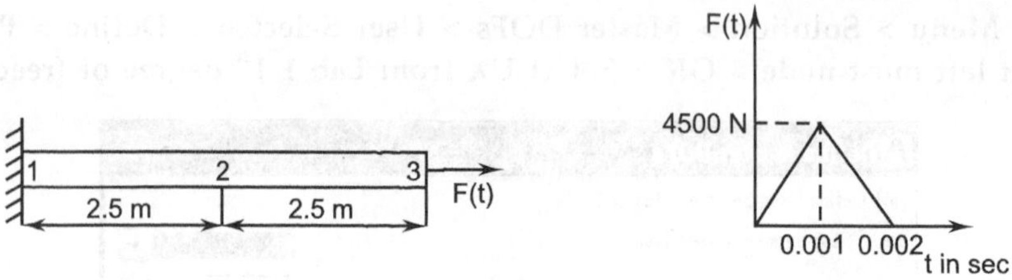

Figure 8.48

SOLUTION. (I) Software results

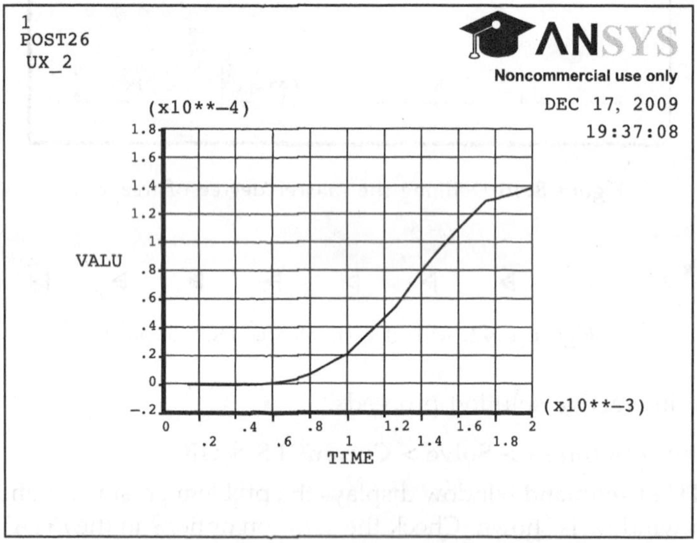

Figure 8.48(a). Displacement response for 0.00025 sec for node 2
(Refer to appendix for colour figures)

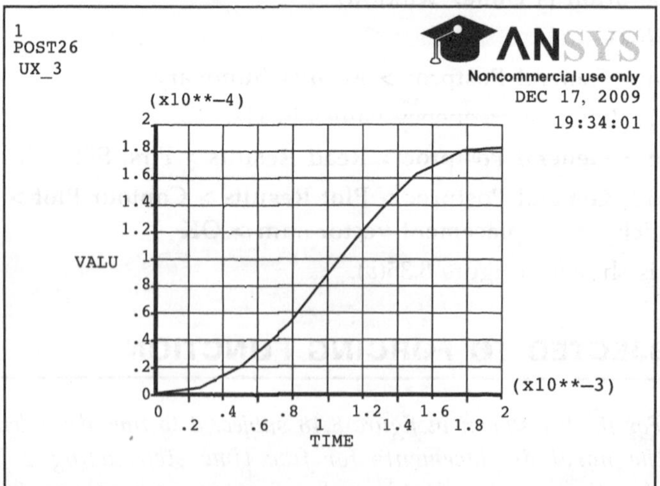

Figure 8.48(b). Displacement response for 0.00025 sec for node 3
(Refer to appendix for colour figures)

Displacement values (in metres) for node 2

TIME	2 UX UX_2
0.0000	0.00000
0.25000E-03	-0.467370E-06
0.50000E-03	-0.821457E-06
0.75000E-03	0.396081E-05
0.10000E-02	0.210563E-04
0.12500E-02	0.535055E-04
0.15000E-02	0.950064E-04
0.17500E-02	0.128841E-03
0.20000E-02	0.138387E-03

Displacement values (in metres) for node 3

TIME	3 UX UX_3
0.0000	0.00000
0.25000E-03	0.375512E-05
0.50000E-03	0.191517E-04
0.75000E-03	0.488709E-04
0.10000E-02	0.889759E-04
0.12500E-02	0.130597E-03
0.15000E-02	0.161991E-03
0.17500E-02	0.179673E-03
0.20000E-02	0.184097E-03

Maximum displacement values (in metres)

Name	Element	Node	Result Item	Minimum	Maximum	X-Axis
TIME			Time	0	0.002	(•)
UX_2		2	X-Component of displacement	-8.21457e-007	0.000138387	( )
UX_3		3	X-Component of displacement	0	0.000184097	( )

Figure 8.48(c). Values of displacement

EXERCISES

1. Determine the first five natural frequencies for the fixed-fixed beam shown in figure 8.49. The beam is made of steel with E = 200 GPa, Poisson's ration = 0.3, length = 2 m, cross-section area = 60 cm^2, mass density ρ = 7800 kg/m^3, moment if inertia I = 200 mm^4.

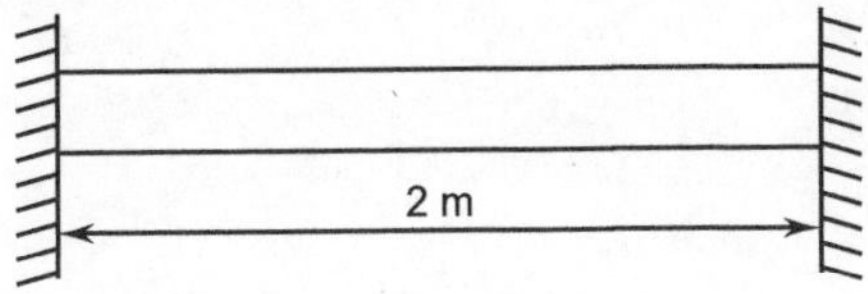

Figure 8.49

2. For the bar shown in figure 8.50, determine nodal displacements for the 5 time steps using two finite elements. Let E = 70 GPa, ρ = 2700 kg/m^3, A = 645 mm^2 and L = 2.5 m.

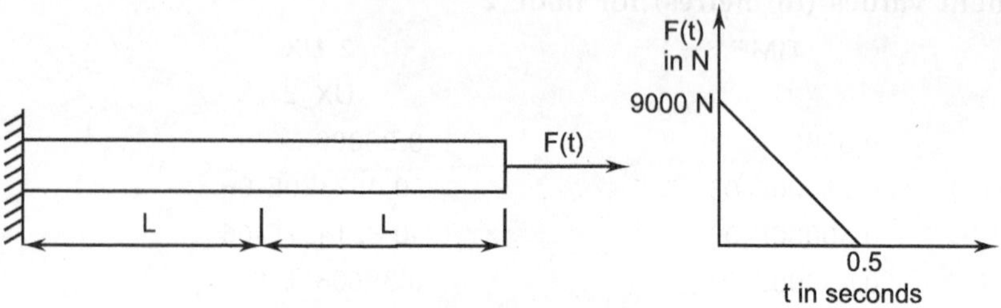

Figure 8.50

3. For the beam shown in figure 8.51, subjected to the forcing functions shown, determine the maximum deflections. Let E = 207 GPa, ρ = 7800 kg/m^3, A = 0.0194 m^2, I = 8.2×10^{-5} m^4, L = 6 m. Take time step of 0.05 seconds.

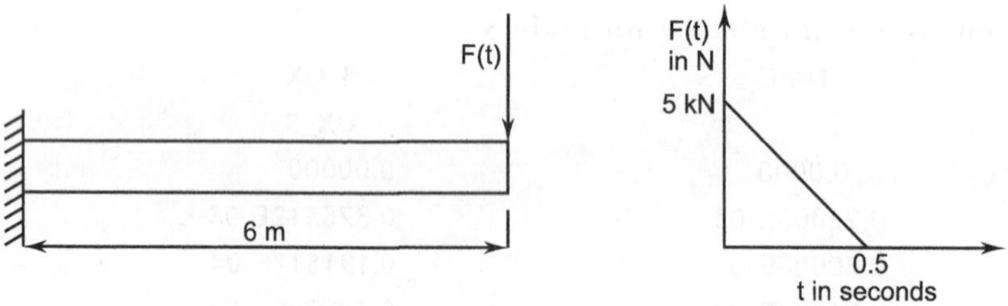

Figure 8.51

Chapter 9

ANSYS WORKBENCH

9.1 INTRODUCTION

ANSYS workbench provides unified data sharing and project file management across the range of ANSYS products.

Workbench is not an application on its own. It is a group of technologies for developing simulating tools. This has been created by ANSYS incorporation to help developers to develop robust tools to meet the needs of designers and analysts. It provides a common interface for the user to access the wealth of technology that ANSYS has with it.

Simulation data is displayed in a hierarchical tree that separates out the various groups of inputs and outputs. Workbench makes use of a Project file storage method to assist with the organization of data, and each can contain all types of information.

Workbench stores the Model information at the highest level and this contains:

- geometry that your mesh is built on,
- contacts (which define how parts interact),
- mesh and
- environment.

The mesh is the core element-based representation of the parts, on which all FEA/CFD tasks are based. Workbench allows you to intelligently refine the mesh.

The Workbench user interface: Tabs based user interface, that is familiar to most users, is used in Workbench. We start with the Project Page tab and this provides us access to the tools we need to read geometry in, set-up the boundary conditions, create results visualisation assets and report on results.

Geometry definition: Data pertaining to geometry can be read from the majority of leading softwares, including Pro/Engineer, Solid Works, NX/Unigraphics, Catia, etc. The system reads the data into the Workbench user interface and preserves part names and material definitions, if any.

Design Modeler: Design Modeler provides us with many tools that allow us to take data from a variety of sources, integrate it into a single pack and use it.

FE Modeler: It allows you to read meshed, element-based data from a wide variety of sources. It performs checks for integrity, extracts the information and places it into the appropriate folders of the simulation tree.

Materials definition: ANSYS Workbench is supplied with a range of standard materials. Materials (if present) can be extracted from the CAD file or may be adapted or enter new materials.

Reporting: Reports can be created based on a wizard style workflow. Various inputs (title, description etc.) may be defined and the system handles rest of the work. Workbench allows you to create diagrams at any point, so the graphics window and a caption is stored and then added to the report.

Design optimisation-DesignXplorer: Workbench has inherent links with CAD system, and bi-directionality can be used to our advantage to carry out optimisation to the maximum extent.

Mechanisms: Workbench also makes it possible to carry out in-depth studies of mechanisms intelligently. It allows us to model both rigid and flexible bodies in one environment.

Lastly, it can be said that ANSYS that provides a single unified environment where user has access to a good functionality within the simulation arena. It helps us to build high quality products that cost less and serve better their intended purpose.

Method of solving problem using workbench is illustrated using a simple example.

PROBLEM 9.1. *For the cantilever beam shown in figure 9.1, determine the total deflection, maximum stress, reaction force and reaction moment. Take Young's modulus E = 210 GPa and Poisson ratio v = 0.3. Force P = 5000 N. Assume plane stress condition.*

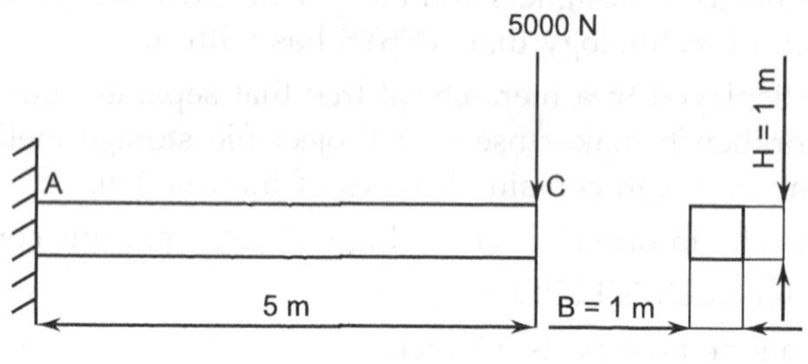

Figure 9.1

SOLUTION. (I) Software method

Procedure for solving the problems using ANSYS®12.0 academic teaching software using ANSYS Workbench.

STEP 1. **Selecting Analysis System and Specifying Material Properties**

Start ANSYS Workbench

Start > Programs > ANSYS 12.0 > Workbench

In **Toolbox Customization** click **Static Structural (ANSYS)**

Figure 9.2. Selecting type of analysis

In **Toolbox**, left click (and hold) on **Static Structural (ANSYS)** and drag the icon to the empty space in the **Project Schematic.** In **Project Schematic** space, right click on **Static Structural (ANSYS)** and **Rename** the project to **Cantilever_beam.**

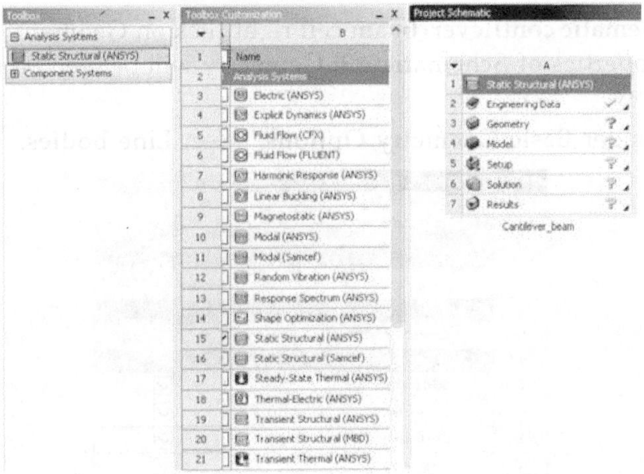

Figure 9.3. Creating cantilever beam cell

Specify Material Properties

In Cantilever beam cell double click on **Engineering Data,** a new page will be opened. The default material given is **Structural Steel**. This will be seen in the **Outline of Schematic A2: Engineering Data**.

In the **Properties of Outline Row 3: Structural Steel** window double click on **Isotropic Elasticity** and change the **Young's Modulus** value. To change the value, double click on the numerical value and enterE = 2.1e11 Pa and **Poisson ratio** = 0.3.

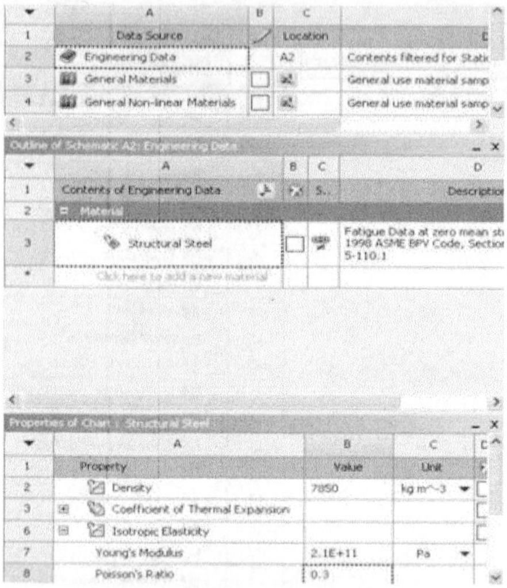

Figure 9.4. Specifying material properties

Next on top, click on **Return to Project** to return to Workbench **Project Schematic** window.

Figure 9.5. Return to Workbench

Step 2. **Geometry Creation**

In the **Project Schematic contilever_beam cell** right click on **Geometry** and select **Properties**. A window named **Properties of Schematic A3: Geometry** will be seen on the right hand side of the Workbench window.

In that window under **Basic Geometry Options**, select **Line bodies**.

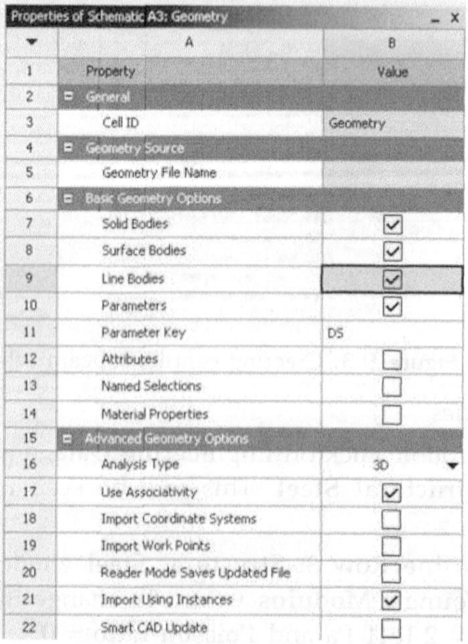

Figure 9.6. Selecting line bodies

Then in the **Project Schematic** window, under the cell, double left click on **Geometry** to start preparing geometry. At this point a new window **ANSYS Design Modeler** will be opened. Select **meter** unit as desired length unit and click **OK.**

Creating a Sketch

Use **XY Plane** for creation of sketch. On the left hand side in **Tree Outline** click on **XY Plane** and then click on **Sketching,** next to **Modeling** tab.

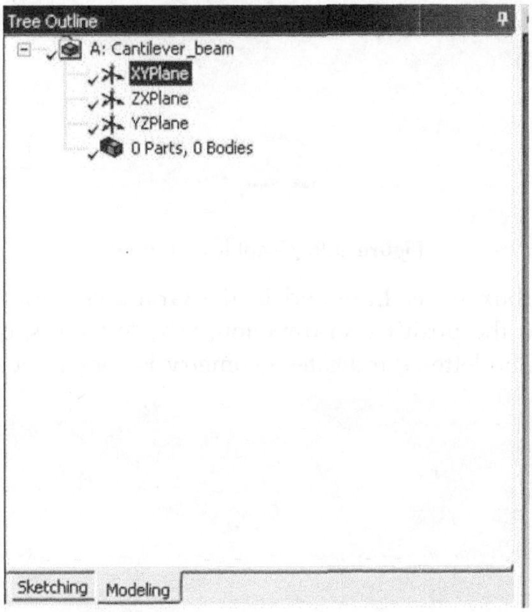

Figure 9.7. Selecting XY plane for sketching

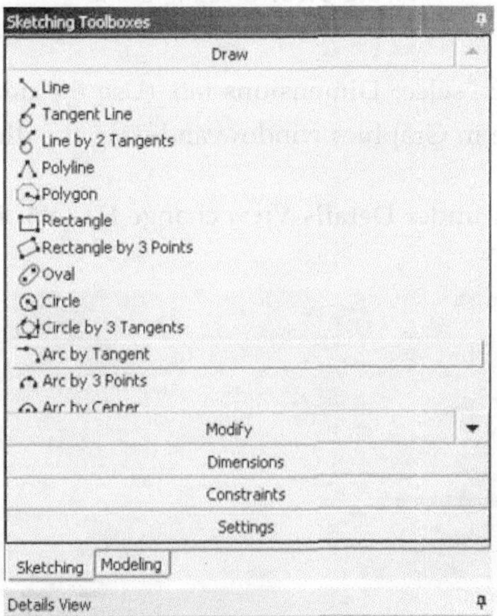

Figure 9.8. Sketching toolbox

Note: In sketching, **undo** features can be used.

On the right hand side, there is a **Graphics** window. At the lower right hand corner of the **Graphics** window, click on **+Z** axis. Then **Graphics** window will appear as shown below.

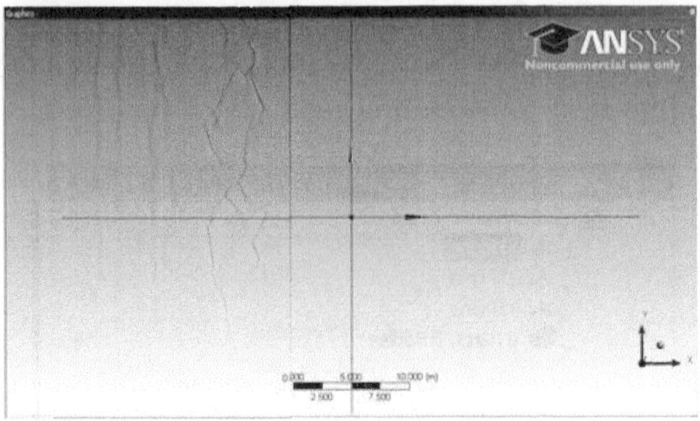

Figure 9.9. Graphics window

In the **Sketching toolbox** select **Line** and in the **Graphics** window create one rough line starting from the origin in the positive XY-direction. (The letter P should be seen at the origin before creating the Line. The letter P indicates geometry is constrained at the origin.)

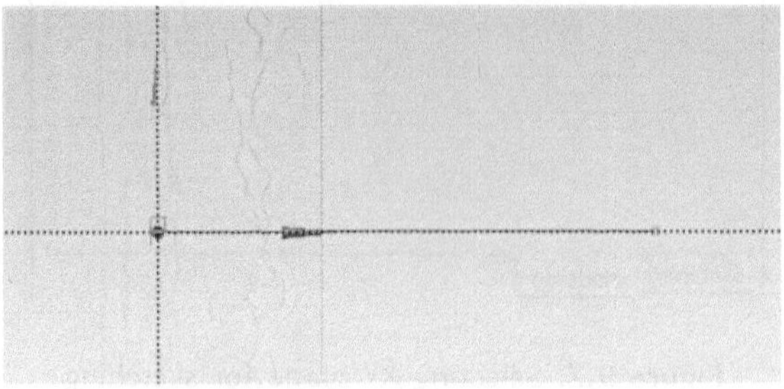

Figure 9.10. Creating a line

Dimensions

Under **Sketching Toolbox**, select **Dimensions** tab. (Use the default dimensioning tools).

Then left click on the line in Graphics window and drag the dimension line. The dimension line shows **H1**

On lower left hand corner under **Details View** change **H1** to **5.** This **5** is the length of the line drawn. Now sketching is done.

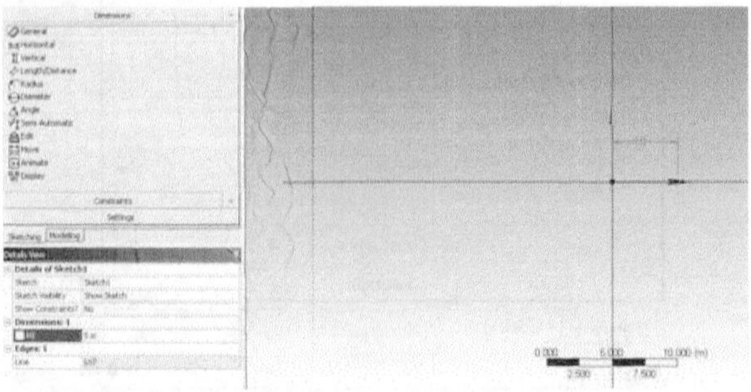

Figure 9.11. Dimensioning the line and entering the length value

Surface Creation

On top menu of the **Design Modeler** window click on

Concept > Lines from Sketches

This will create a new line **Line1**. In **Tree Outline** double click on **XY Plane** then select **Sketch 1**, then under **Details View**, in **Base Objects** click **Apply**. Finally on the top, click **Generate** to generate the surface.

Create Cross-Section

Concept > Cross-Section > Rectangular

Under **Details View**, input B = 1 m and H = 1 m.

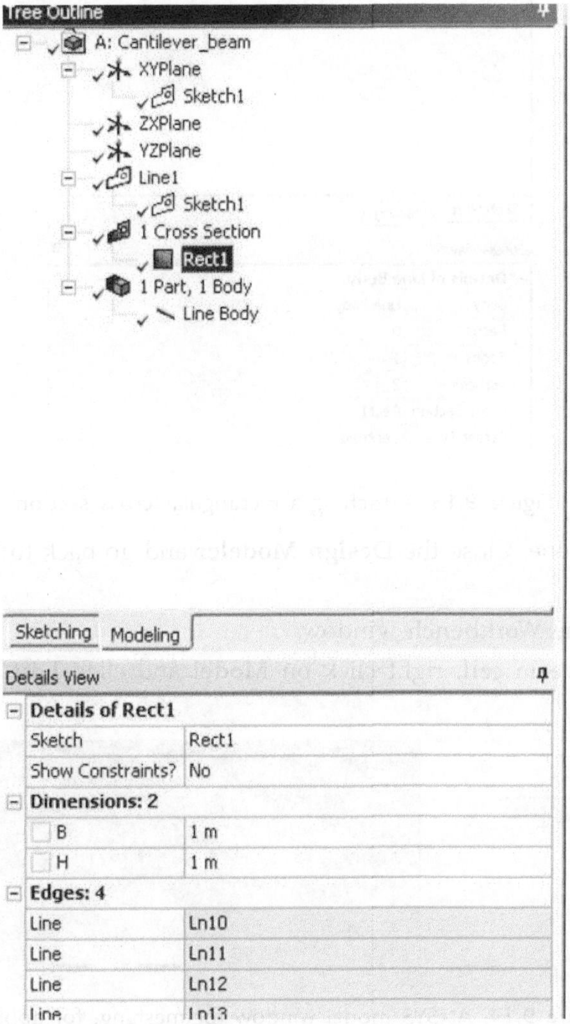

Figure 9.12. Entering the breadth and height values of the beam

Now under **Tree Outline > 1 Part, 1 Body > Line Body,**

Attach **Rect1** to **Cross-Section** under **Details View** as shown below.

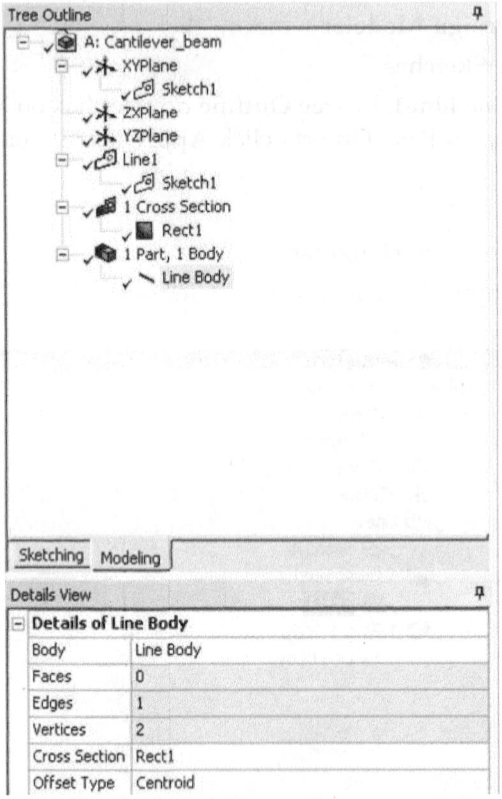

Figure 9.13. Attaching a rectangular cross-section

Now geometry is done. Close the **Design Modeler** and go back to **Workbench**.

Step 3. **Meshing**

Save the work in the **Workbench** window.

In the **Cantilever_beam cell**, right click on **Model** and click **Edit.** A new **ANSYS Model** window will open.

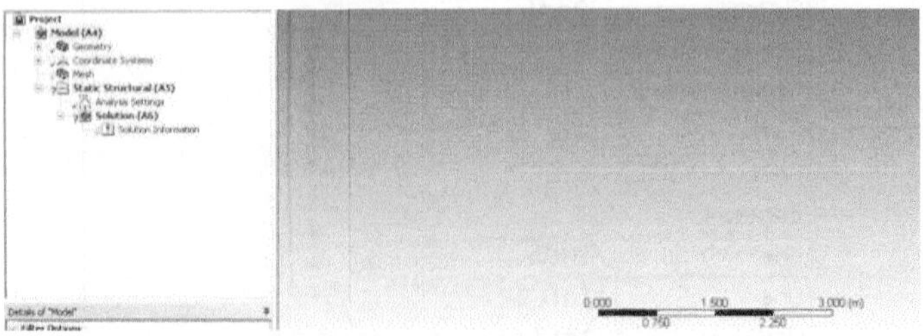

Figure 9.14. ANSYS model window for meshing, for applying boundary conditions and for solution

In **Outline** heading right click on **Mesh** and click **Generate Mesh**. Then Meshed model will appear in Graphics window.

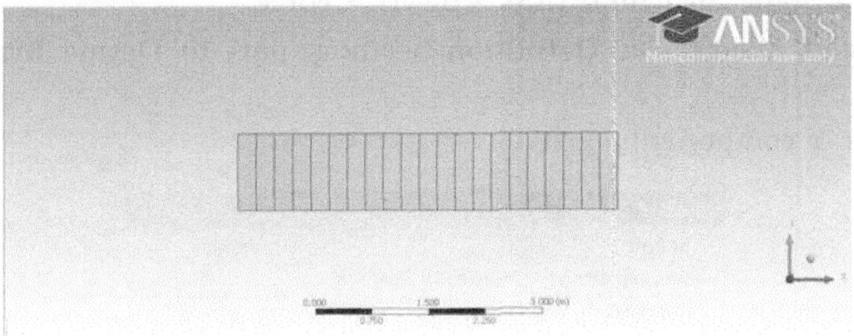

Figure 9.15. Meshed model

STEP 4. **Boundary Conditions**

Start the setting up of boundary condition at A.

In **Outline** right click **Static Structural (A5) > Insert** then select **Remote Displacement.**

In **Graphics** window right click and change **Cursor Mode** to **Vertex.**

Select point A in the **Graphics** window and click **Apply** next to **Geometry** under Details of **"Remote Displacement"**

Details of "Remote Displacement"

Scope	
Scoping Method	Geometry Selection
Geometry	1 Vertex
Coordinate System	Global Coordinate System
X Coordinate	0. m
Y Coordinate	0. m
Z Coordinate	0. m
Location	Click to Change

Figure 9.16. Selecting point A

Now under the heading of **Details of "Remote Displacement"** and under **Definition** heading change **UX, UY, UZ, ROTX, ROTY and ROTZ to zero.**

Definition	
Type	Remote Displacement
X Component	0. m (ramped)
Y Component	0. m (ramped)
Z Component	0. m (ramped)
Rotation X	0. ° (ramped)
Rotation Y	0. ° (ramped)
Rotation Z	0. ° (ramped)
Suppressed	No
Behaviour	Deformable

Figure 9.17. Applying displacement boundary conditions

At point C Force component needs to be applied.

In **Graphics** window select point C.

Then **Outline > Static Structural (A5) > Insert > Force**

Under **Details of "Force"**, in **Definition** heading, next to **Define By** change **Vector** to **Components.**

Enter–**5000** for **Y component**.

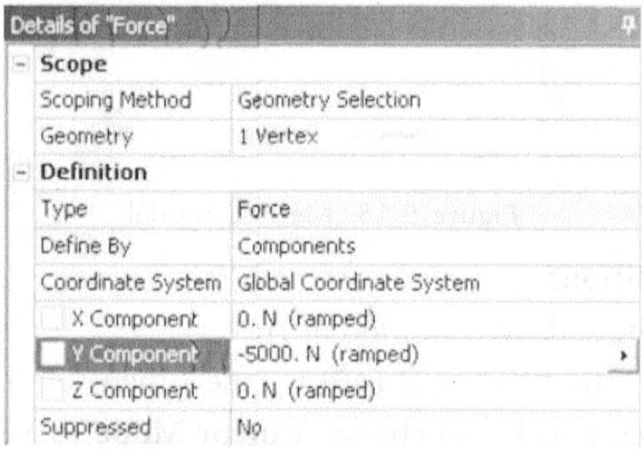

Figure 9.18. Applying force

Click on **Static Structural (A5)** to view the model with boundary conditions in **Graphics** window shown below.

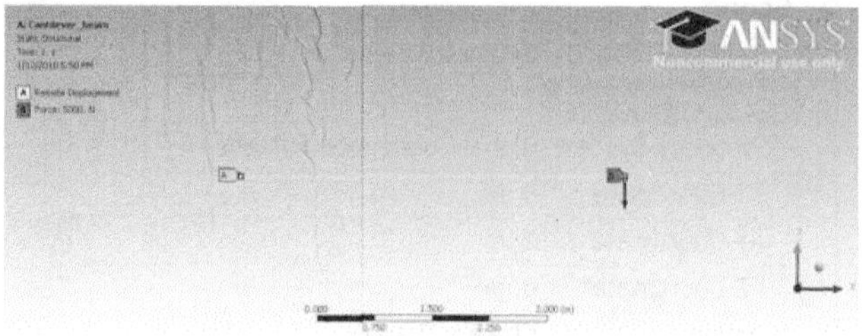

Figure 9.19. Model with boundary conditions

Step 5. **Solution**

For finding deflection,

Under **Outline** heading right click **Solution (A6), Insert > Deformation > Total**

For finding Stresses in beam,

Outline > Solution (A6) > Insert > Beam Tool > Beam Tool

For finding reaction force,

Outline > Solution (A6) > Insert > Probe > Force Reaction

Select point A and Under **Details of "Force Reaction"**, in **Definition** heading next to **Boundary Condition** select **Remote Displacement.**

For finding reaction moment,

Outline > Solution (A6) > Insert > Probe > Moment Reaction

Select point A and Under **Details of "Moment Reaction"**, in **Definition** heading next to **Boundary Condition** select **Remote Displacement.**

Then on top click Solve button.

***Step* 6. Results**

Under **Outline** and under **Solution (A6)** click on **Total Deformation**. The result is as shown below.

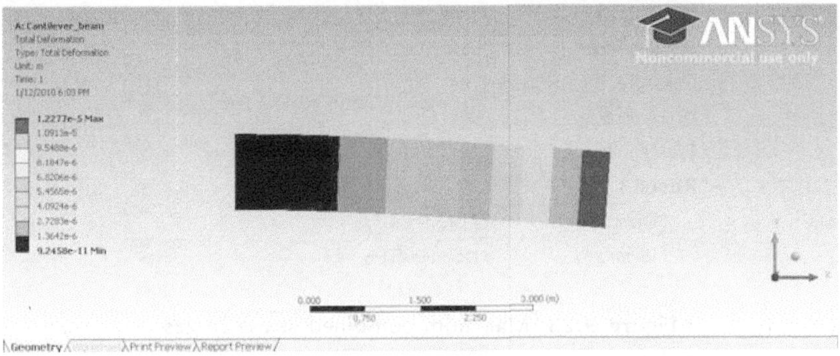

Figure 9.20. Deflection pattern for a cantilever beam

Details of "Total Deformation"	
Scope	
Scoping Method	Geometry Selection
Geometry	All Bodies
Definition	
Type	Total Deformation
By	Time
Display Time	Last
Calculate Time History	Yes
Identifier	
Results	
Minimum	9.2458e-011 m
Maximum	1.2277e-005 m
Information	

Figure 9.21. Deflection values for a cantilever beam

Then double click on **Beam Tool** click on **Maximum combined stress**. This result is as shown below.

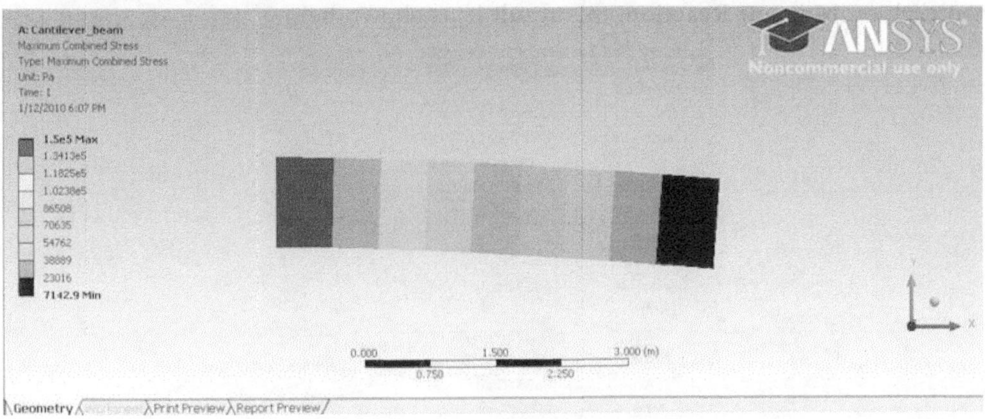

Figure 9.22. Maximum combined stress on a cantilever beam

Details of "Maximum Combined Stress"	
Definition	
Type	Maximum Combined Stress
By	Time
Display Time	Last
Calculate Time History	Yes
Use Average	Yes
Identifier	
Results	
Minimum	7142.9 Pa
Maximum	1.5e+005 Pa

Figure 9.23. Maximum combined stress values

Then click on **Force Reaction**, this result is as shown below

Details of "Force Reaction"	
Definition	
Type	Force Reaction
Location Method	Boundary Condition
Boundary Condition	Remote Displacement
Orientation	Global Coordinate System
Options	
Result Selection	All
Display Time	End Time
Results	
X Axis	8.5265e-014 N
Y Axis	5000. N
Z Axis	-7.9581e-013 N
Total	5000. N

Figure 9.24. Values of reaction force

Then click on **Moment Reaction,** this result is as shown below

Details of "Moment Reaction"	
Definition	
Type	Moment Reaction
Location Method	Boundary Condition
Boundary Condition	Remote Displacement
Orientation	Global Coordinate System
Summation	Centroid
Options	
Result Selection	All
Display Time	End Time
Results	
X Axis	-2.1032e-012 N·m
Y Axis	-5.1159e-013 N·m
Z Axis	25000 N·m
Total	25000 N·m

Figure 9.25. Values of reaction moment

(II) Analytical method (Refer figure 9.1)

Maximum bending stress occurs at the support

Bending moment at support M = P × L = 5 × 10^3 × 5 = 25000 N-m

$$\text{Section modulus } s = \frac{B \times H^2}{6} = \frac{1 \times 1^2}{6} = 0.1667 \text{ m}^3$$

$$\text{Maximum stress } \sigma_{max} = \frac{M}{s} = \frac{25000}{0.1667} = 150000 \text{ Pa} = 0.15 \text{ MPa}$$

$$\text{Moment of inertia} = I = \frac{B \times H^3}{12} = \frac{1 \times 1^3}{12} = 0.08333 \text{ m}^4$$

$$\text{Maximum deflection } \delta_{max} = \frac{PL^3}{3EI} = \frac{5 \times 10^3 \times (5)^3}{3 \times 210 \times 10^9 \times 0.08333} = 1.2 \times 10^{-5} \text{ m} = 0.012 \text{ mm}.$$

ANSWERS OF PROBLEM 9.1

Parameter	Software results	Analytical method
Maximum deflection	0.012277 mm	0.012 mm
Maximum stress	0.15 MPa	0.15 MPa
Reaction force	5000 N	5000 N
Reaction moment	25000 N-m	25000 N-m

THANKS NOTE

We are very much indebted to ANSYS Incorporation USA and ANSYS Incorporation India for giving permission to use photoshots of ANSYS menus, dialog boxes and plots in our book.

REFERENCES

1. Singiresu S. Rao, The Finite Element Method in Engineering, Buutterworth–Heinemann, An Imprint of Elsevier, New Delhi, Fourth Edition, 2005.
2. Daryl L. Logan, A First Course in the Finite Element Method Using Algor, PWS Publishing Company, Boston, First Edition 1997.
3. Tirupathi R. Chandrapatla and Ashok D. Belegundu, Introduction to Finite Elements in Engineering, Pearson Education, New Delhi, Third Edition, 2006.
4. J. N. Reddy, An Introduction to the Finite Element Method, Tata McGraw-Hill Publishing Company Limited, New Delhi, Third Edition, 2005.
5. P. Seshu, Textbook of Finite Element Analysis, Prentice-Hall of India Private Limited, New Delhi, Fifth Edition, 2007.
6. Nitin S. Gokhale, Sanjay S. Deshpande, Sanjeev V. Bedekar and Anand N. Thite, Practical Finite Element Analysis, Finite to Infinite, Pune, First Edition, 2008.
7. S. M. Murigendrappa, Fundamentals of Finite Element Method, Interline Publishing, Bangalore, Second Edition, 2009.
8. S. Ramamrutham, Strength of Materials, Dhanpat Rai Publishing Company (P) Ltd. New Delhi. Fifteenth Edition 2004.
9. ANSYS Release 9.0 Tutorials.
10. ANSYS Tutorials of University of Alberta, http://www.mece.ualberta.ca
11. ANSYS Tutorials , http://mae.uta.edu/~lawrence
12. ANSYS Tutorials, http://www.andrew.cmu.edu
13. ANSYS Tutorials, https://confluence.cornell.edu/display/SIMULATION

APPENDIX

Appendix
Colour Figures

Chapter 2

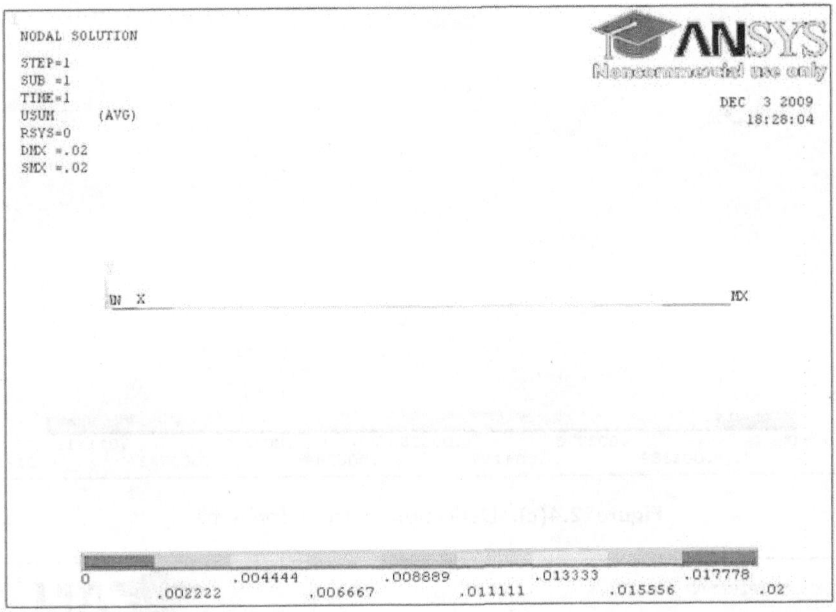

Figure 2.3(c). Deflection pattern for a bar

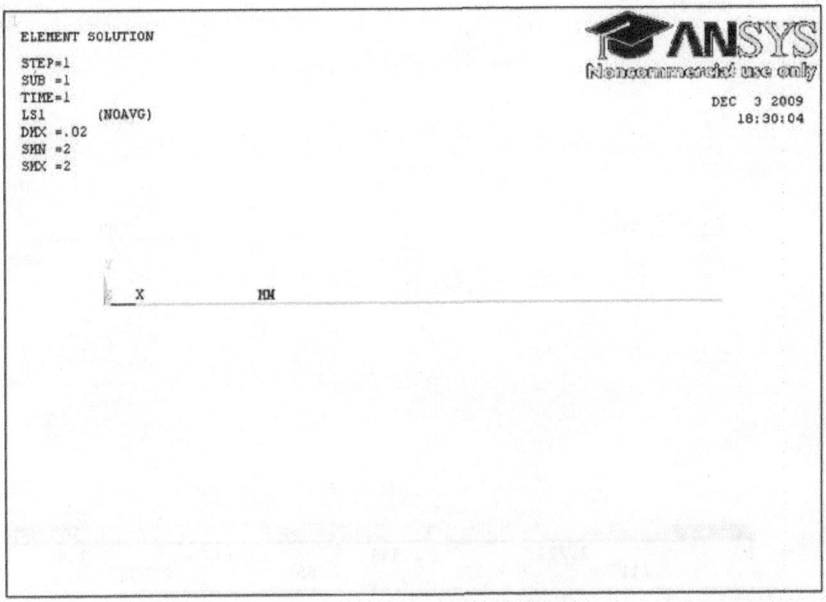

Figure 2.3(d). Stress pattern for a bar

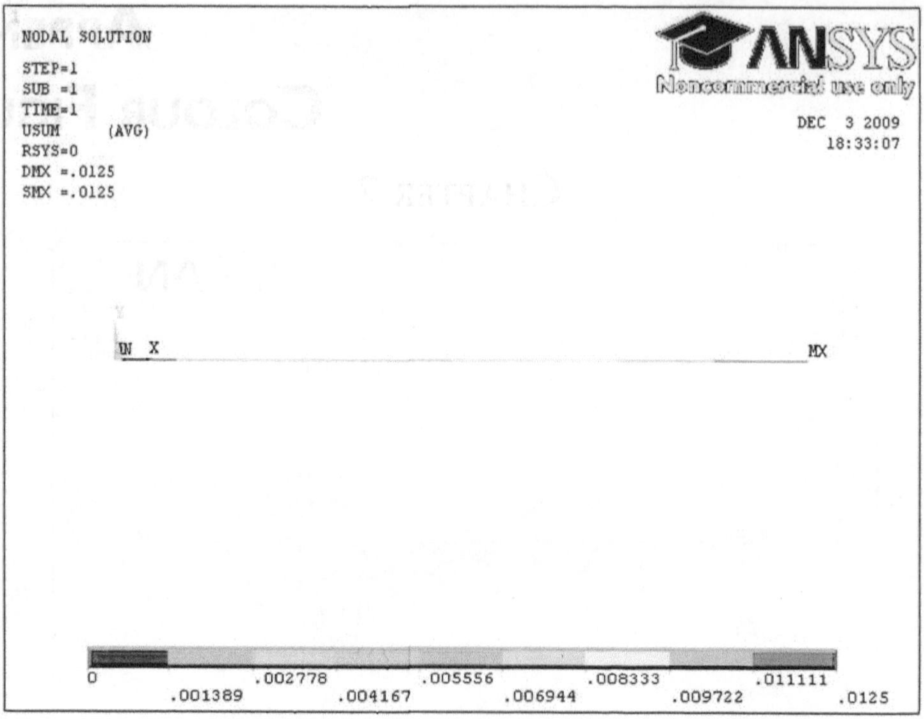

Figure 2.4(c). Deflection pattern for a bar

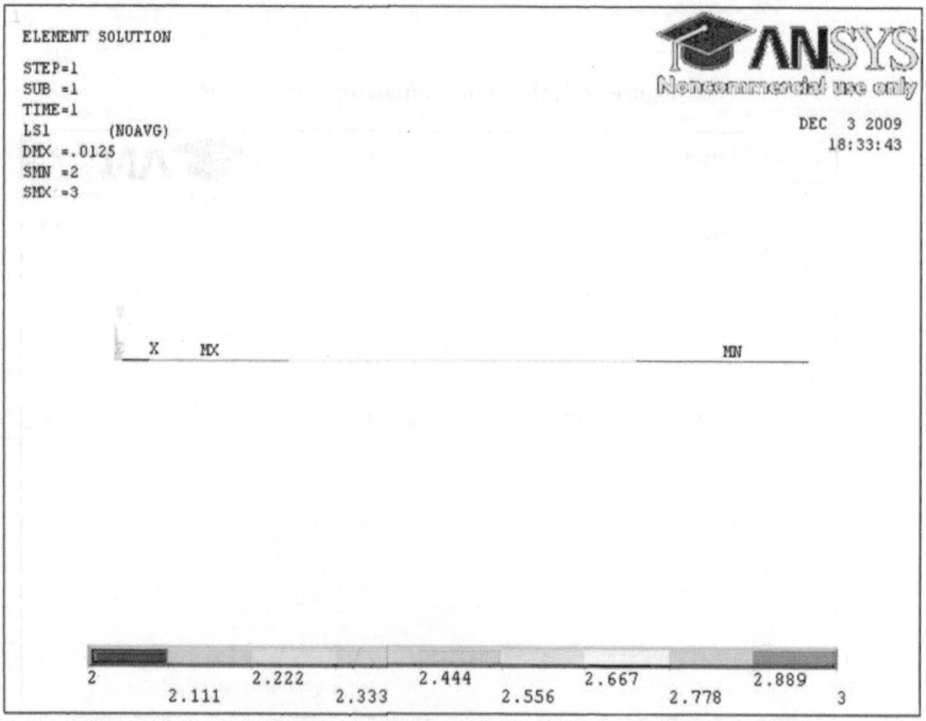

Figure 2.4(d). Stress pattern for a bar

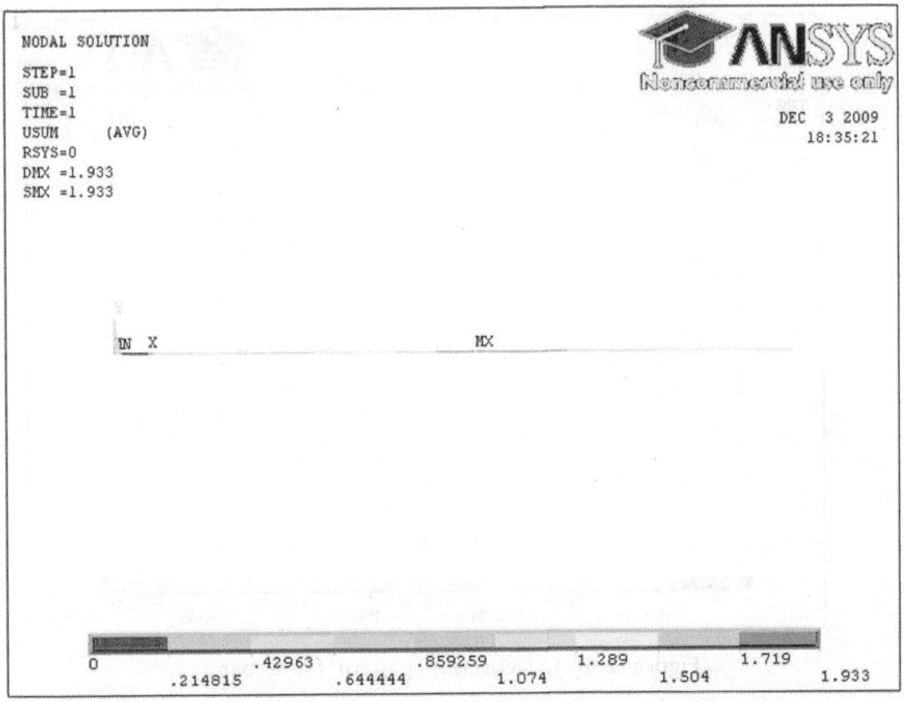

Figure 2.5(c). Deflection pattern for a bar

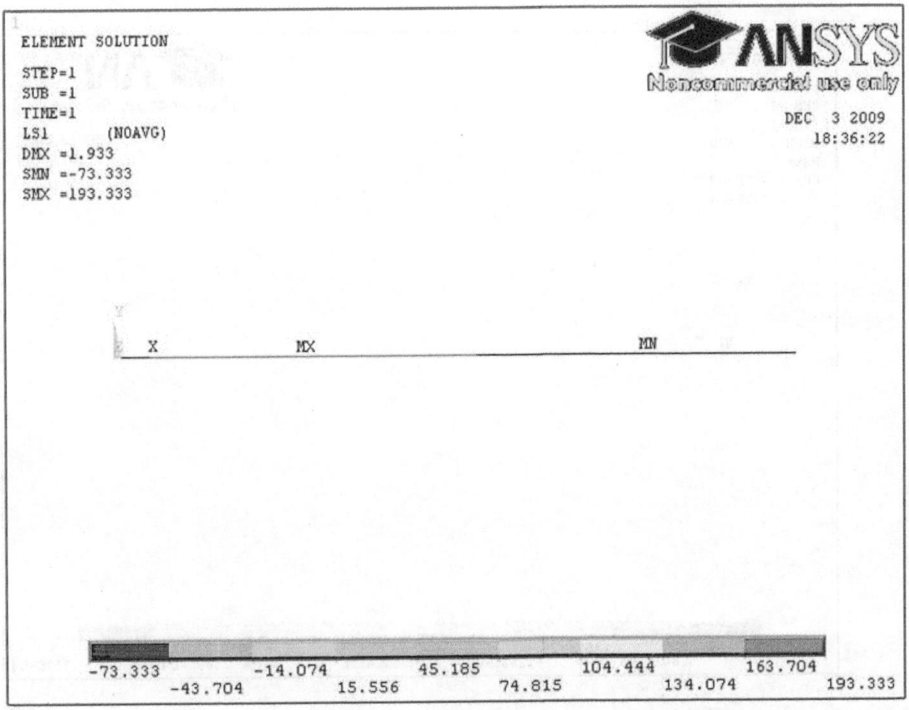

Figure 2.5(d). Stress pattern for a bar

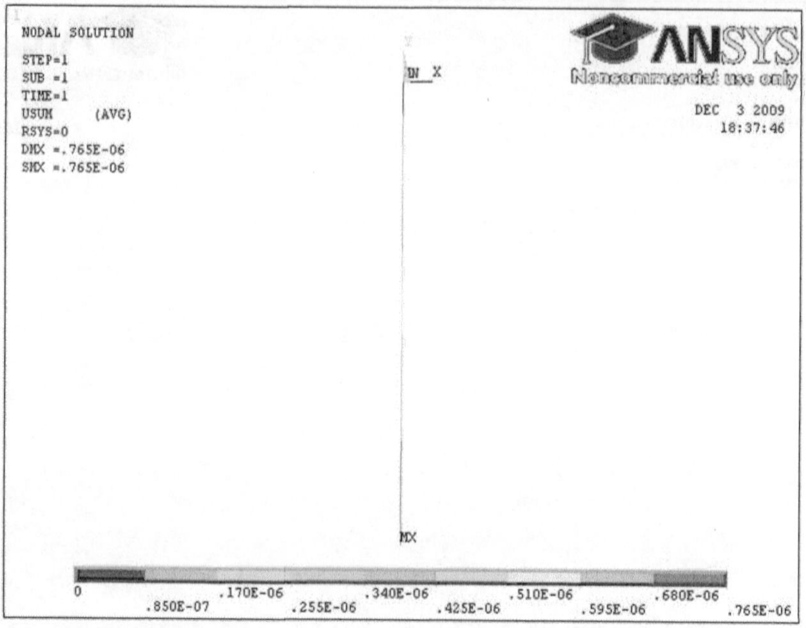

Figure 2.6(c). Deflection pattern for a bar

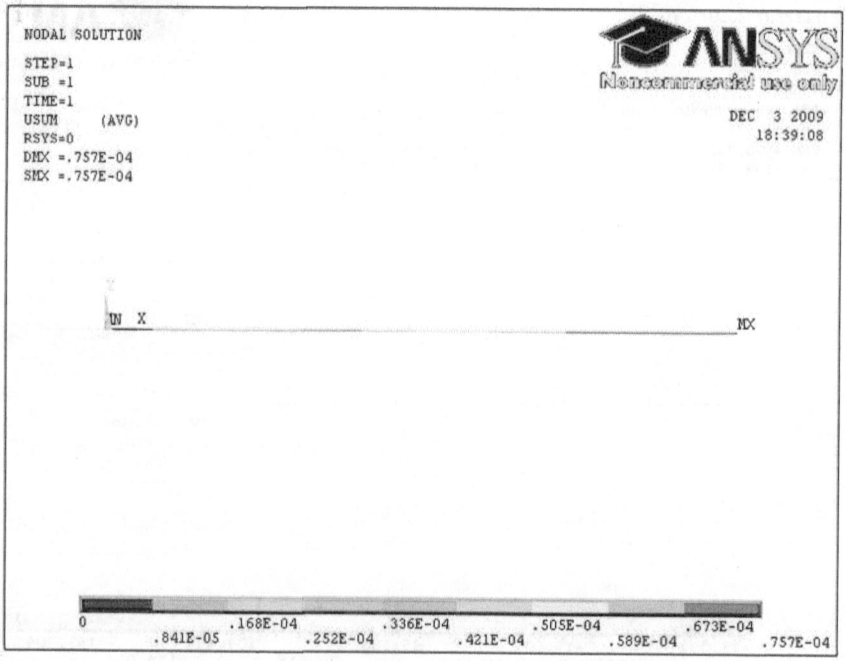

Figure 2.7(c). Deflection pattern for a rod

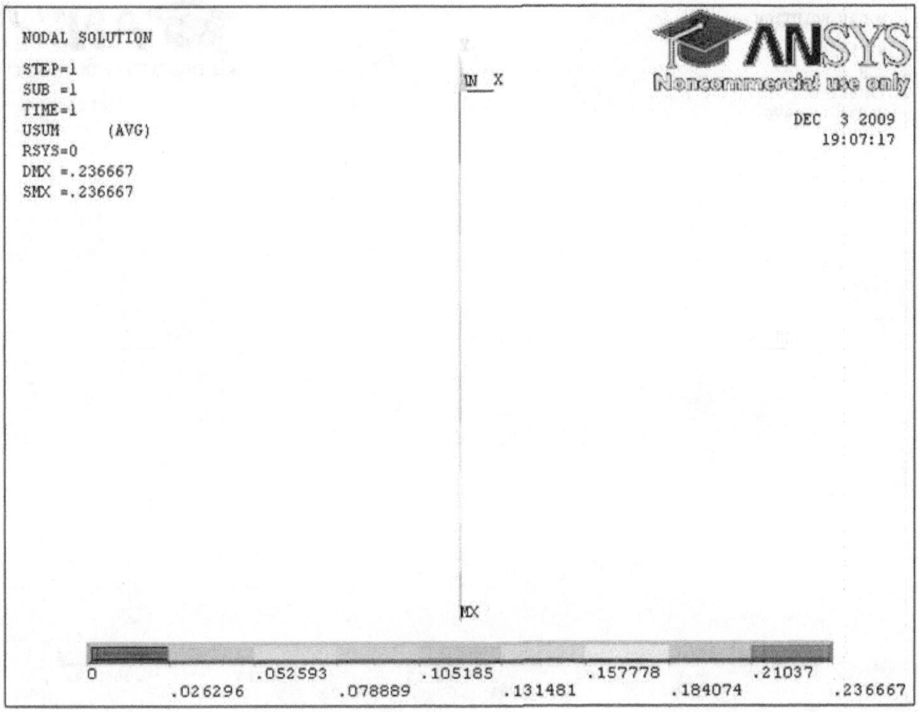

Figure 2.16(d). Deflection pattern for a tapered bar

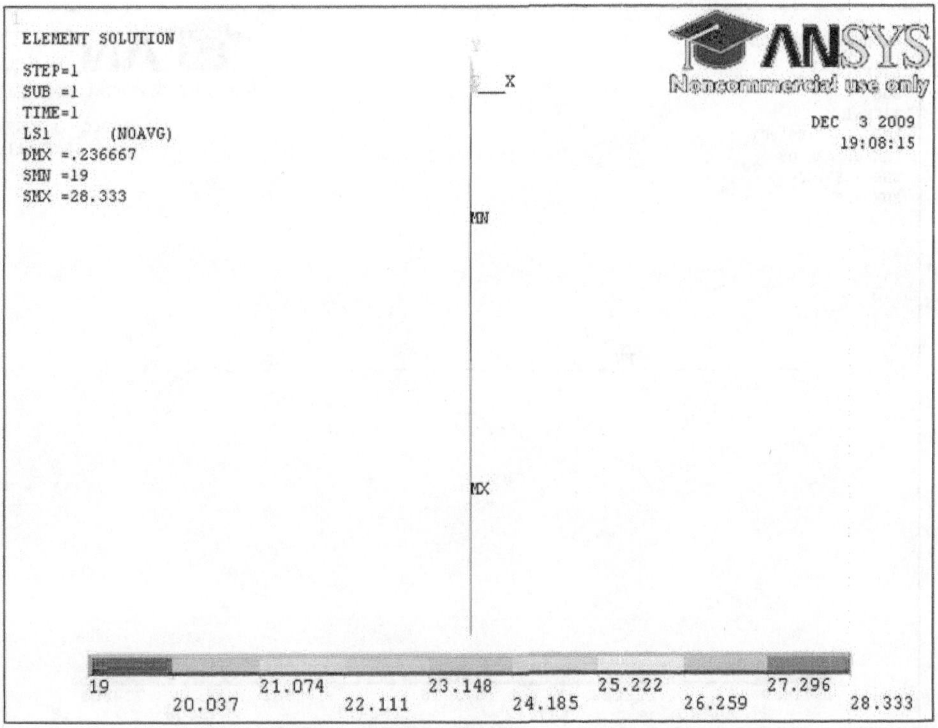

Figure 2.16(e). Stress pattern for a tapered bar

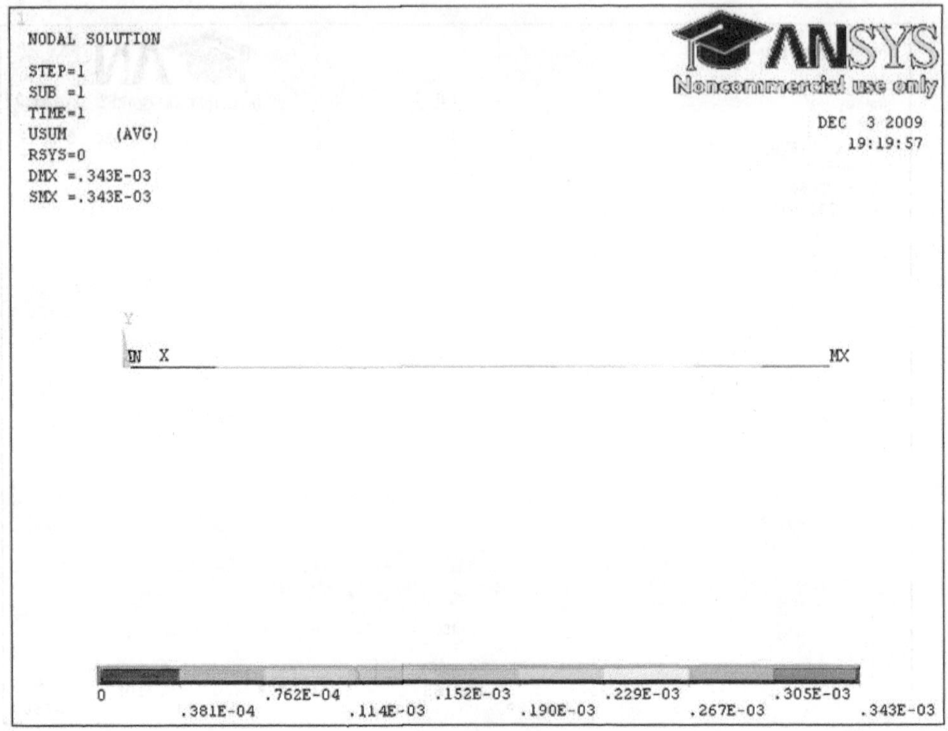

Figure 2.17(d). Deflection pattern for a tapered bar

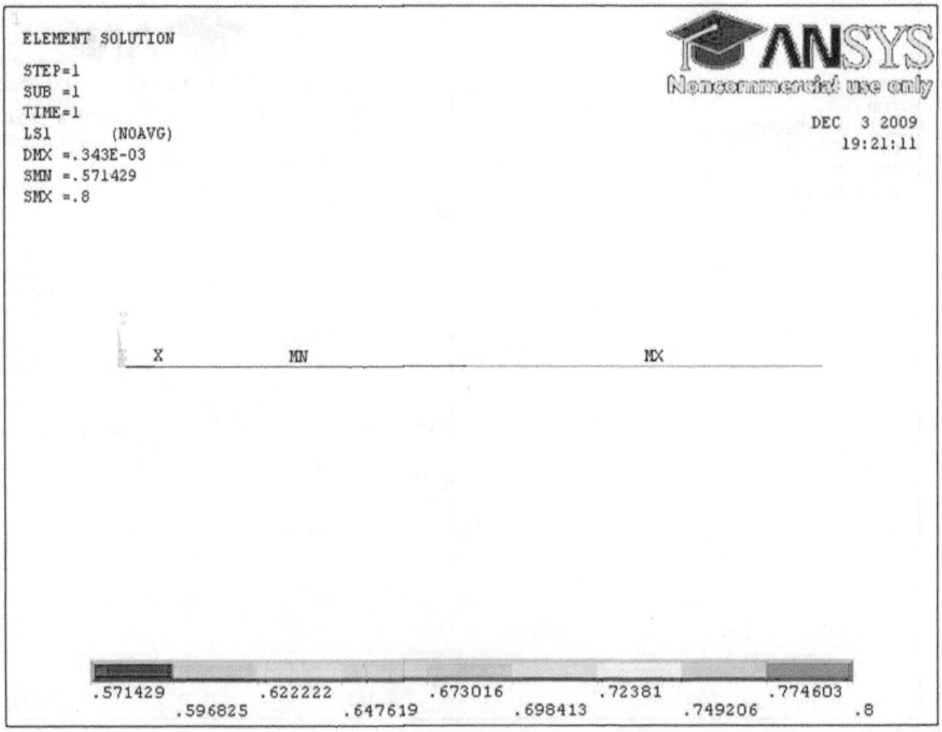

Figure 2.17(e). Stress pattern for a tapered bar

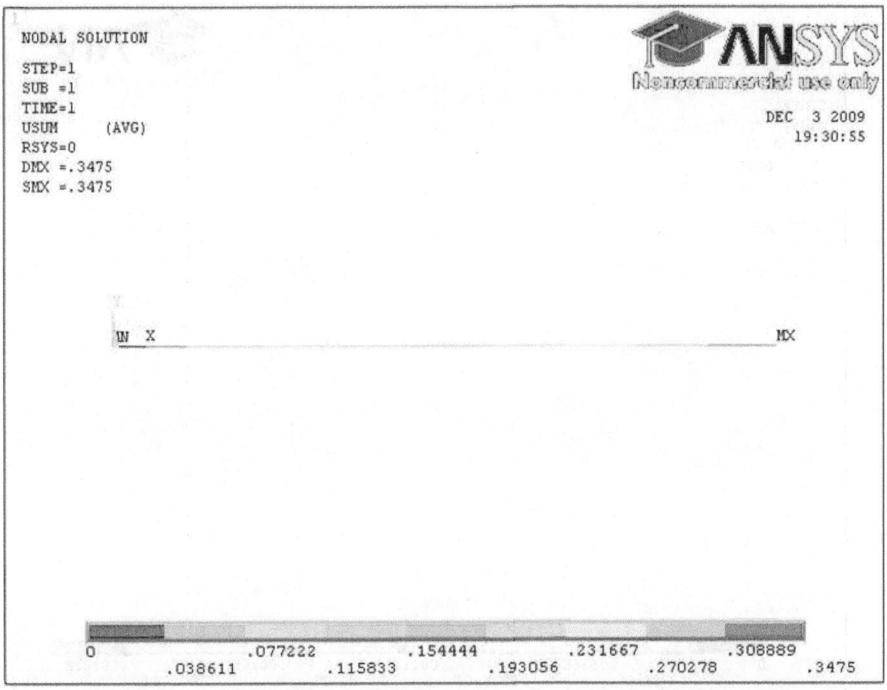

Figure 2.18(c). Deflection pattern for a tapered bar

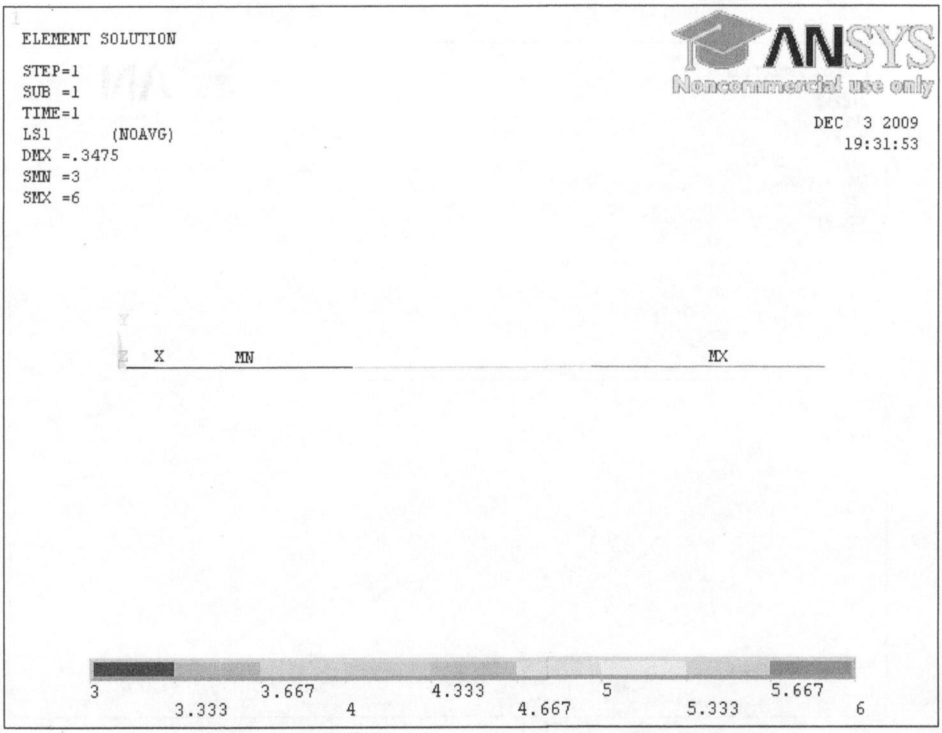

Figure 2.18(d). Stress pattern for a tapered bar

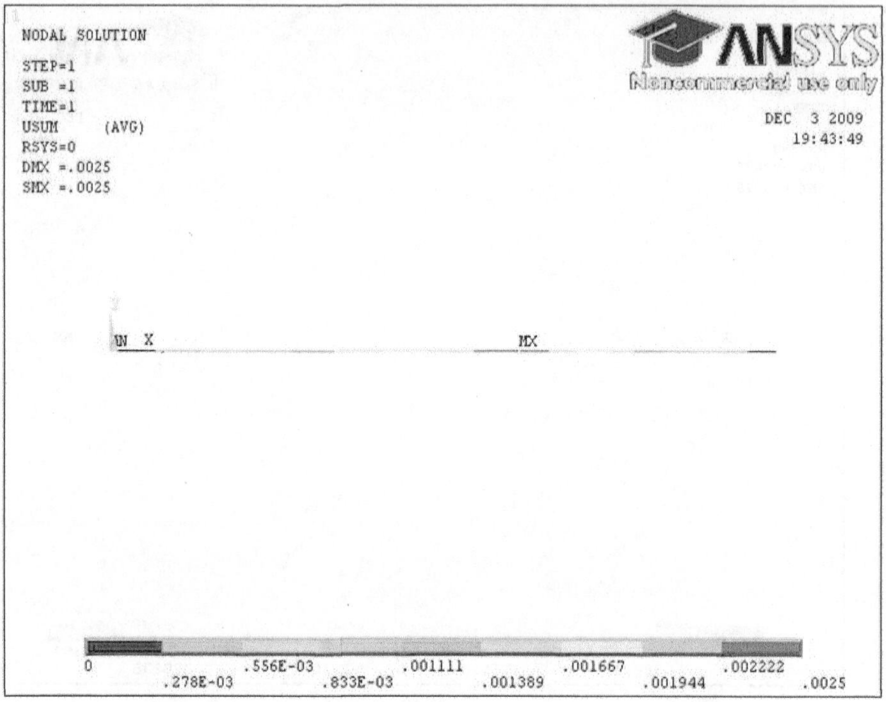

Figure 2.29(b). Deflection pattern for a stepped bar

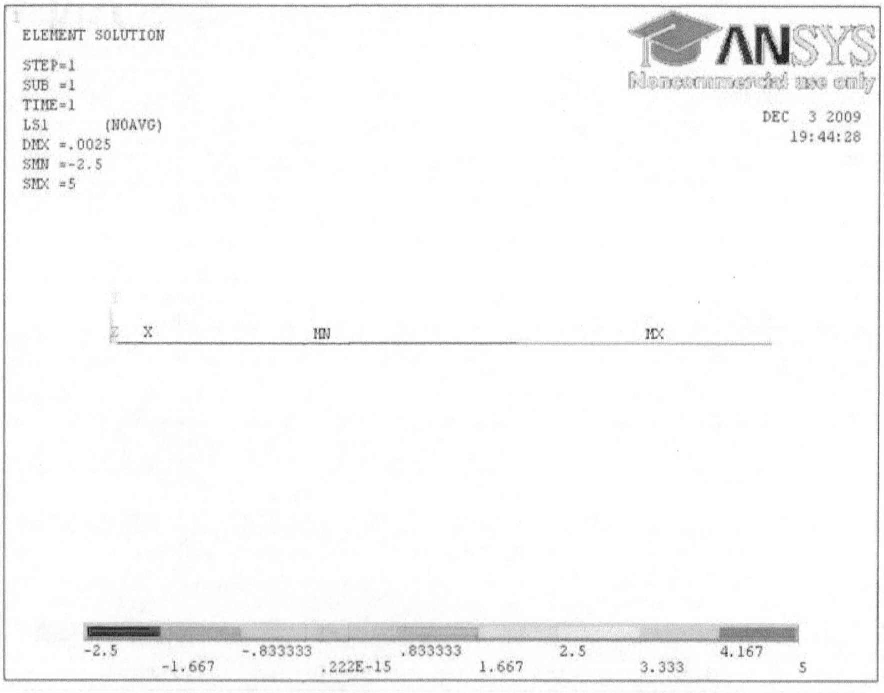

Figure 2.29(c). Stress pattern for a stepped bar

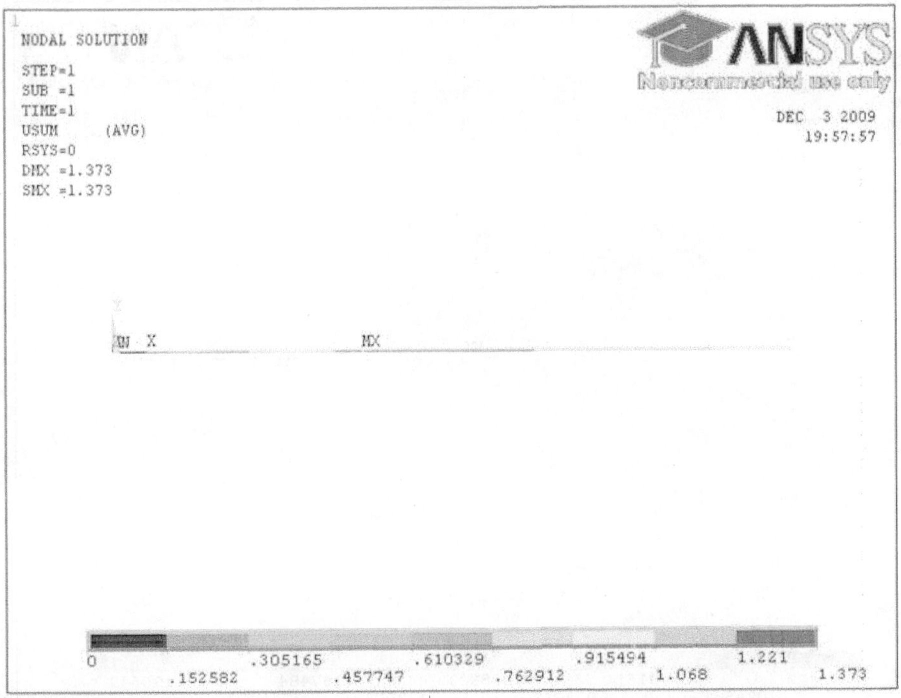

Figure 2.30(b). Deflection pattern for a stepped bar

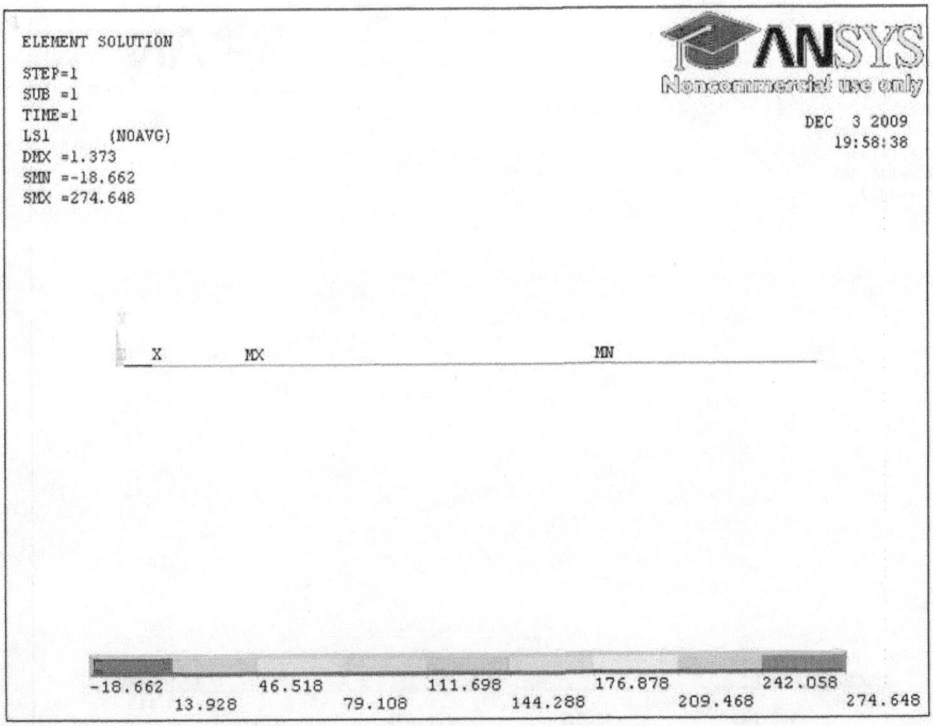

Figure 2.30(c). Stress pattern for a stepped bar

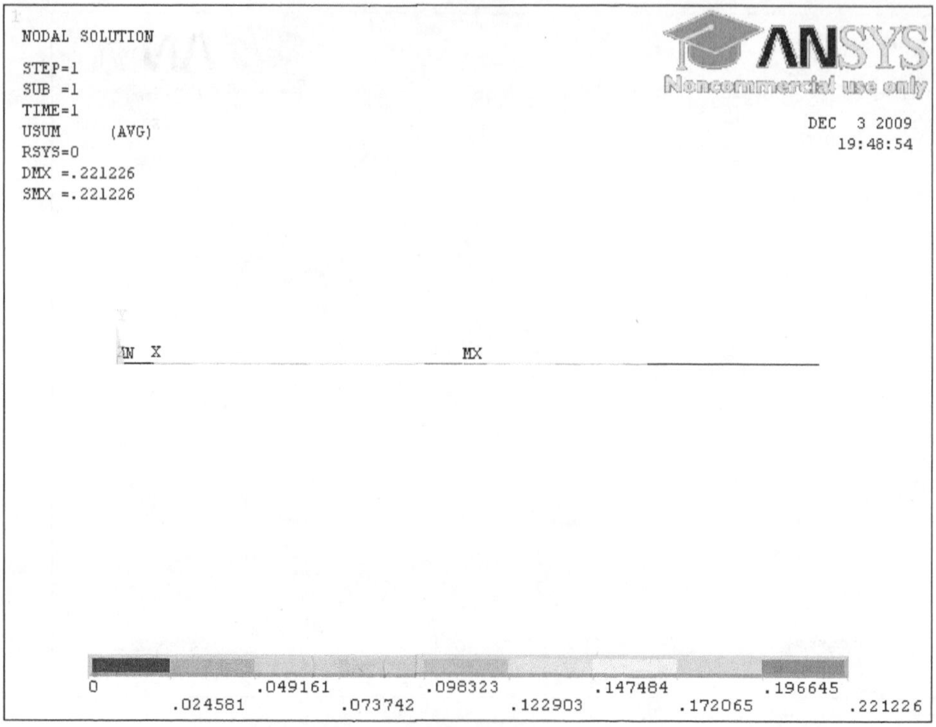

Figure 2.31(b). Deflection pattern for a stepped bar

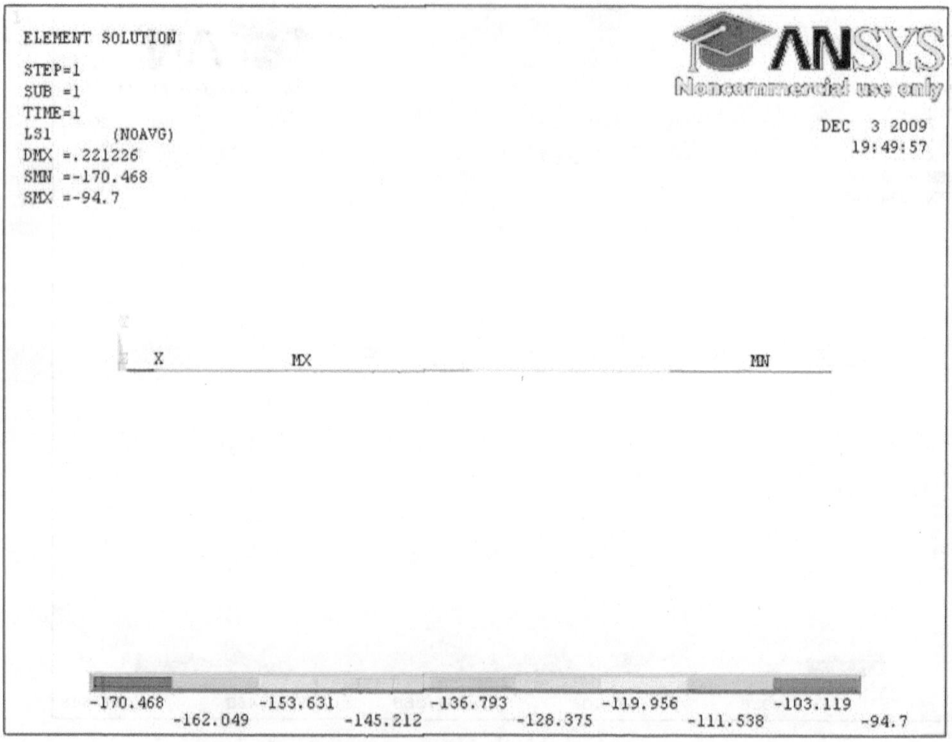

Figure 2.31(c). Stress pattern for a stepped bar

CHAPTER 3

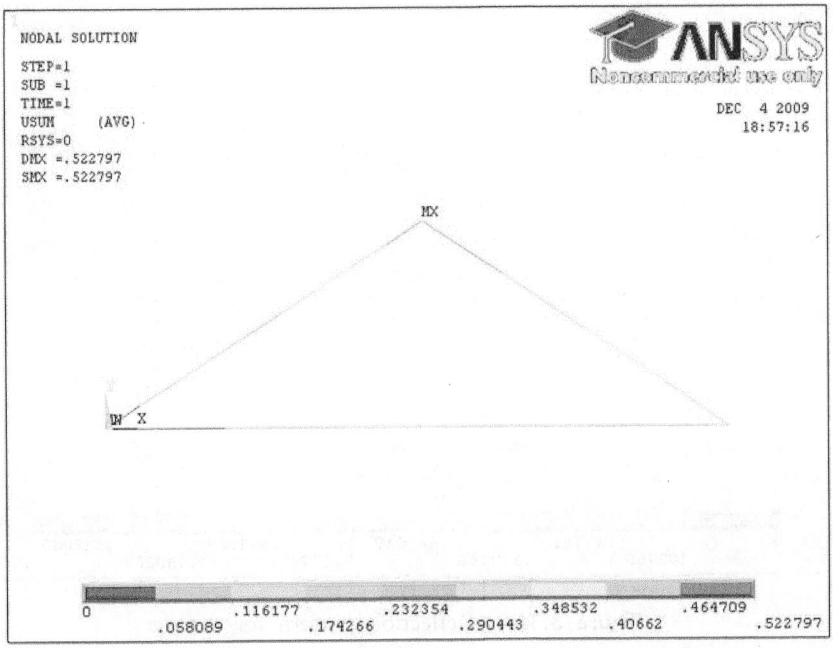

Figure 3.2(g). Deflection pattern for a truss

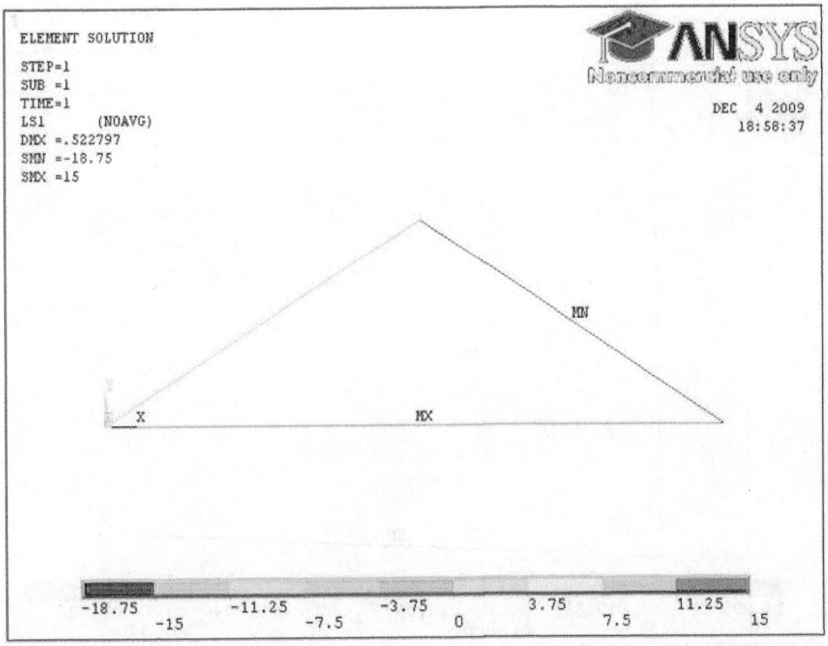

Figure 3.2(h). Stress pattern for a truss

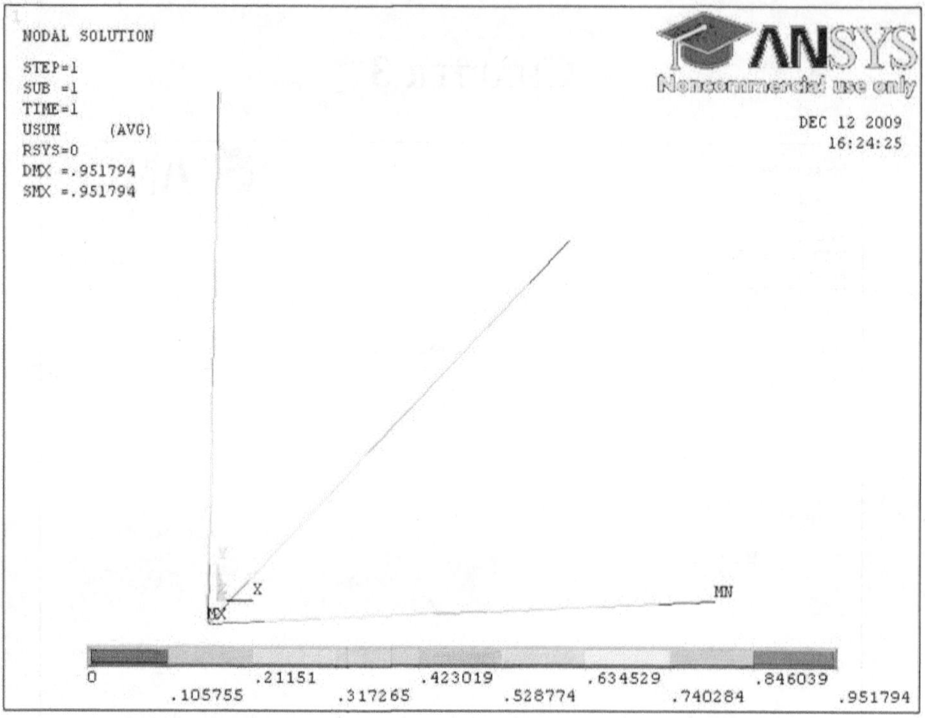

Figure 3.3(a). Deflection pattern for a truss

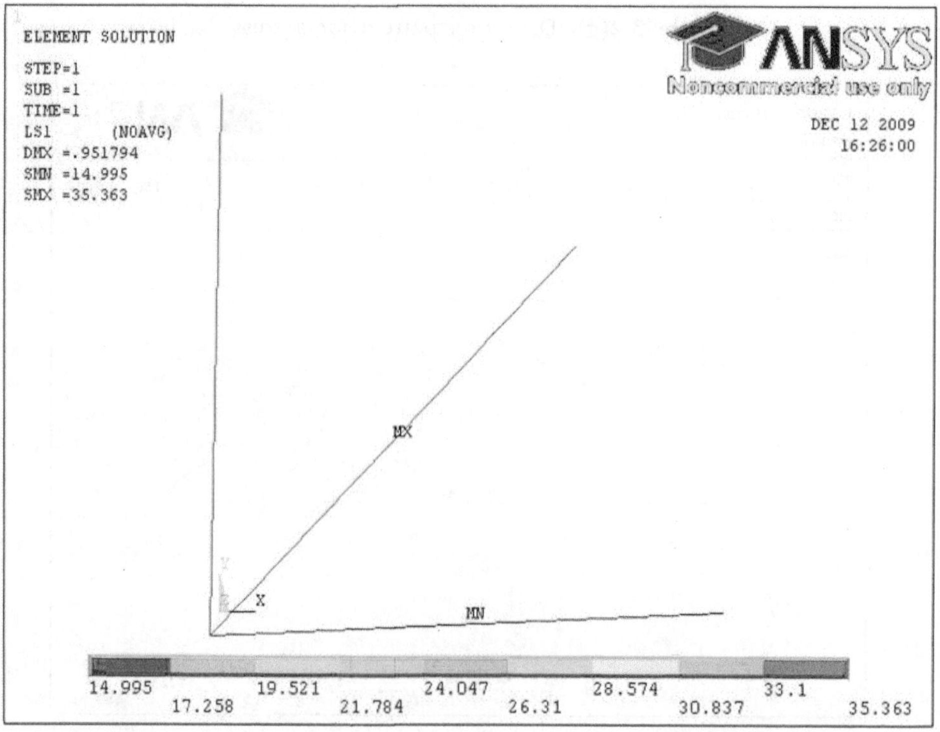

Figure 3.3(b). Stress pattern for a truss

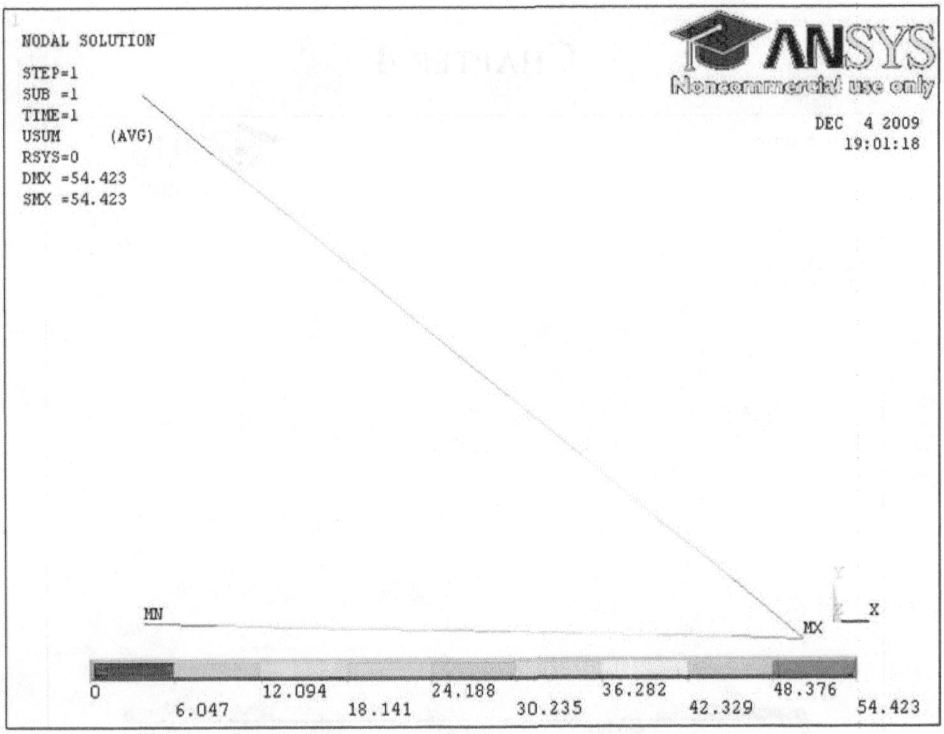

Figure 3.4(b). Deflection pattern for a truss

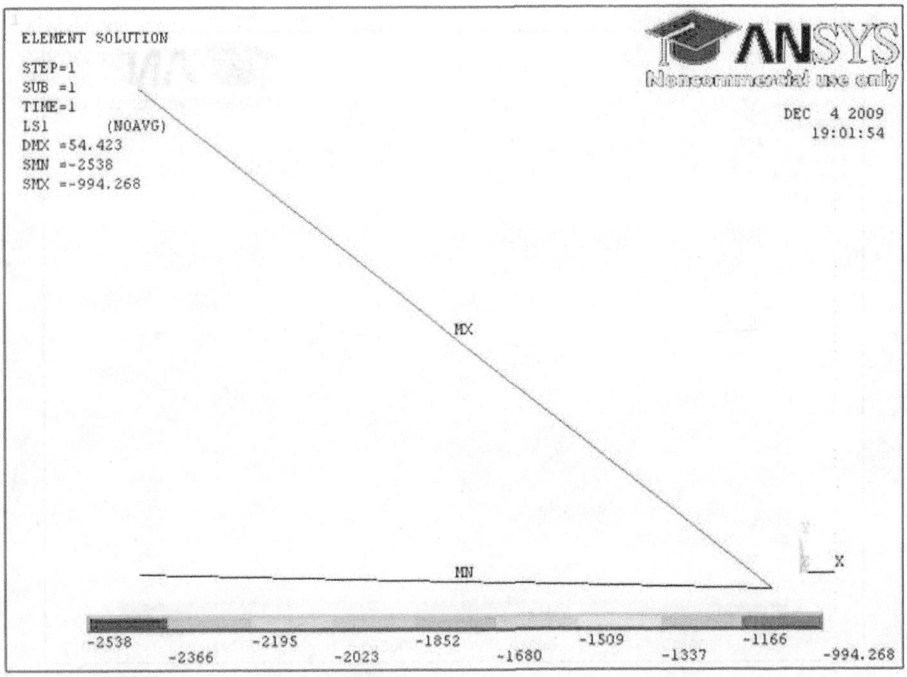

Figure 3.4(c). Stress pattern for a truss

Chapter 4

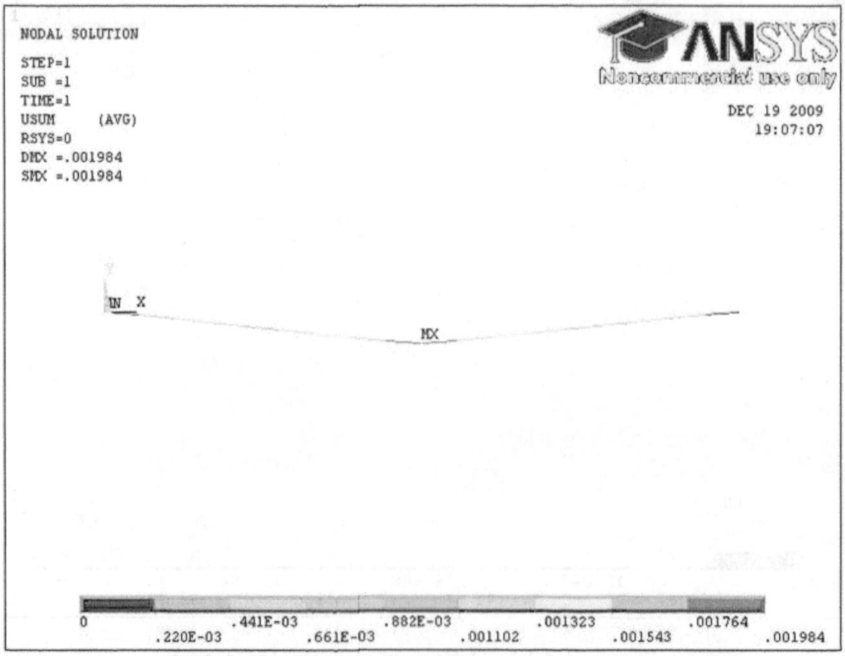

Figure 4.2(b). Deflection pattern for a simply supported beam

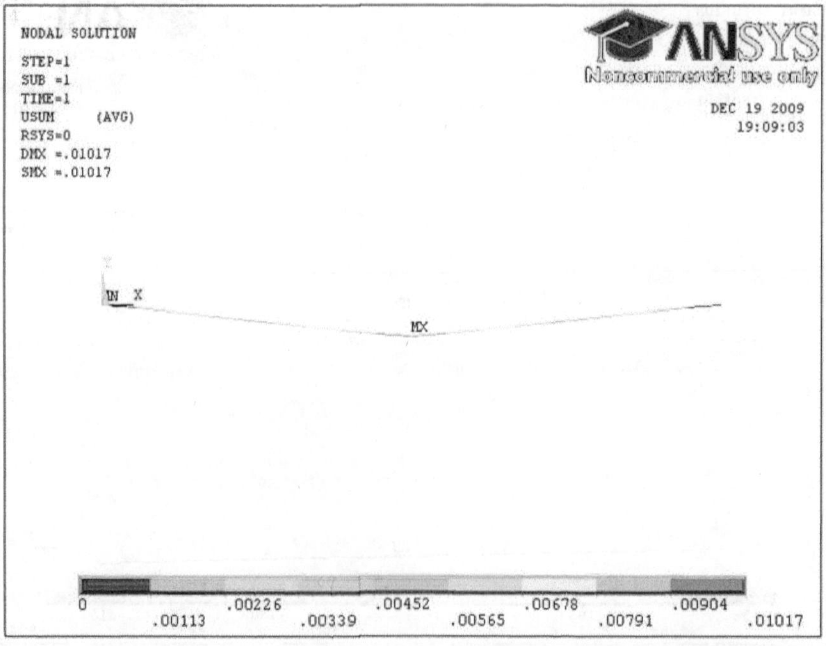

Figure 4.3(f). Deflection pattern for a simply supported beam

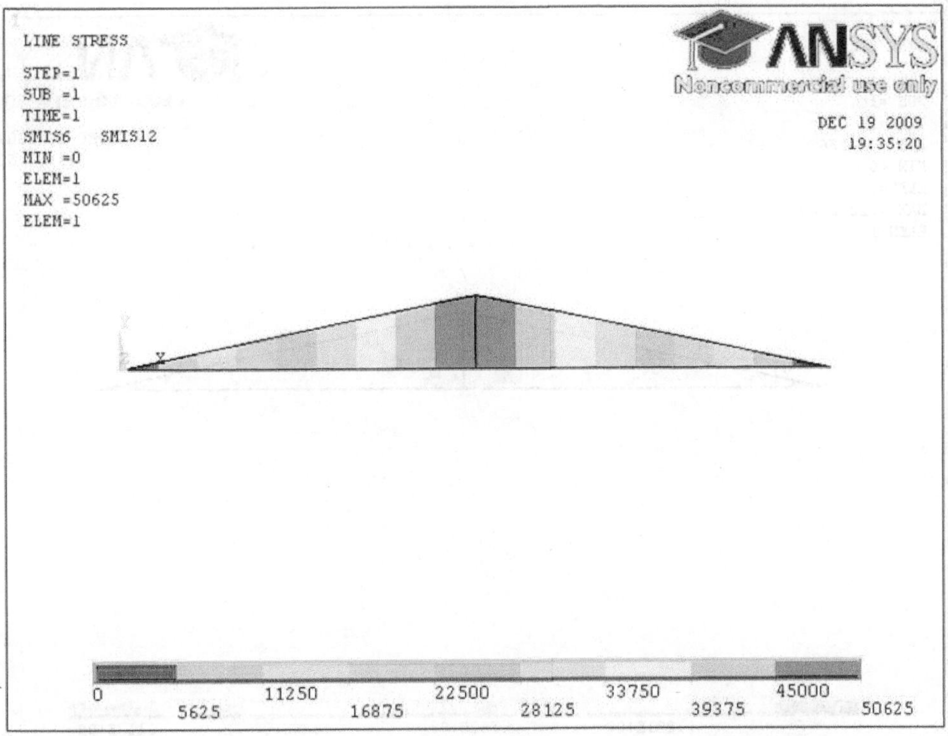

Figure 4.3(g). Bending moment diagram for a simply supported beam

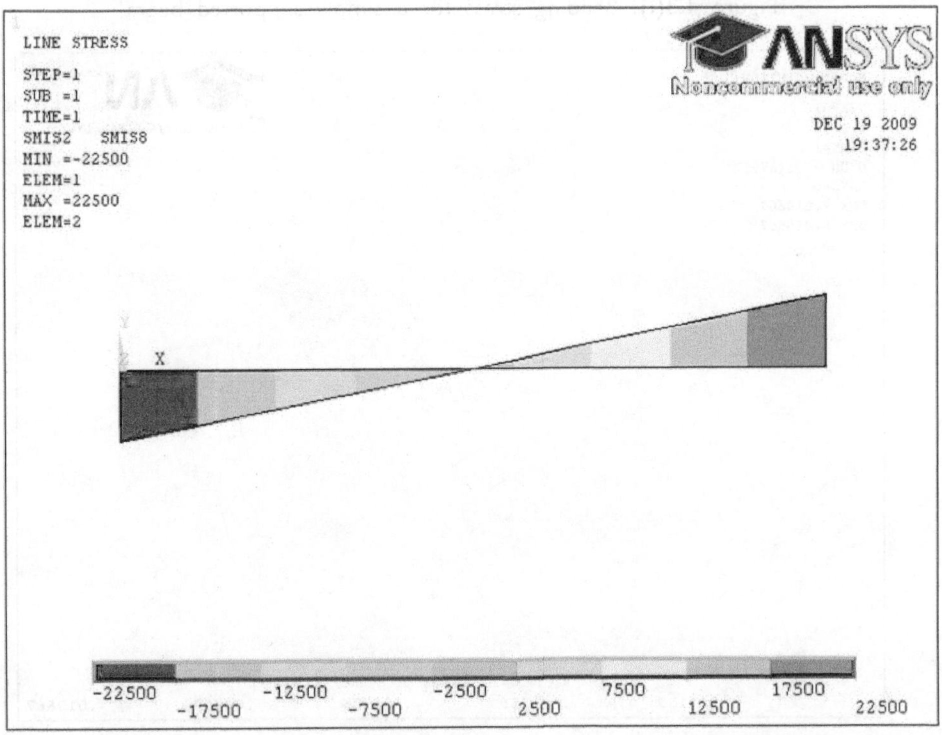

Figure 4.3(h). Shear force diagram for a simply supported beam

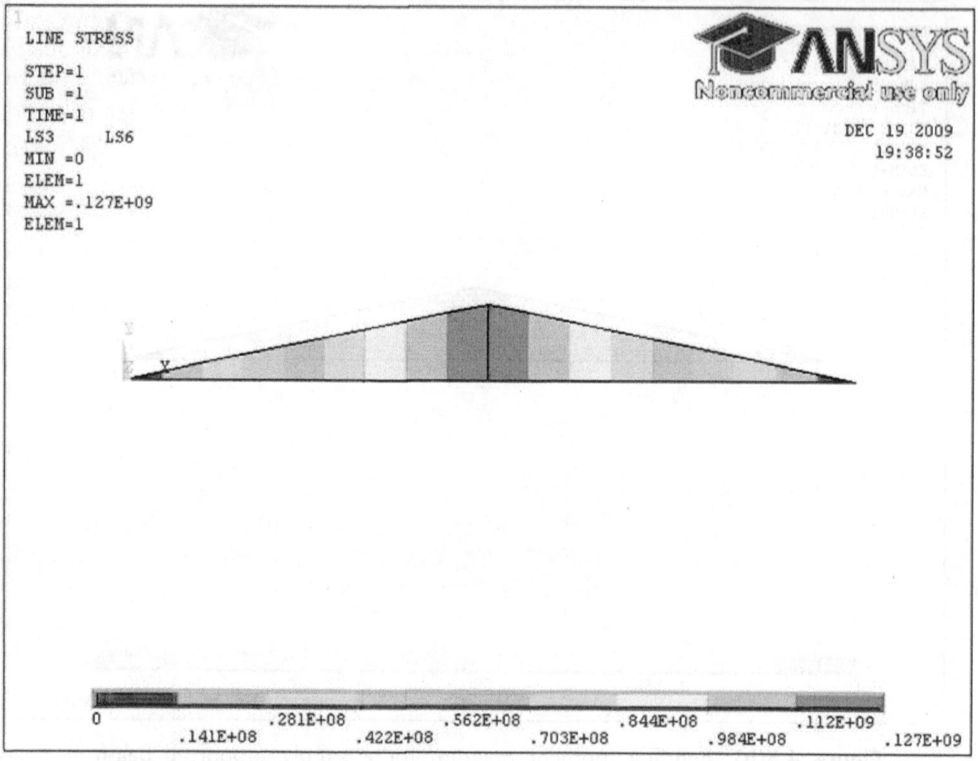

Figure 4.3(i). Bending stress for a simply supported beam

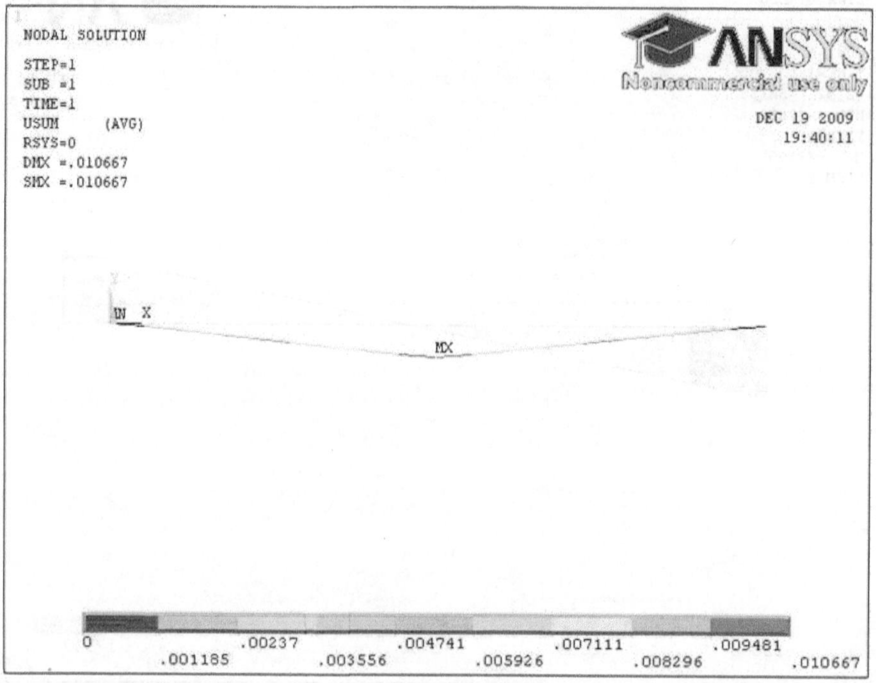

Figure 4.4(b). Deflection pattern for a simply supported beam

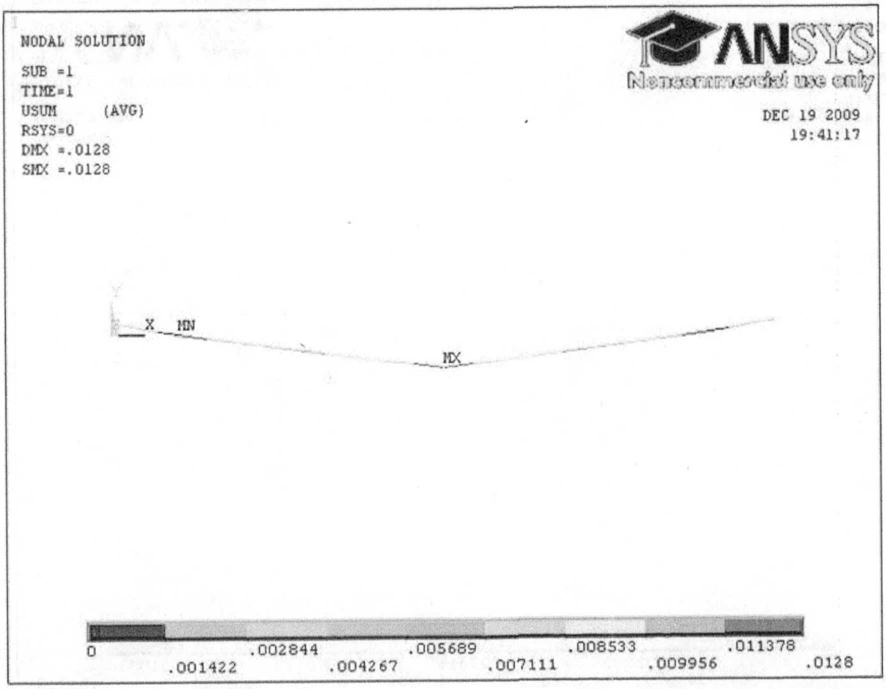

Figure 4.5(c). Deflection pattern for a simply supported beam

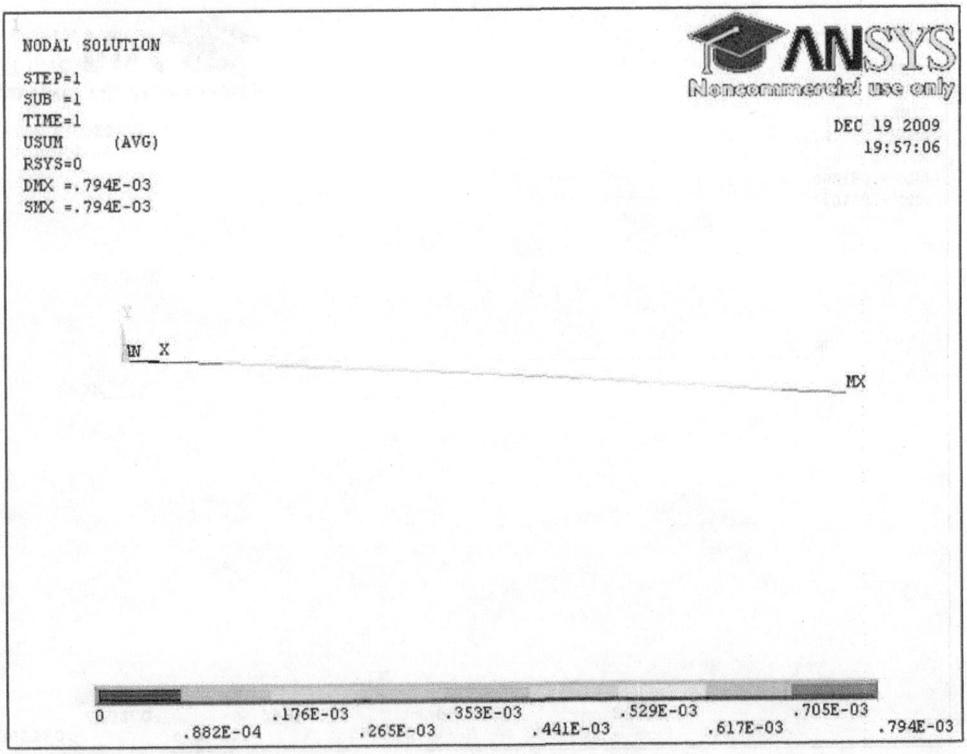

Figure 4.16(c). Deflection pattern for a cantilever beam

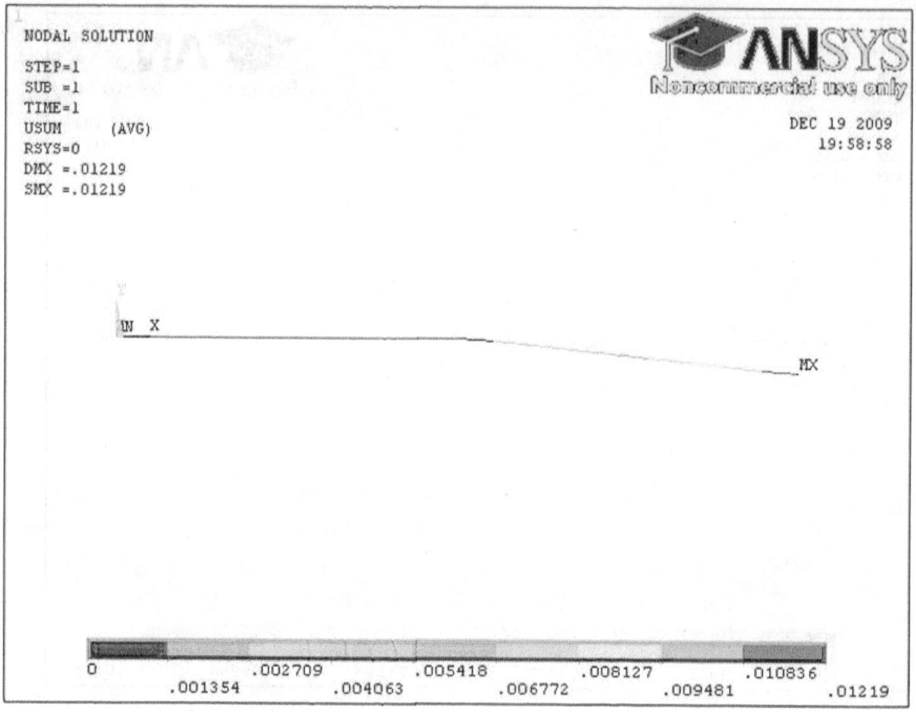

Figure 4.17(d). Deflection pattern for a cantilever beam

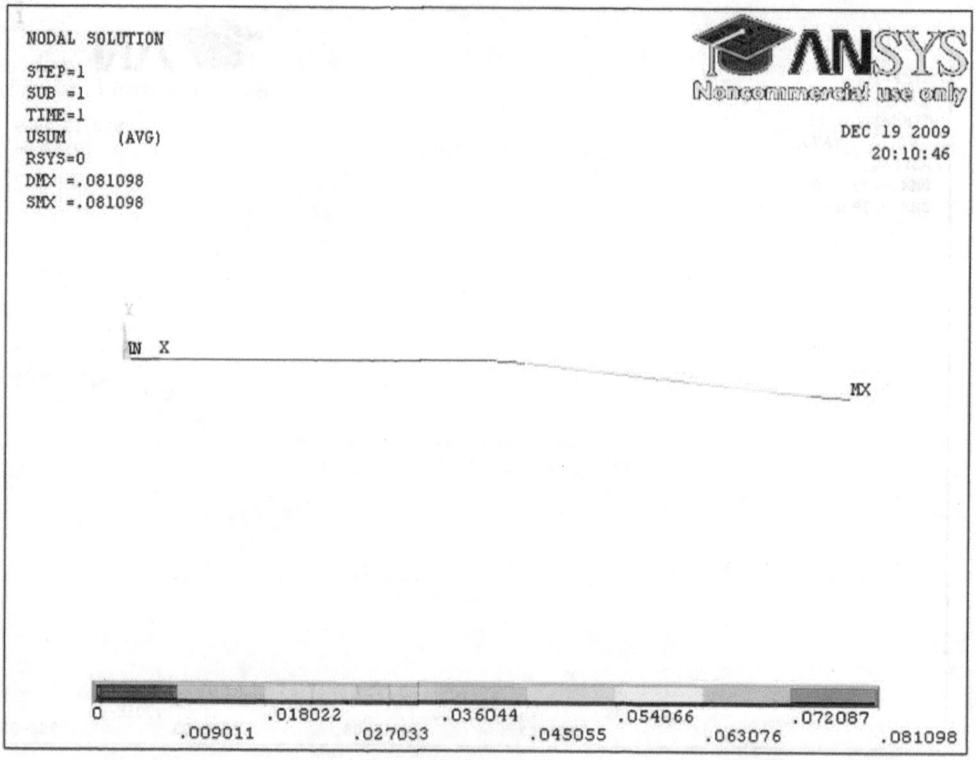

Figure 4.18(d). Deflection pattern for a cantilever beam

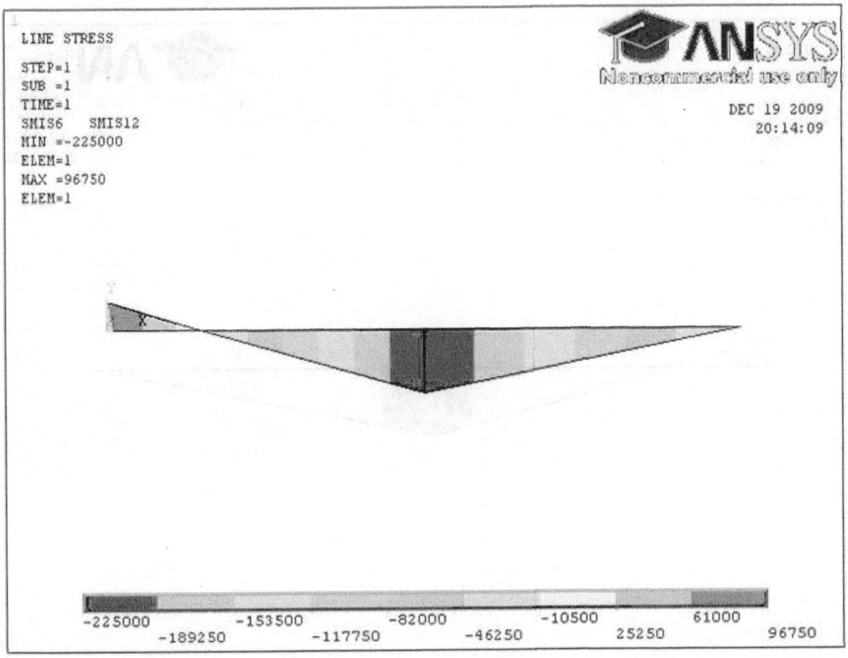

Figure 4.18(e) Bending moment diagram for a propped cantilever beam

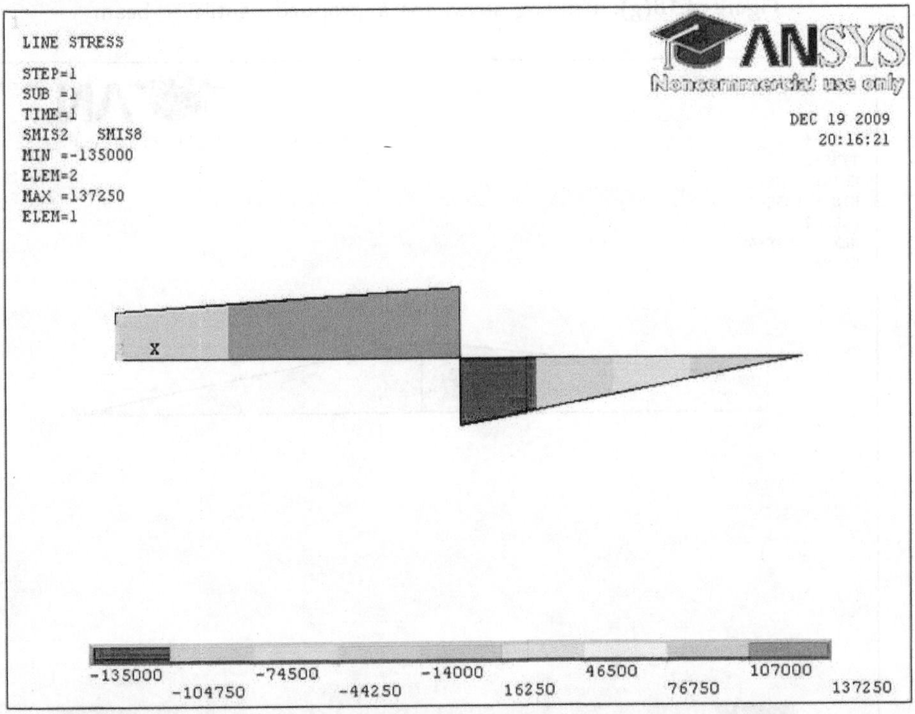

Figure 4.18(f). Shear force diagram for a propped cantilever beam

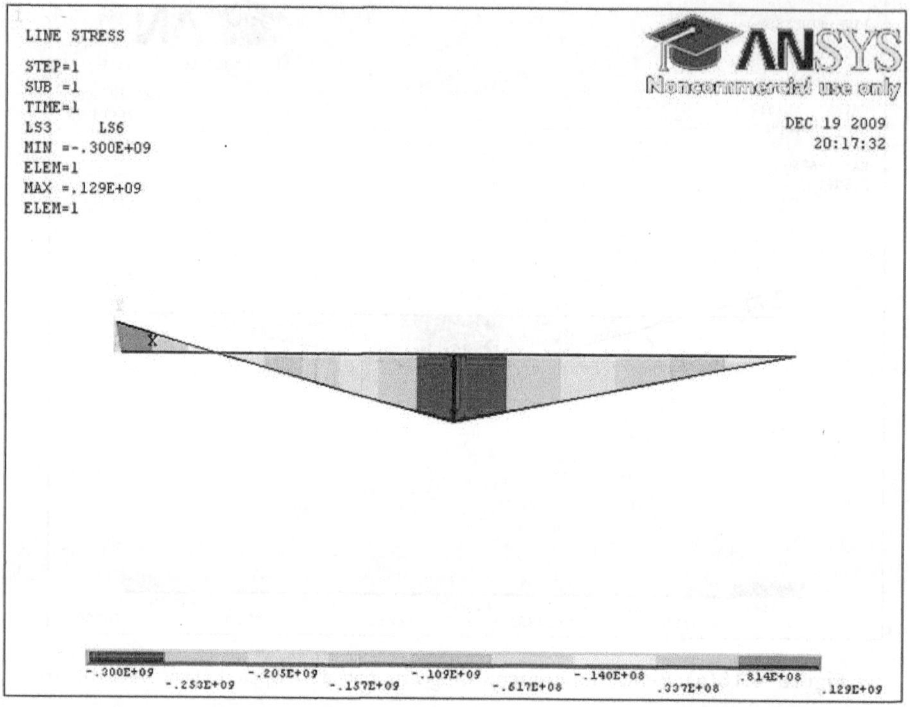

Figure 4.18(g). Bending stress for a propped cantilever beam

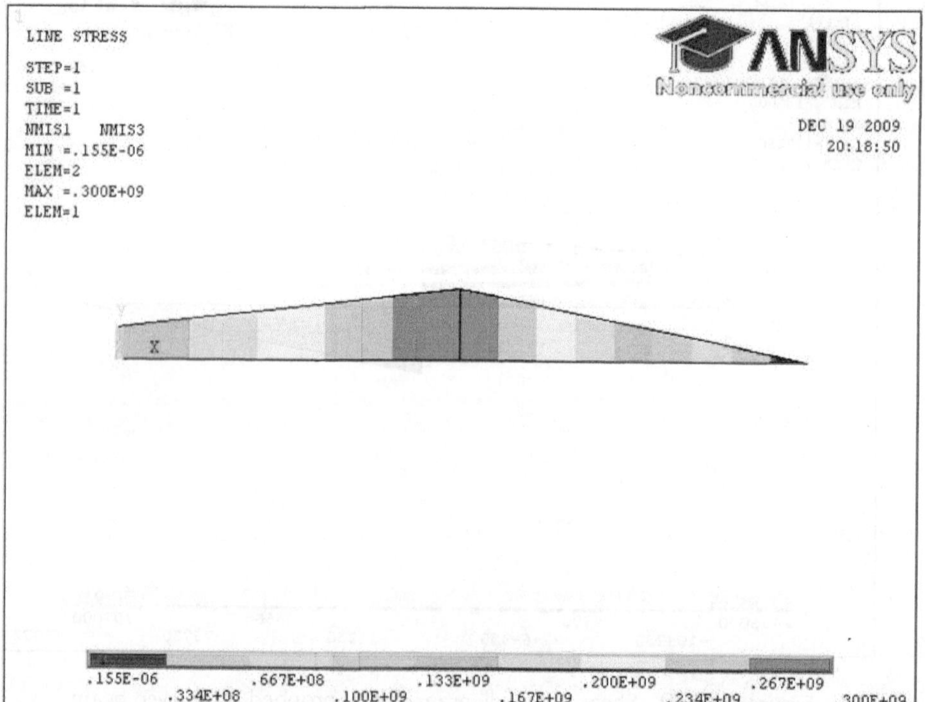

Figure 4.18(h). Maximum stress for a propped cantilever beam

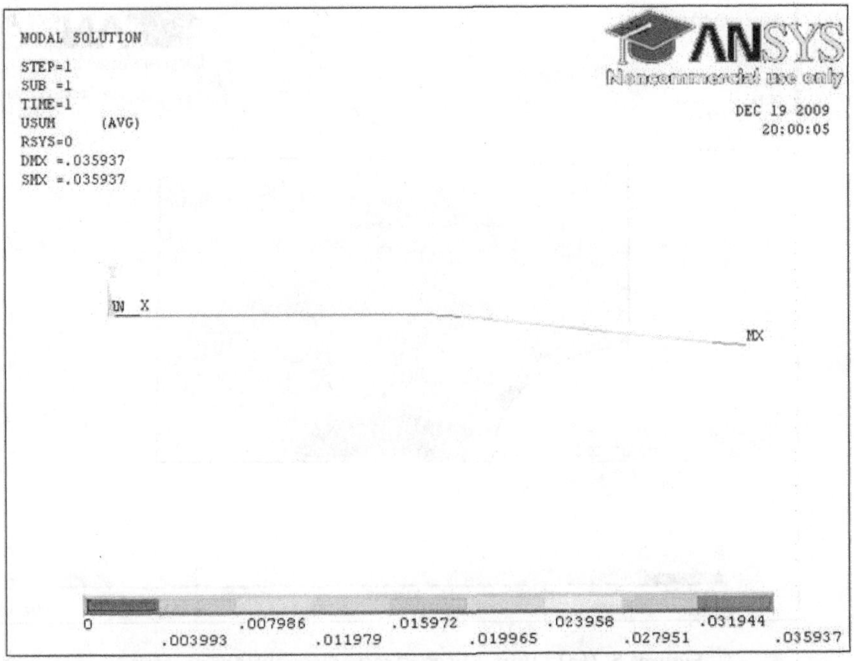

Figure 4.19(d). Deflection pattern for a cantilever beam

CHAPTER 5

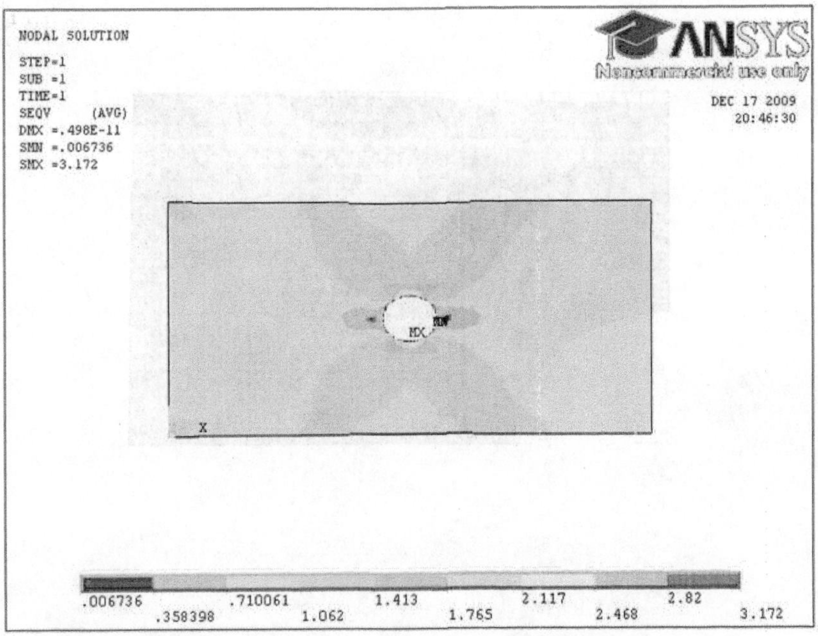

Figure 5.6(a). Von Mises stress distribution pattern

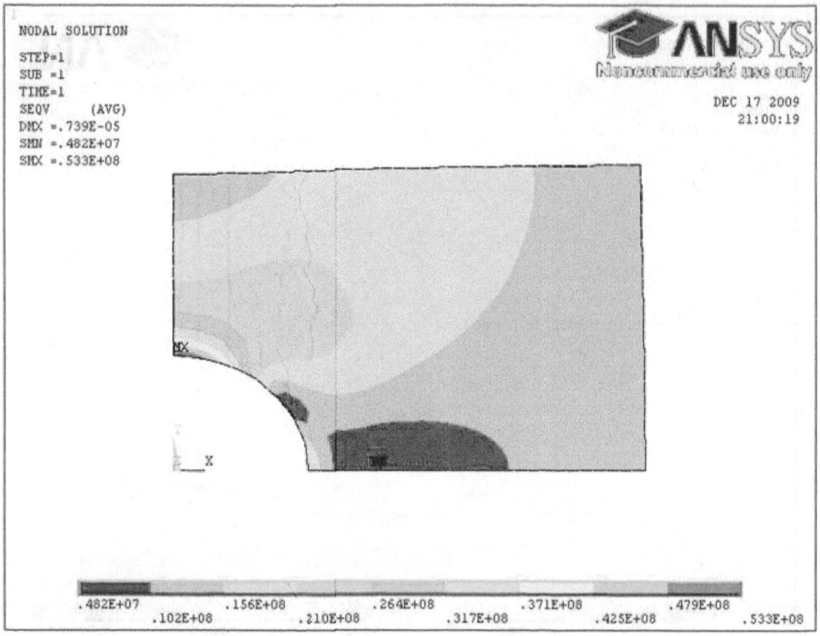

Figure 5.7(a). Von Mises stress distribution pattern

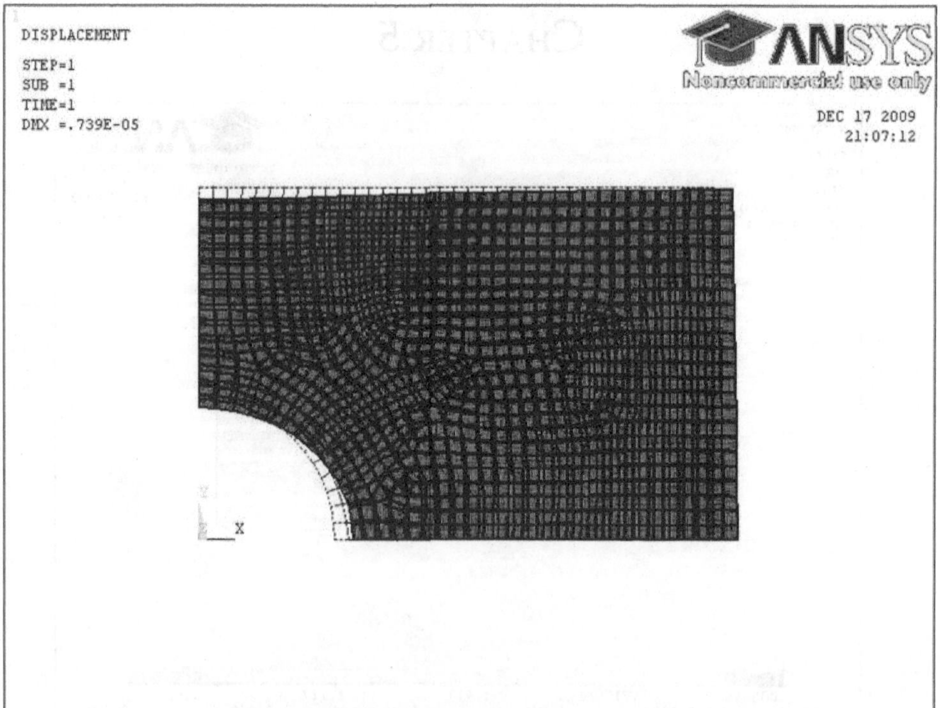

Figure 5.7(b). Deformed shape of the hole

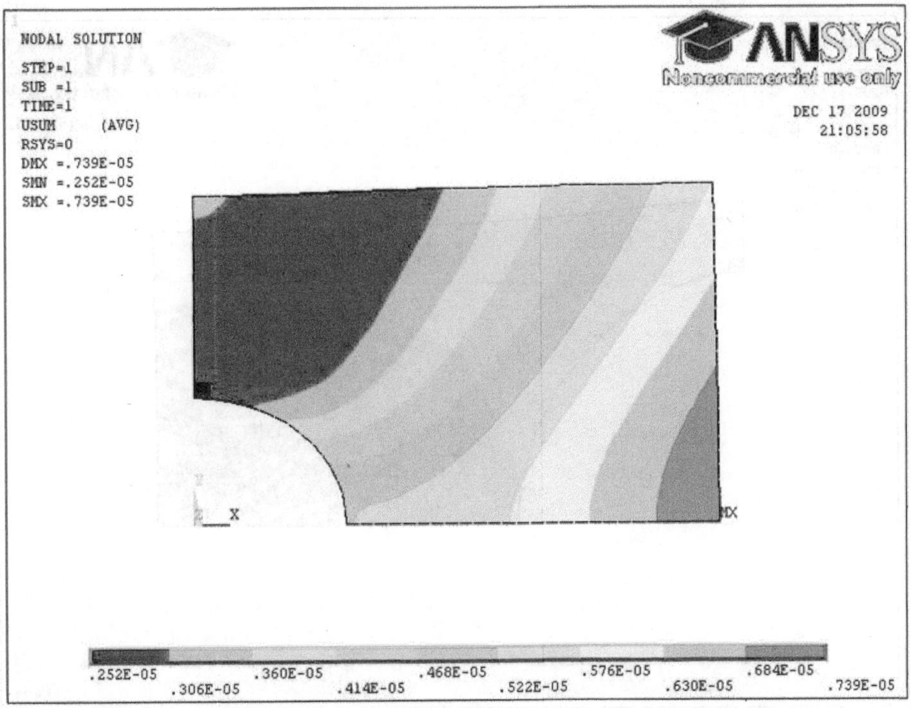

Figure 5.7(c). Deflection pattern

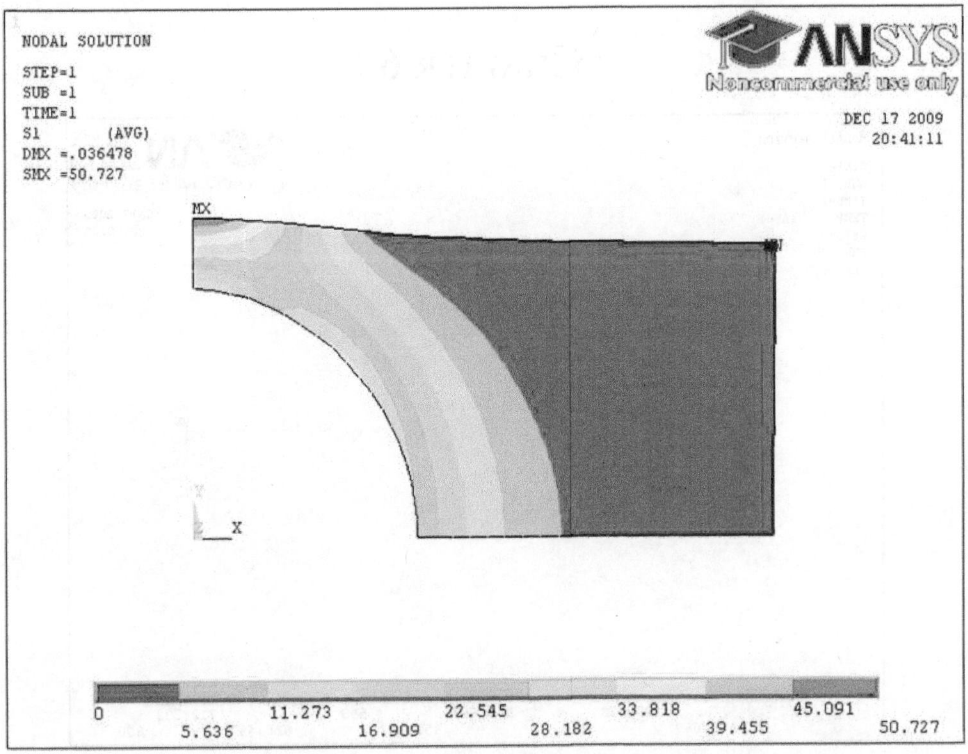

Figure 5.8(a). 1st principal stress distribution pattern

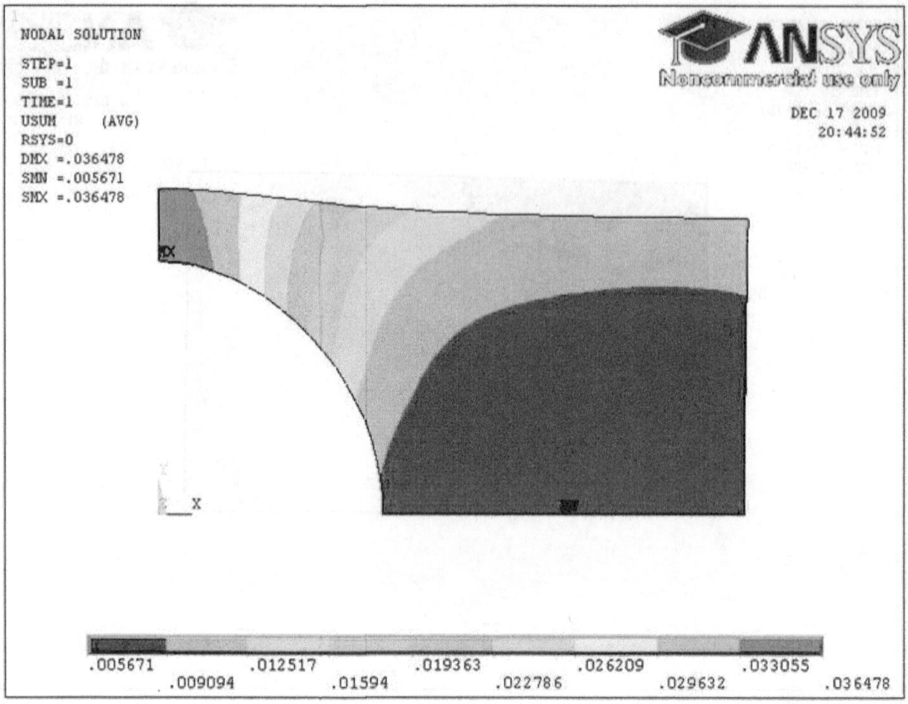

Figure 5.8(b). Deflection pattern

CHAPTER 6

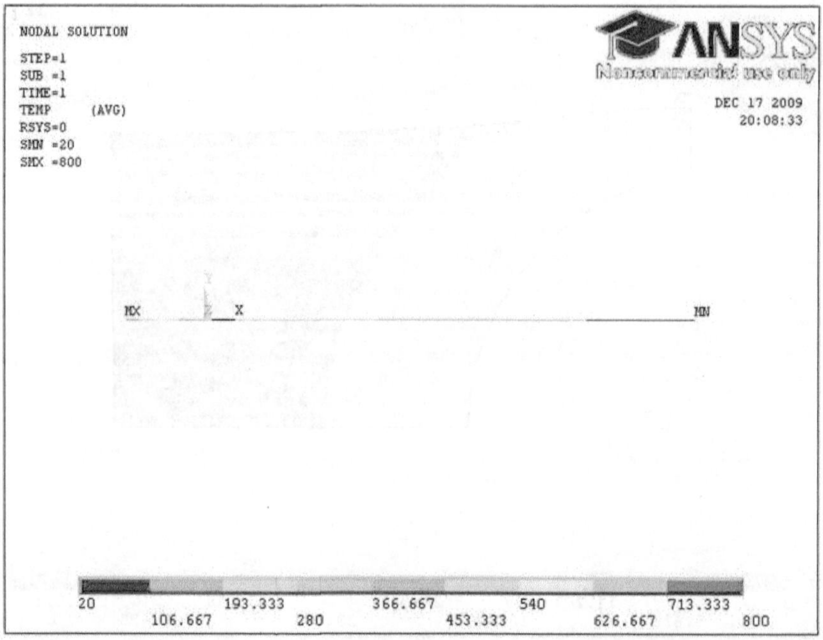

Figure 6.1(c). Temperature distribution in a composite wall

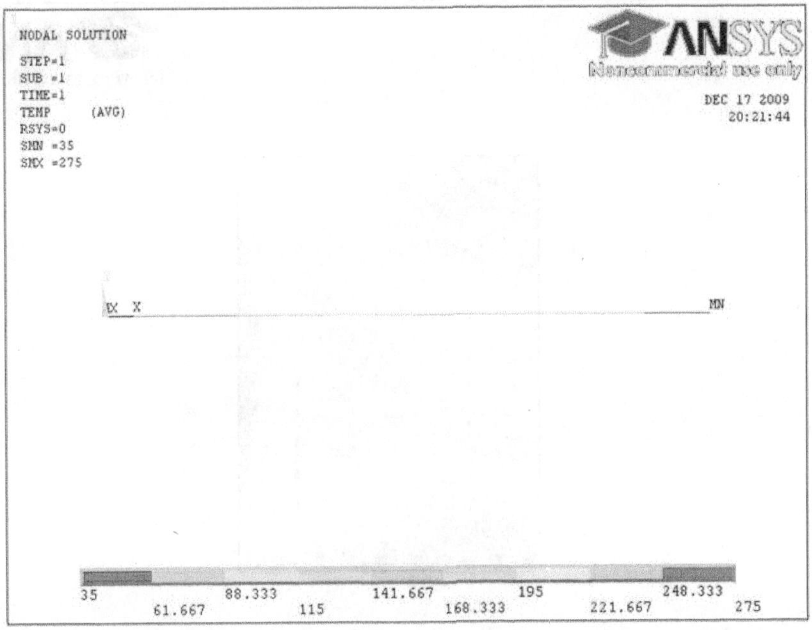

Figure 6.14(b). Temperature distribution in a large plate

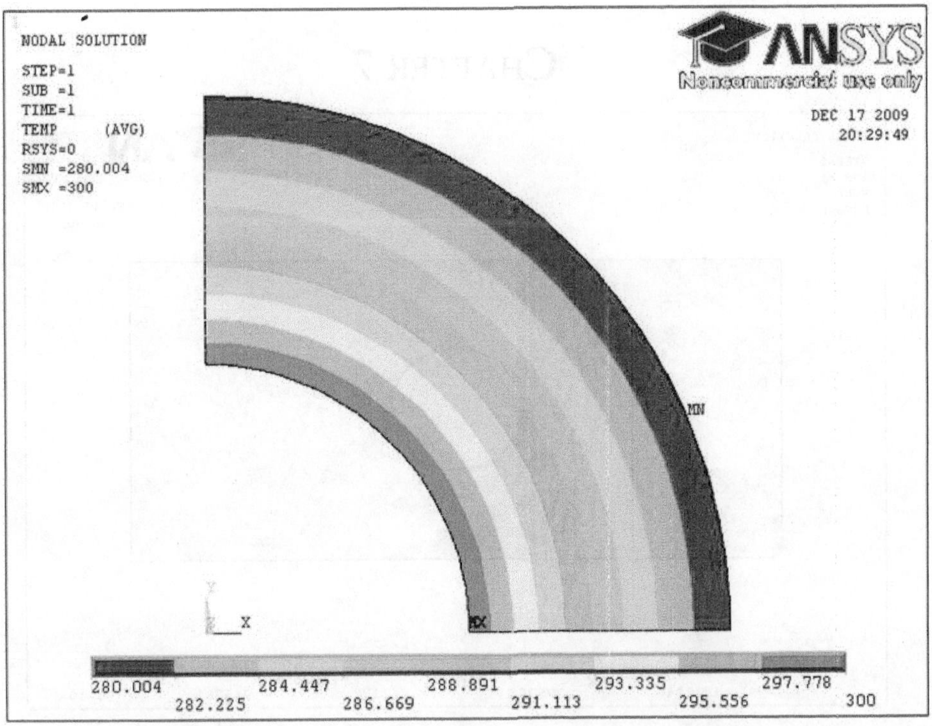

Figure 6.25(c). Temperature distribution in a long cylinder

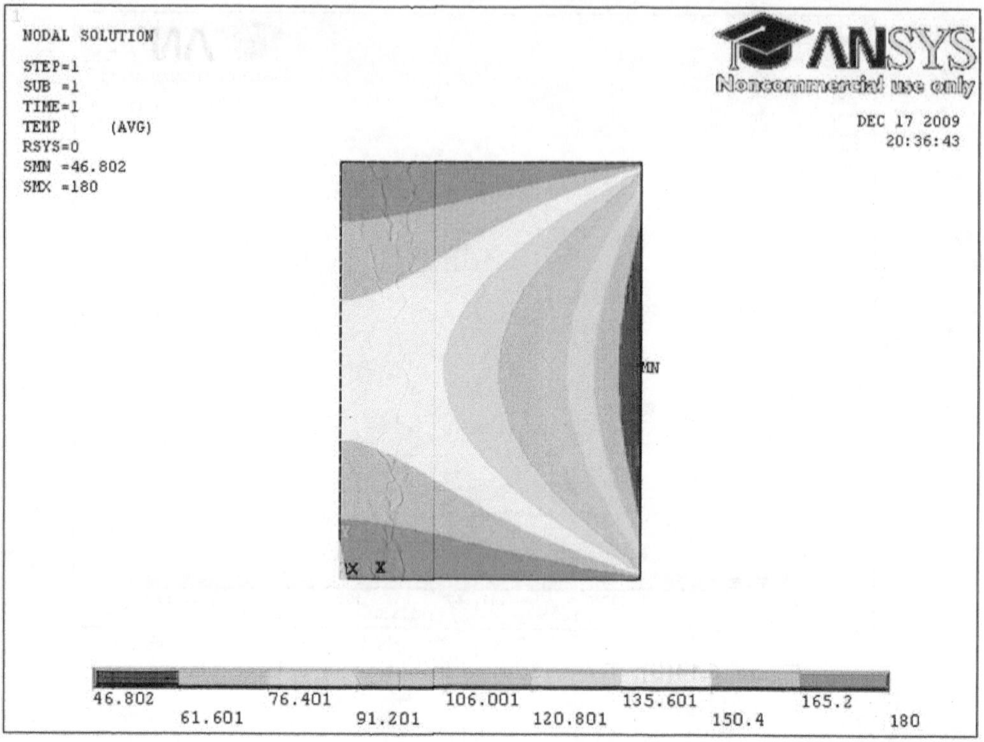

Figure 6.36(a). Temperature distribution in a body of rectangular cross-section

CHAPTER 7

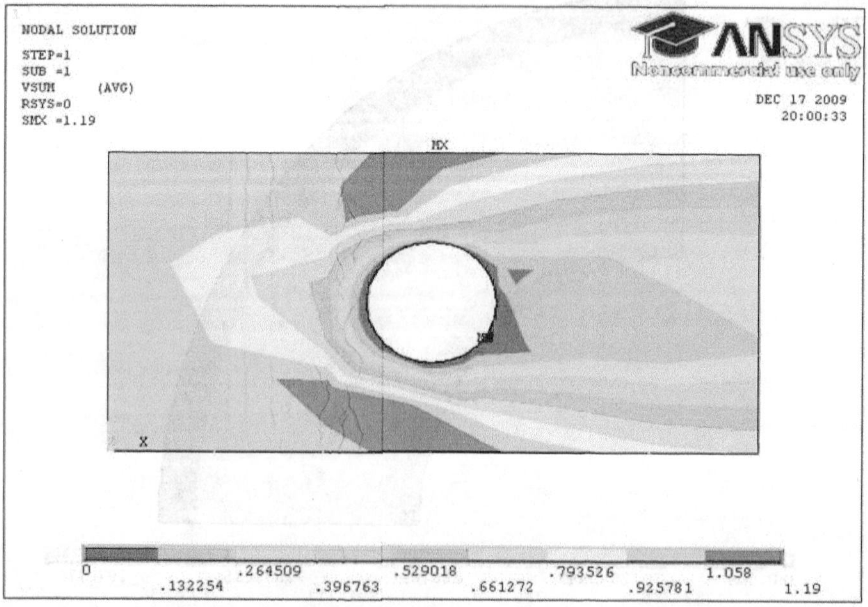

Figure 7.13. Velocity distribution over a cylinder

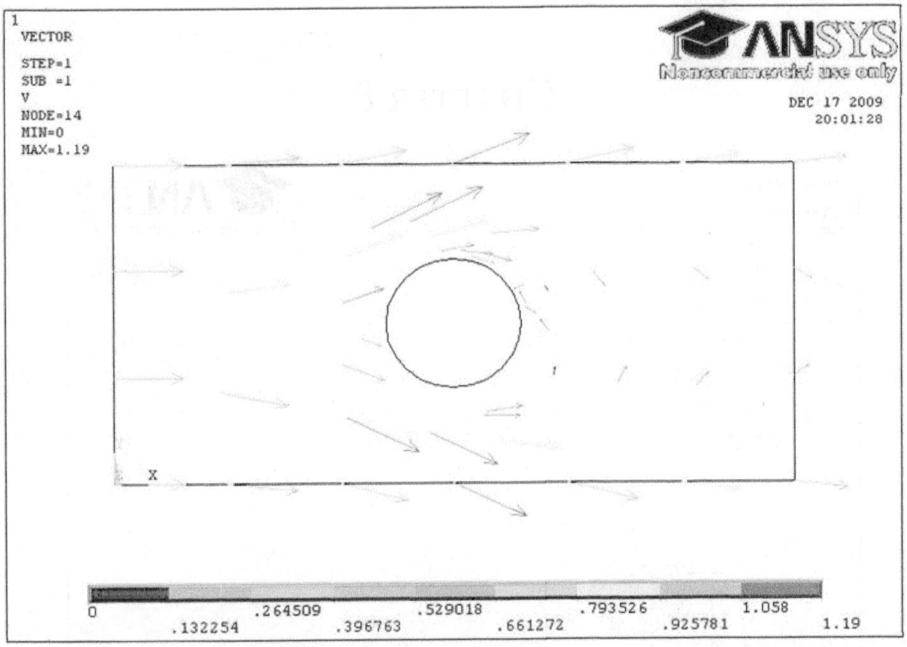

Figure 7.14. Vector plot of the fluid velocity

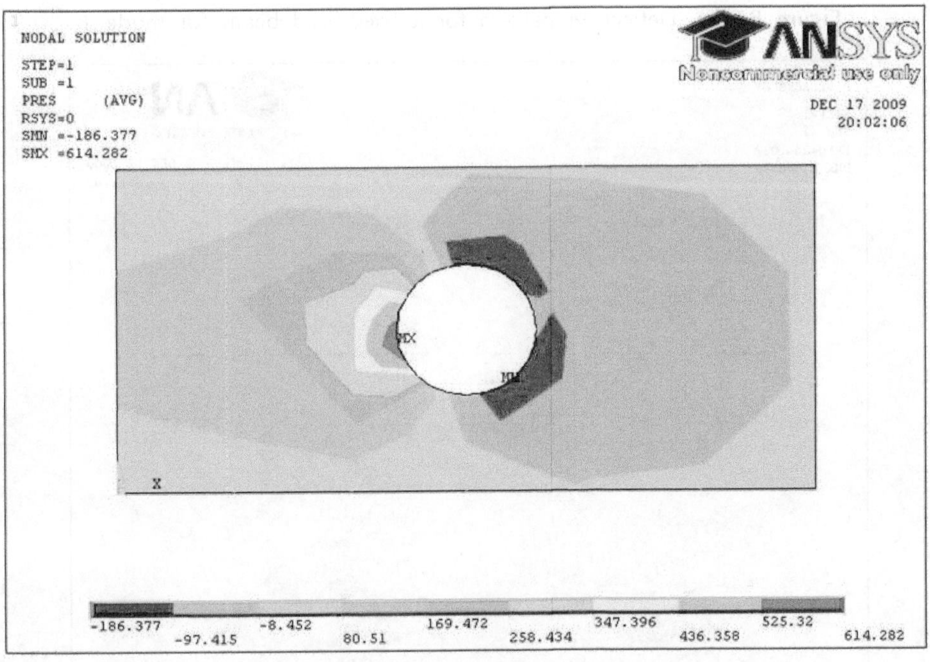

Figure 7.15. Pressure distribution over a cylinder

CHAPTER 8

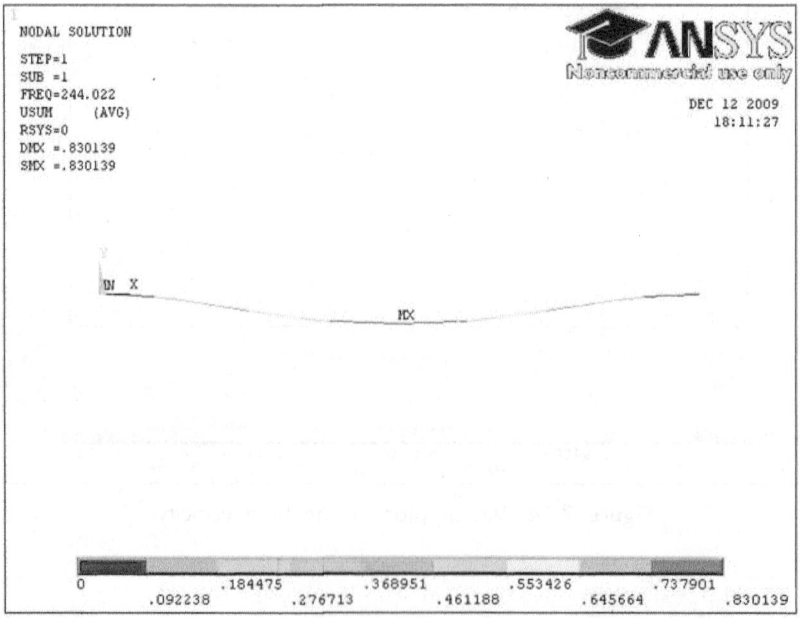

Figure 8.1(b). Deflection pattern for a fixed-fixed beam for mode 1

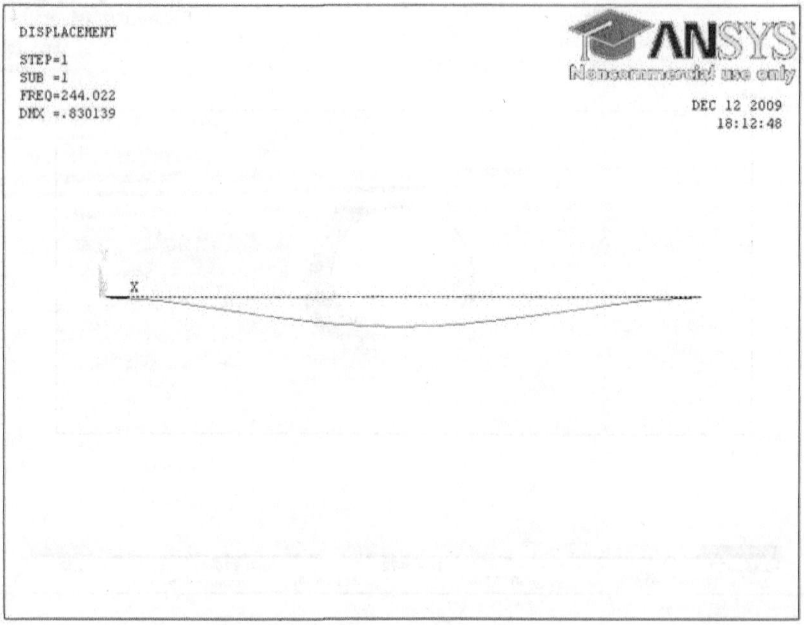

Figure 8.1(c). Mode 1 for fixed-fixed beam

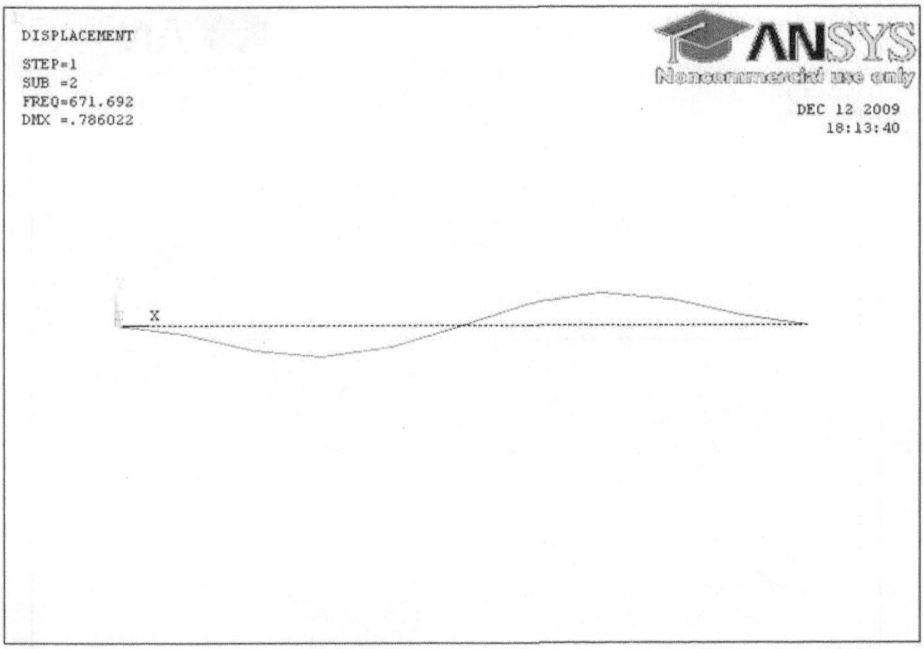

Figure 8.1(d). Mode 2 for fixed-fixed beam

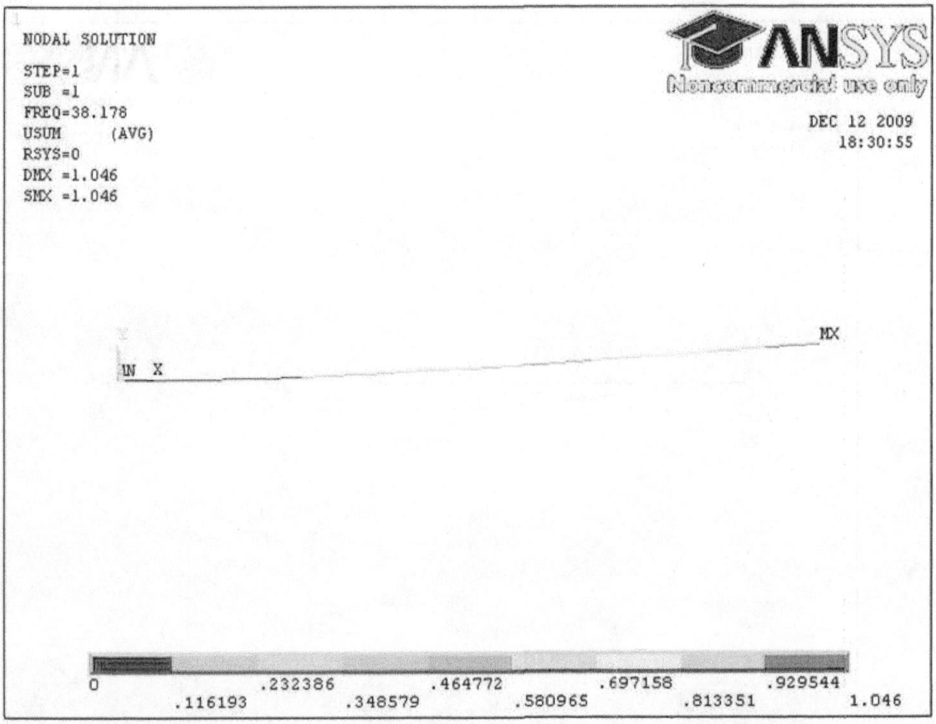

Figure 8.11(b). Deflection pattern for a cantilever beam for mode 1

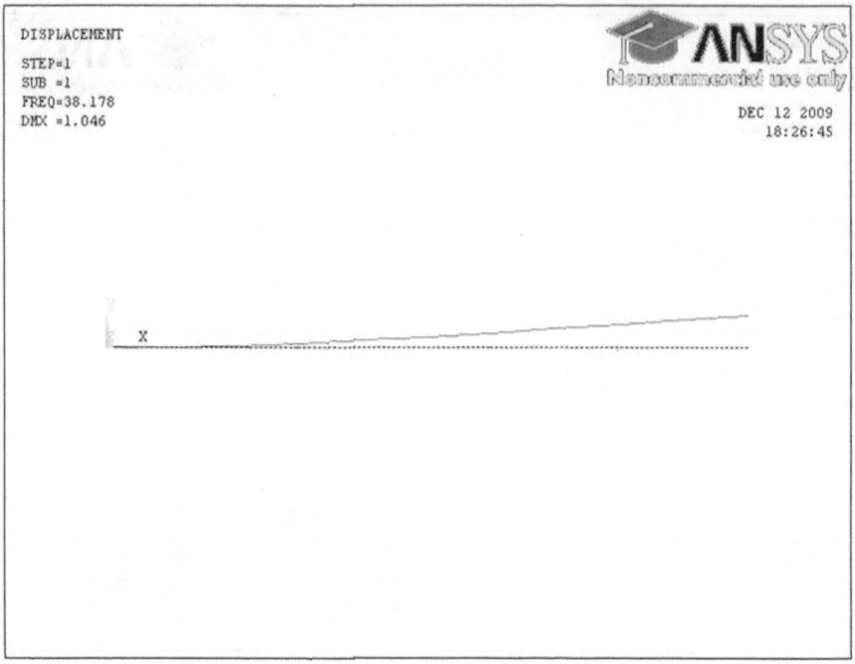

Figure 8.11(c). Mode 1 for cantilever beam

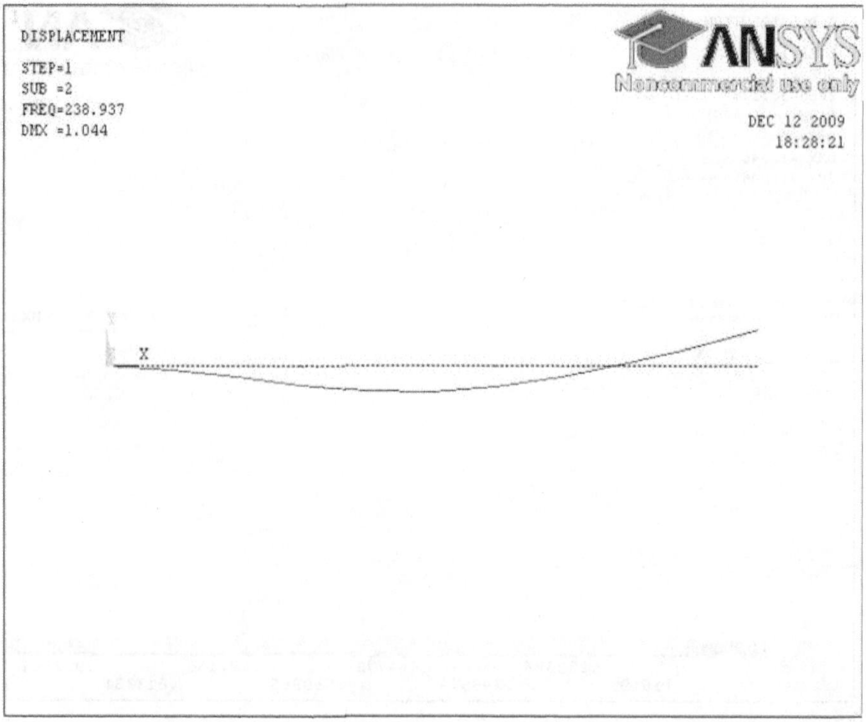

Figure 8.11(d). Mode 2 for cantilever beam

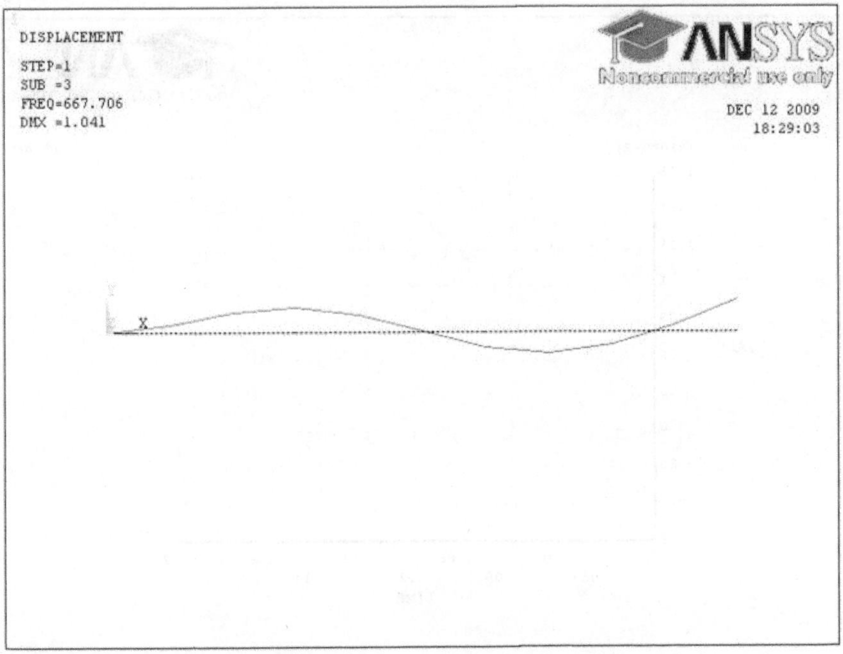

Figure 8.11(e). Mode 3 for cantilever beam

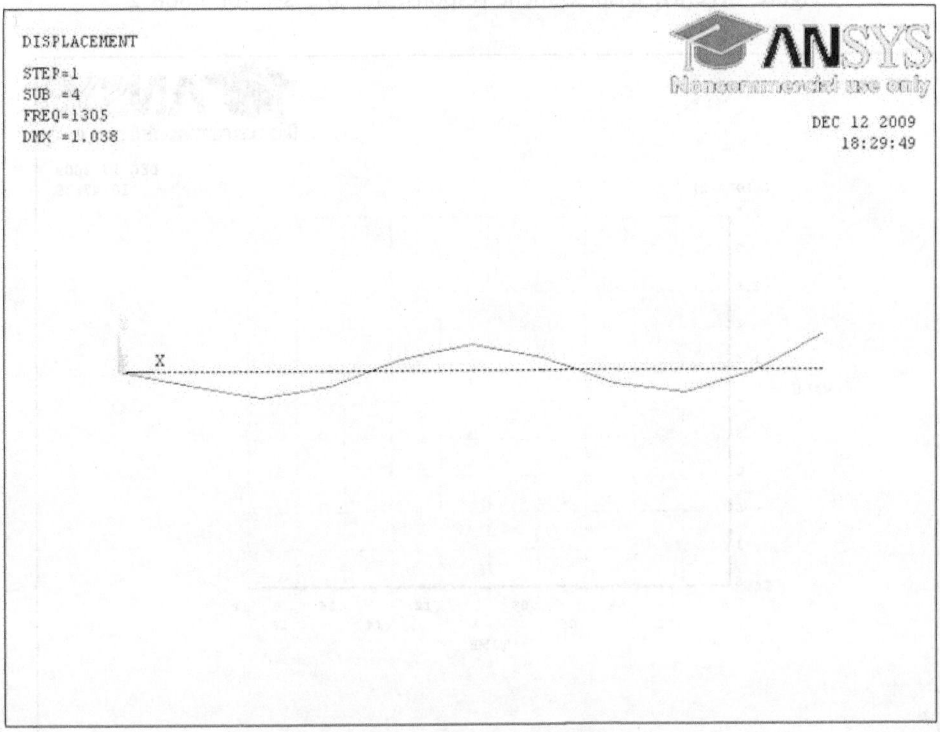

Figure 8.11(f). Mode 4 for cantilever beam

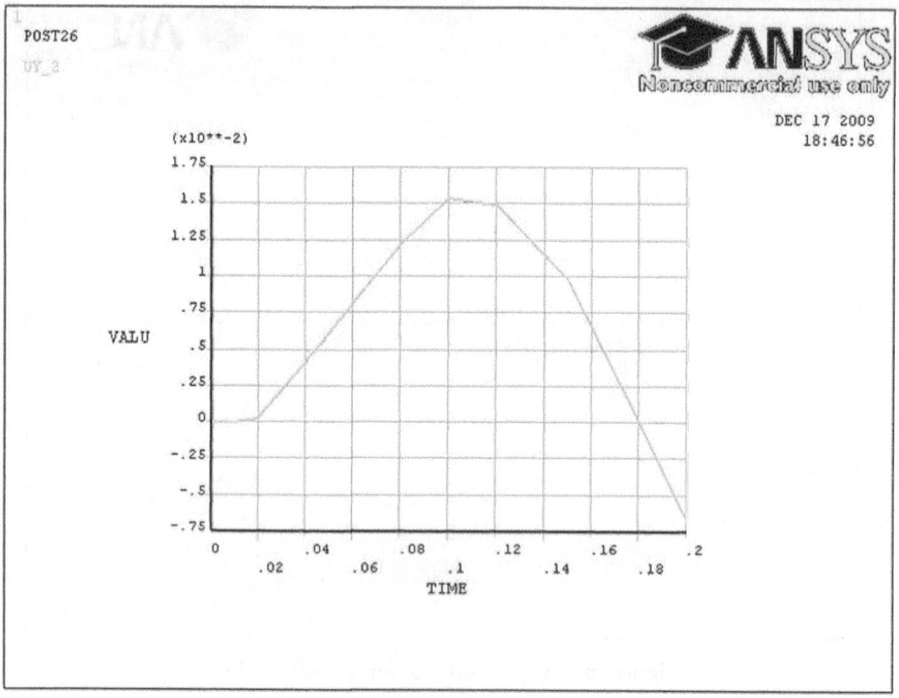

Figure 8.12(a). Displacement response for 0.2 sec for node 2

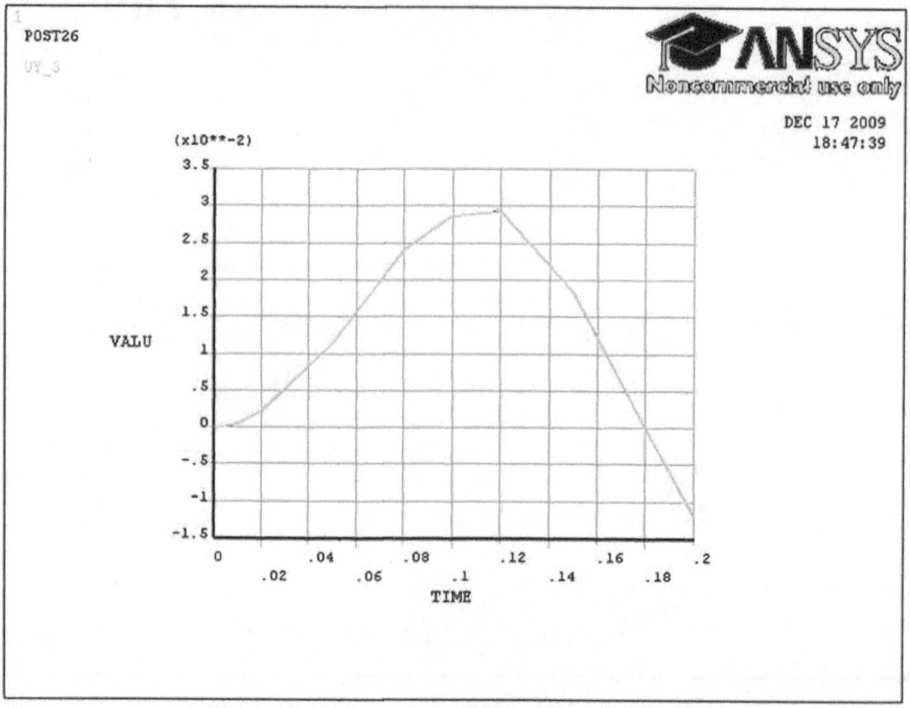

Figure 8.12(b). Displacement response for 0.2 sec for node 3

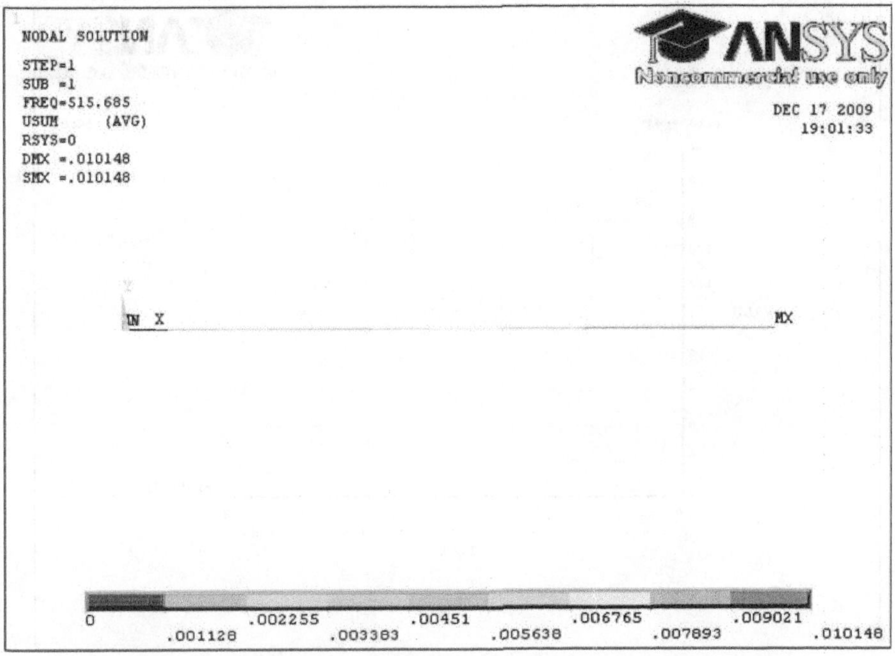

Figure 8.35(b). Deflection pattern for a bar

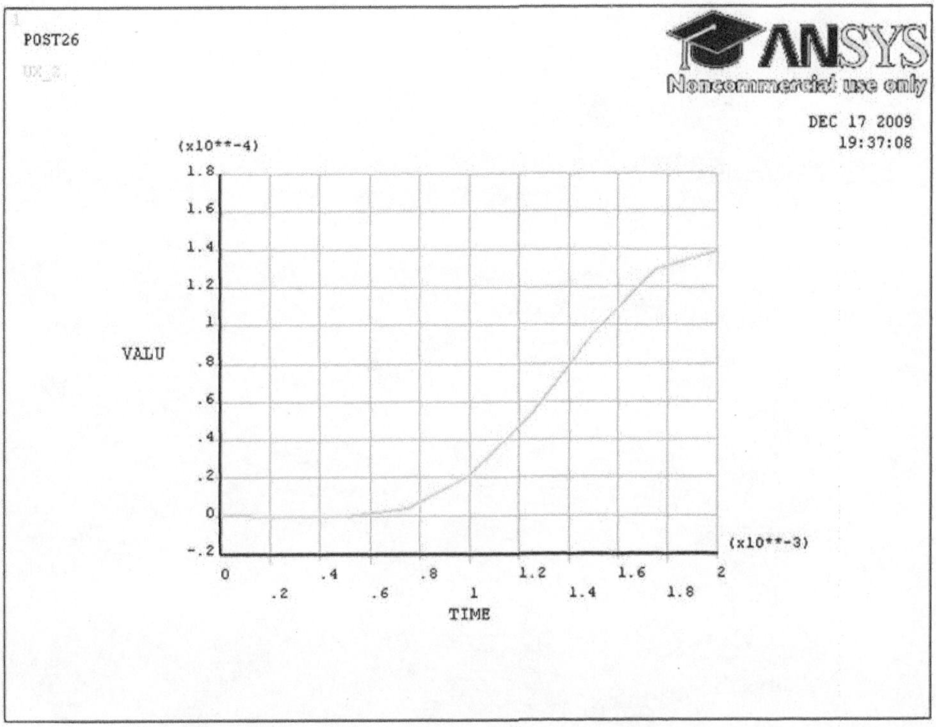

Figure 8.48(a). Displacement response for 0.00025 sec for node 2

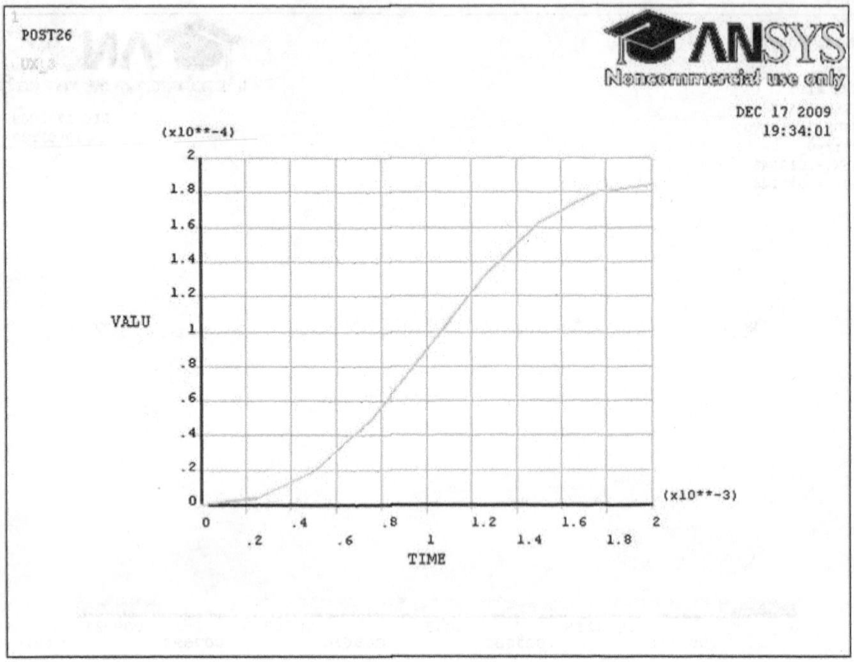

Figure 8.48(b). Displacement response for 0.00025 sec for node 3